Handbook of Hydraulics

About the Authors

James E. Lindell is Senior Vice President Emeritus at MWH, now part of Stantec. He was formerly Chief Hydraulic Engineer at Harza Engineering and Editor in Chief of the seventh edition of *Handbook of Hydraulics*. He has been a member of the Civil Engineering faculty at the University of Notre Dame and a visiting lecturer at the University of Illinois at Chicago and Northwestern University.

Wade P. Moore is a Senior Hydraulic Engineer at Stantec. He specializes in the application of advanced techniques, including computational fluid dynamics (CFD), to the analysis and design of hydraulic structures. He made significant contributions to the seventh edition of *Handbook of Hydraulics*.

Horace W. King (deceased) was the author of the original edition of *Handbook of Hydraulics*, published in 1918, and was also Professor of Hydraulics at the University of Michigan.

Handbook of Hydraulics

James E. Lindell
Wade P. Moore
Horace W. King

Eighth Edition

McGraw Hill Education

New York Chicago San Francisco
Athens London Madrid
Mexico City Milan New Delhi
Singapore Sydney Toronto

Library of Congress Control Number: 2017947815

Handbook of Hydraulics, Eighth Edition

1 2 3 4 5 6 7 8 9 QVS 22 21 20 19 18 17

ISBN 978-1-259-85968-7
MHID 1-259-85968-1

Sponsoring Editor
Lauren Poplawski

Editing Supervisor
Stephen M. Smith

Production Supervisor
Pamela A. Pelton

Acquisitions Coordinator
Lauren Rogers

Project Manager
Anubhav Singh,
Cenveo® Publisher Services

Copy Editor
Surendra Shivam,
Cenveo Publisher Services

Proofreader
Amit Kumar,
Cenveo Publisher Services

Art Director, Cover
Jeff Weeks

Composition
Cenveo Publisher Services

Contents

Preface to the Eighth Edition

Horace Williams King, as Professor of Engineering Hydraulics at the University of Michigan, was the creator of this *Handbook of Hydraulics*, which became such a standard for hydraulic engineers that it became known as *King's Handbook*.

Work on this handbook has been continued over the years by Professor Ernest F. Brater and, most recently in the seventh edition, by James E. Lindell and C. Y. Wei, editors, and Wade P. Moore, a contributor.

Now, the eighth edition of the *Handbook of Hydraulics* takes further steps to keep abreast of developments in hydraulic engineering and advanced methods of problem-solving. Recognizing the increased capability of computers and other electronic devices, tables have been replaced, where appropriate, by methods to solve equations that can be easily accomplished using today's advanced technology. To make the handbook more useful to a wider audience, equations, examples, and coefficients are presented in both U.S. customary and metric units.

It is the objective of this handbook to present the fundamentals needed to solve common hydraulic problems and to provide tables, graphs, and computer techniques to facilitate their solution. Many numerical examples are included. The book has been organized to provide maximum assistance to engineers engaged in hydraulic engineering projects. In addition, it is suitable as a text, especially for courses that deal with nonuniform flow in open channels or coastal engineering. It is only by solving many numerical problems that the student or the young engineer will obtain a sound working knowledge of hydraulic engineering. The tools provided in this book will facilitate the solution of these problems.

The authors wish to acknowledge the vast contributions made by individual researchers, practicing engineers, and governmental organizations, whose published material provides the essential information for the solution of hydraulic problems. References to these many contributions are made throughout the text.

Special acknowledgment is made to Justin Bartels and James C. Lindell, who made valuable contributions to the preparation of figures and tables, and to Research Professor Steve Barfuss of Utah State University, who contributed Sec. 14.

James E. Lindell
Wade P. Moore

Preface to the First Edition

In applied hydraulics rational theory must give place to experimental knowledge. Though every particle of flowing water moves in accordance with definite fixed laws, such laws are intricate and imperfectly understood. In many instances the basic formulas used in hydraulic computations are derived from theoretical considerations, but they must invariably be corrected by experimental coefficients and frequently they become thereby so transformed as to bear but a slight resemblance to the original formulas.

Many thousands of experiments on flowing water have been performed during the last two centuries, the results of which form the basis of our present science of hydraulics. These experiments present many incongruities and as they do not cover the range of conditions required in practice, it is difficult to devise from them accurate working rules and formulas. The hydraulic engineer is therefore confronted with the task of making what appears to be the most reasonable applications of the available data to each problem that he encounters.

A great number of empirical formulas have been devised, which provide an indirect method of transferring experimental results to practical problems. In using such formulas, however, the engineer should not lose sight of the fact that results obtained by them will be subject to errors corresponding to the discrepancies in the experiments on which the formulas are based.

The active interest in experimental research during recent years has been productive of such a rapidly increasing number of hydraulic formulas that engineers generally are not in a position to make critical comparisons and select those that possess the greatest merit. The result has been a tendency to cling to the old and accepted formulas. The author believes that unless the newer formulas have apparent advantages over the old, the latter are preferable inasmuch as their peculiarities are known and it is easier to select coefficients for them, but they should be discarded as soon as more accurate or simpler formulas become available.

In this book, the older and commonly accepted formulas are given preference except where a gain in accuracy or simplicity or both will result from the adoption of new formulas or methods. The author departs from standard American practice in advocating the use of the Manning formula in place of the Kutter formula. He has not done this, however, until he has been able to prove that the two formulas give practically identical results by using the same coefficient. New weir formulas are also submitted which are shown to be simpler and to conform to existing experimental data more consistently than other formulas. Exponential formulas are advocated for pipes but a simplified method of using them is given in detail.

This book is intended primarily to assist in the solution of hydraulic problems. In preparing the manuscript the author has continually kept in mind the twofold purpose, of securing an accuracy consistent with the best experiments and of simplifying calculations. This has necessitated an examination of a vast amount of data and has resulted in the preparation of a great many tables. A knowledge of the fundamental principles of hydraulics is presupposed and derivations have been omitted except where they have appeared necessary in explaining new methods. It is believed that the book will be useful to practising engineers and to students.

In the preparation of tables care has been taken to make them correct to the last figure and all computations and formulas have been independently checked. The author will be grateful to those who may call to his attention any errors or omissions.

A work of this kind is, in a large measure, a recompilation of the results of others, and a great many books and publications have necessarily been consulted. Reference to such use has been made at the proper place in the text. In the preparation of this volume, the author acknowledges assistance from the following:

Mr. Robert E. Horton reviewed the manuscript and proof and made many valuable criticisms and suggestions relative to the character of material and scope of the book. He gave the author free access to all of the records in his office and many of the data contained herein were obtained from this source. For being able to present the book in its present form the author is, in a large measure, indebted to Mr. Horton's helpful suggestions, and he takes this opportunity to express his grateful appreciation.

Professor Theodore R. Running rendered valuable assistance in mathematical computations, especially in checking the author's weir formula by the method of least squares and in suggesting the method employed in the construction of the Manning formula diagrams.

Mr. Chester O. Wisler assisted in checking formulas and tables and in reading proof, and gave many valuable suggestions which were made use of in preparing this book.

Mr. Harry R. Leach, Mr. Floyd A. Nagler, and Mr. Russell A. Dodge shared with the author the bulk of the labor of computing and checking tables, reading proof, and other details. It has only been through the hearty cooperation, loyalty, and active interest of these men that the completion of this volume at the present time has been made possible.

Messrs. M. J. Orbeck, J. B. Jewell, C. N. Ward, R. B. Sleight, and W. O'B. Henderson rendered valuable assistance in computing and checking.

Horace W. King
Ann Arbor, Michigan
January 1918

SECTION 1

Fluid Properties and Hydraulic Units

This section provides information on the fluid properties and quantities used to solve hydraulic engineering problems. Table 1.1 lists the quantities together with the commonly used symbols, dimensions, and basic units. The dimensions are expressed in terms of three fundamental quantities: mass M, length L, and time T. The basic units are given in both U.S. customary units (English units) and the metric system, also known as the International System (SI). Some numerical values of fluid properties are presented in Table 1.2. Table 1.3 is devoted to the conversion of units from the U.S. customary system to the metric system, and Table 1.4 lists the unit weights of materials used in hydraulic structures.

Fluid Properties

Before discussing fluid properties, it is necessary to define the unit of mass. To convert *weight* W to *mass* M, it is necessary to divide by the *gravitational acceleration* g,

$$M = \frac{W}{g} \tag{1.1}$$

The standard units of force and mass are slugs and pounds (U.S. customary) and newtons and kilograms (metric), respectively. In the U.S. customary system a force of 1 pound (lb) will accelerate 1 slug at a rate of 1 foot per second per second (ft/s^2). Likewise, in the metric system a force of 1 newton (N) will accelerate a mass of 1 kilogram (kg) at the rate of 1 meter per second per second (m/s^2). The standard gravitational acceleration is 32.2 ft/s^2 (9.81 m/s^2).

Table 1.1 Fluid Properties and Hydraulic Quantities

Quantity	Symbols	Dimensions	Basic units (English system)	Basic units (Metric system)
Length	l, d, D, h, H	L	Feet (ft)	Meters (m)
Area	a	L^2	Square feet (ft^2)	Square meters (m^2)
Volume	$\forall$	L^3	Cubic feet (ft^3)	Cubic meters (m^3)
Time	T, t	T	Seconds (s)	
Velocity	V, υ, u	L/T	Feet per second (ft/s)	Meters per second (m/s)
Kinematic viscosity	υ	L^2/T	Square feet per second (ft^2/s)	Square meters per second (m^2/s)
Discharge	Q	L^3/T	Cubic feet per second (ft^3/s)	Cubic meters per second (m^3/s)
Mass	M	M	Slug ($lb \cdot s^2/ft$)	Kilograms (kg)
Density	ρ	M/L^3	Slugs per cubic foot ($lb{\cdot}s^2/ft^4$)	Kilograms per cubic meter (kg/m^3)
Force, weight	F, W	ML/T^2	Pound (lb)	Newton (N)
Specific weight (unit weight)	γ	M/L^2T^2	Pounds per cubic foot (lb/ft^3)	Newtons per cubic meter (N/m^3)
Viscosity	μ	M/LT	Slugs per foot second ($lb \cdot s/ft^2$)	Newton-seconds per square meter ($N \cdot s/m^2$)
Surface tension	σ	M/T^2	Pounds per foot (lb/ft)	Newtons per meter (N/m)
Modulus of elasticity (bulk)	E	M/LT^2	Pounds per square foot (lb/ft^2)	Newtons per square meter (N/m^2)
Pressure intensity	ρ	M/LT^2	Pounds per square foot (lb/ft^2)	Newtons per square meter (N/m^2)
Shear stress	τ	M/LT^2	Pounds per square foot (lb/ft^2)	Newtons per square meter (N/m^2)
Momentum or impulse		ML/T	Slug-feet per second ($lb \cdot s$)	Kilogram-meters per second ($kg \cdot m/s$)
Energy or work		ML^2/T^2	Foot pound ($ft \cdot lb$)	Meter-newtons ($m \cdot N$)
Power		ML^2/T^3	Foot pound per second ($ft \cdot lb/s$)	Meter-newtons per second ($m \cdot N/s$)
Specific gravity	sp. gr.			
Vapor pressure	p_v	M/LT^2	Pound per square foot (lb/ft^2)	Newtons per square meter (N/m^2)

TABLE 1.2a Numerical Values of Some Fluid Properties,* U.S. Customary System

Fluid	Temperature T, °F	Specific gravity, sp. gr.	Specific weight w, lb/ft³	Absolute viscosity μ, slugs/ft·s	Kinematic viscosity υ, ft²/s	Vapor pressure p_v, lb/ft²	Surface tension σ, lb/ft
Water	32	0.9999	62.42	0.00003746	0.00001931	12.7	0.518
	39.2	1.0000	62.427	0.00003274	0.00001687	16.9	0.514
	50	0.9997	62.41	0.00002735	0.00001410	25.6	0.509
	60	0.9990	62.37	0.00002359	0.00001217	36.8	0.504
	70	0.9980	62.30	0.00002050	0.00001059	52.3	0.498
	80	0.9966	62.22	0.00001799	0.00000930	73	0.492
	100	0.9931	62.00	0.00001424	0.00000739	136	0.480
	200	0.9630	60.12	0.00000637	0.00000341	1650	0.412
Seawater	—	—	64.00	—			
Mercury	32		—	0.00000.355			
	68		—	0.00000328			
	80.6		—	—			
Castor oil	50	—	—	0.0505			
	59		—	0.0315			
	68	—	—	0.0206			
Dry air at standard atmospheric pressure	32	0.00129	—	$0.355(10^{-6})$			
	50	0.00125	—	$0.372(10^{-6})$			
	59	0.00122	—	$0.376(10^{-6})$			
	68	0.00121	—	$0.380(10^{-6})$			

*Values are based on *Handbook of Chemistry and Physics*, *International Critical Tables*, and *ASCE Manual for Hydraulic Laboratory Studies*. Many are averages of those obtained from various sources. Other sources of similar data are *Smithsonian Physical Tables* and *Standard Density and Volumetric Tables*, U.S. Bureau of Standards.

Table 1.2b Numerical Values of Some Fluid Properties,* Metric System

Fluid	Temperature T, °C	Specific gravity, sp. gr.	Specific weight w, N/m³	Absolute viscosity μ, N·s/m²	Kinematic viscosity υ, m²/s	Vapor pressure p_υ, N/m²	Surface tension σ, N/m
Water	0	0.9999	9805	0.00179	1.795×10^{-6}	608	0.0756
	4	1.0000	9806	0.00157	1.568×10^{-6}	809	0.0750
	10	0.9997	9804	0.00131	1.310×10^{-6}	1226	0.0743
	15	0.9990	9798	0.00113	1.131×10^{-6}	1762	0.0735
	21	0.9980	9787	0.00098	0.984×10^{-6}	2504	0.0727
	27	0.9966	9774	0.00086	0.864×10^{-6}	3495	0.0718
	38	0.9931	9739	0.00068	0.687×10^{-6}	6512	0.0700
	93	0.9630	9444	0.00030	0.317×10^{-6}	79,002	0.0601
Seawater	—	—	10,054	—			
Mercury	0	13.60	—	0.00170			
	20	13.55	—	0.00157			
	27	13.53	—	—			
Castor oil	10	—	—	2.418			
	15	0.969	—	1.508			
	20	—	—	0.986			
Dry air at standard atmospheric pressure	0	0.00129	—	0.000017			
	10	0.00125	—	0.000018			
	15	0.00122	—	0.000018			
	20	0.00121	—	0.000018			

*Values are based on *Handbook of Chemistry and Physics*, *International Critical Tables*, and *ASCE Manual for Hydraulic Laboratory Studies*. Many are averages of those obtained from various sources. Other sources of similar data are *Smithsonian Physical Tables* and *Standard Density and Volumetric Tables*, U.S. Bureau of Standards.

Table 1.3 Factors for Converting USCS Units to SI Units

Length:	
Feet (ft) × 0.3048*	= meters (m)
Inches (in.) × 0.0254*	= meters (m)
Miles (mi) × 1609.34	= meters (m)
Miles (mi) × 1.60934	= kilometers (km)
Nautical miles (nm) × 1852.0*	= meters (m)
Yards (yd) × 0.9144*	= meters (m)
Area:	
Feet^2 (ft^2) × 0.0929030	= meters^2 (m^2)
Acres × 4046.86	= meters^2 (m^2)
Miles^2 (mi^2) × 2.58999	= kilometers^2 (km^2)
Volume:	
Feet^3 (ft^3) × 0.02831685	= meters^3 (m^3)
Feet^3 (ft^3) × 28.31685	= liters†
Yards^3 (yd^3) × 0.764555	= meters^3 (m^3)
Acre-feet × 1233.48	= meters^3 (m^3)
Gallons (gal) × 0.00378541	= meters^3 (m^3)
Gallons (gal) × 3.78541	= liters†
Velocity:	
Feet per second (ft/s) × 0.3048*	= meters per second (m/s)
Miles per hour (mi/h) × 0.44704*	= meters per second (m/s)
Miles per hour (mi/h) × 1.60934	= kilometers per hour (km/h)
Knots × 0.514444	= meters per second (m/s)
Knots × 1.852*	= kilometers per hour (km/h)
Discharge:	
Feet^3 per second (ft^3/s) × 0.02831685	= meters^3 per second(m^3/s)
Millions of gallons per day (mgd) × 0.0438126	= meters^3 per second (m^3/s)
Acre-feet per day × 0.0142764	= meters^3 per second (m^3/s)
Gallons per minute (gal/min) × 0.0000630902	= meters^3 per second (m^3/s)
Force:	
Pounds (lb) × 0.453592	= kilograms force (kgf)
Pounds (lb) × 453.592	= grams (g)
Pounds (lb) × 4.44822	= newtons† (N)
Tons × 0.907185	= metric tons† (t)

(*Continued*)

TABLE 1.3 Factors for Converting USCS Units to SI Units (*Continued*)

Pressure:		
Pounds per foot² (lb/ft²) × 47.8803	=	newtons per meter² (N/m²)
Pounds per foot² (lb/ft²) × 4.88243	=	kilograms force per meter² (kgf/m²)
Pounds per inch² (lb/in.²) × 6894.76	=	newtons per meter² (N/m²)
Pounds per inch² (lb/in.²) × 703.070	=	kilograms force per meter² (kgf/m²
Millibars (mb) × 100.0*	=	newtons per meter² (N/m²)
Unit weight:		
Pounds per foot³ (lb/ft³) × 157.0875	=	newtons per meter³ (N/m³)
Pounds per foot³ (lb/ft³) × 16.0185	=	kilograms force per meter³ (kgf/m³)
Pounds per foot³ (lb/ft³) × 0.0160185	=	grams per centimeter³ (g/cm³)
Mass and density:		
Slugs × 14.5939	=	kilograms (kg)
Slugs per foot³ × 515.379	=	kilograms per meter³ (kg/m³)
Viscosity:		
Pound-seconds per foot² (lb·s/ft²) or	=	newton-seconds per meter²
slugs per foot-second × 47.8803	=	(N · s/m²)†
Feet² per second (ft²/s) × 0.0929031	=	meters² per second (m²/s)

*Exact values.

†Liters × 1000.0 = centimeters³

Liters × 0.001 = meters³

Metric tons × 1000.0 = kilograms force

Kilograms force × 9.80665 = newtons

Newtons × 100,000.0 = dynes

Newton-seconds per meter² × 0.1 = poises

Specific weight, or *unit weight*, w is defined as the weight per unit of volume of a substance. The mass per unit of volume is called *density* ρ. It follows from the preceding discussion that the specific weight may be converted to density by dividing by the gravitational acceleration,

$$\rho = \frac{W}{g} \tag{1.2}$$

By solving either Eq. (1.1) or Eq. (1.2), it may be seen that g is the weight per unit of mass. In other words, 1 slug or kg of mass weighs 32.2 lb or 9.81 N. The value of g varies slightly with both latitude and altitude. The value 32.2 ft/s² or 9.81 m/s² is sufficiently accurate for hydraulic computations. The densities and specific weights of water and some other fluids are given in Table 1.2. The specific weights of various materials are given in Table 1.4.

Table 1.4 Average Specific Weights of Various Materials Used in Construction

Substance	lb/ft³	kN/m³
Clay, earth, and mud:		
Clay	122–162	19.2–25.5
Earth, dry and loose	72–80	11.2–12.6
Earth, dry and shaken	82–92	13.0–14.5
Earth, dry and moderately rammed	90–100	14.2–15.7
Earth, slightly moist, loose	70–76	11.0–12.0
Earth, more moist, loose	66–68	10.4–10.6
Earth, more moist, shaken	75–90	12.0–14.2
Earth, more moist, moderately rammed	90–100	14.2–15.7
Earth, as soft flowing mud	104–112	16.4–17.8
Earth, as soft mud well pressed into a box	110–120	17.3–19.0
Mud, dry, close	80–110	12.6–17.3
Mud, wet, moderately pressed	110–130	17.3–20.6
Mud, wet, fluid	104–120	16.4–19.0
Masonry and its materials:		
Brick, best, pressed	150	23.8
Brick, common, hard	125	19.9
Brick, soft, inferior	100	15.7
Brickwork, pressed brick, fine joints	140	22.1
Brickwork, medium quality	125	19.9
Brickwork, coarse, inferior soft bricks	100	15.7
Cement, pulverized, loose	72–105	11.4
Cement, pressed	115	18.3
Cement, set	168–187	26.4–29.4
Concrete, 1:3:6	140	22.1
Gravel, loose	82–125	13.0–19.9
Gravel, rammed	90–145	14.2–23.0
Masonry of granite or stone of like weight:		
Well dressed	165	25.9
Well-scabbled rubble, 20% mortar	154	24.4
Roughly scabbled rubble, 25–35% mortar	150	23.9
Well-scabbled dry rubble	138	21.8
Roughly scabbled dry rubble	125	19.9

(*Continued*)

TABLE 1.4 Average Specific Weights of Various Materials Used in Construction (*Continued*)

Substance	lb/ft³	kN/m³
Masonry and its materials (cont.):		
Masonry of sandstone or stone of like weight weighs about seven-eighths of the above		
Mortar, hardened	96–115	14.2–18.3
Sand, pure quartz, dry, loose	87–106	13.8–16.8
Sand, pure quartz, dry, slightly shaken	92–110	14.5–17.3
Sand, pure quartz, dry, rammed	100–120	15.7–19.0
Sand, natural, dry, loose	80–110	12.6–17.3
Sand, natural, dry, shaken	85–125	13.7–19.9
Sand, wet, voids full of water	118–128	18.7–20.2
Stone	135–195	21.6–30.6
Stone, quarried, loosely piled	80–110	12.6–17.3
Stone, broken, loose	77–112	12.2–17.8
Stone, broken, rammed	79–121	12.4–19.1
Metal and alloys:		
Brass (copper and zinc)	487–524	77–83
Bronze (copper and tin)	524–537	83–85
Copper, cast	537–548	85–87
Copper, rolled	548–562	87–89
Iron and steel, cast	438–483	70–77
Average	450	71
Iron and steel, wrought	475–494	75–78
Average	480	76
Spelter or zinc	425–450	68–71
Tin, cast	450–470	71–75
Steel	490	77
Tin	459	72
Zinc	438	69
Mercury (32°F, 0°C)	849	133
Woods (dry)*:		
White oak	46.4	7.3
White pine	25.6	4.0
Southern long-leaf pine	38.1	6.0
Douglas fir	32.1	5.1
Short-leaf yellow pine	38.4	6.0

(*Continued*)

Table 1.4 Average Specific Weights of Various Materials Used in Construction (*Continued*)

Substance	lb/ft³	kN/m³
Woods (dry)* (cont.):		
Norway pine	30.2	4.8
Spruce and eastern fir	25.0	4.0
Hemlock	26–32	4.1–5.1
Cypress	29.8	4.7
Cedar	23.1	3.7
Chestnut	41.0	6.5
California redwood	26.2	4.1
California spruce	25.0	4.0

*The weights of green or unseasoned timbers are 20–40% greater.

The *specific gravity* (sp. gr.) of a substance is obtained by dividing the weight of a certain volume of the substance by the weight of an equal volume of water at 39.2°F or 4.2°C. Values of the specific gravity of several fluids are given in Table 1.2.

The *viscosity* μ of a fluid is a measure of the relative ease or difficulty with which a particle of the fluid may be deformed. For example, heavy oil has a greater viscosity than water, and water is more viscous than air. The viscosity of a liquid decreases as the temperature increases, but the opposite is true of a gas. Assume that the space between the two horizontal plates, shown in Fig. 1.1, is filled with a fluid, that the upper plate is in motion with velocity V and that the lower plate is stationary. Those particles of fluid that are in contact with a boundary surface will have the same velocity as the boundary. It follows that the velocity of the fluid will vary from V at the top plate to zero at the bottom one and that the velocity of any layer of fluid will be different from that of the adjacent layers. As a result, all particles of fluid will be continuously deformed by the shear stresses between adjacent layers, as illustrated in Fig. 1.1 by successive positions of a

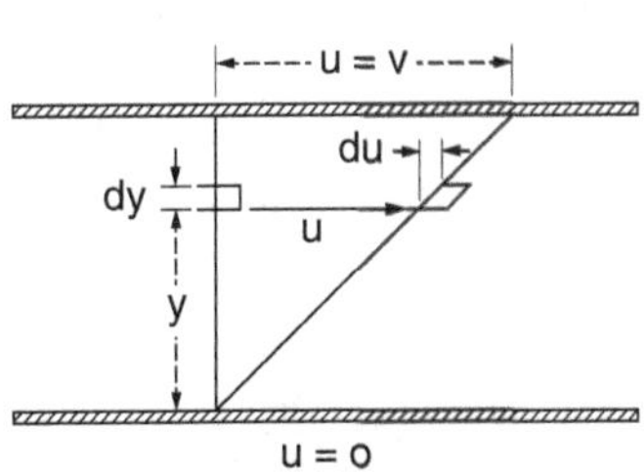

Figure 1.1 Velocity distribution in Newtonian fluids.

particle at a distance y from the lower plate. It has been shown by laboratory tests that for a particular fluid at a given temperature the ratio of the shear stress τ to the rate of deformation du/dy is constant. This constant is called coefficient of viscosity, dynamic viscosity, absolute viscosity, or simply viscosity. The preceding definition is expressed mathematically by the equation

$$\mu = \frac{\tau}{du/dy} \tag{1.3}$$

This relationship is called Newton's law of viscosity, and fluids which behave in this manner are called newtonian fluids. Most fluids dealt with by engineers are newtonian fluids. Some thick, highly viscous fluids are nonnewtonian, meaning that the shear stress is not directly proportional to du/dy for the entire range of shear stresses.

By inserting the units for the terms in the right side of Eq. (1.3), the units of viscosity are found to be pound-seconds per square foot or newton-seconds per square meter. The viscosities of a number of fluids are given in Table 1.2. The *kinematic viscosity* ν is obtained by dividing viscosity by density,

$$\nu = \frac{\mu}{\rho} \tag{1.4}$$

The unit of kinematic viscosity, obtained by inserting the units of μ and ρ into Eq. (1.4), is square feet per second or square meters per second.

The *surface tension* σ is the property of liquids that causes capillary action in small tubes. If the liquid *wets* the surface of the tube, the liquid surface will be concave upward, and the liquid will rise, as shown in Fig. 1.2*a*. If the liquid does not *wet* the tube, the surface will be concave downward, causing the liquid to be depressed, as shown in Fig. 1.2*b*. When pressures are measured by manometers, care must be exercised to avoid errors resulting from capillarity. A number of experimental values of capillary rise or fall for glass tubes of various sizes, as determined by Folsom,[1] are plotted in Fig. 1.3. The straight line shown on the graph indicates the rise that might be expected with very pure water in contact with air. Some values of surface tension of water are given in Table 1.2.

The *modulus of elasticity* E of liquids is defined as the change in pressure intensity divided by the corresponding change in volume per unit volume,

$$E = \frac{\Delta p}{\Delta \forall / \forall} \tag{1.5}$$

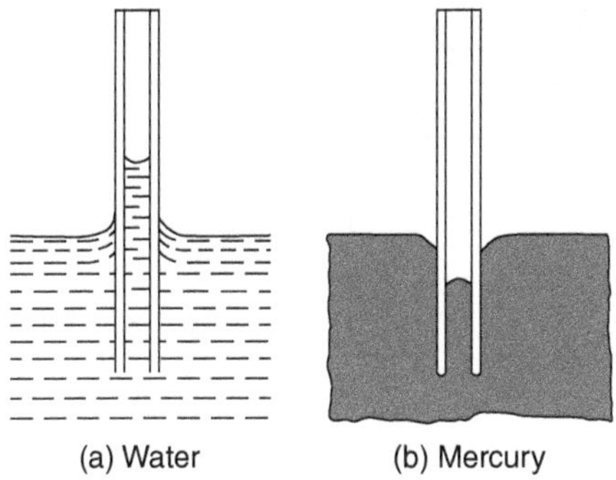

Figure 1.2 Capillary tubes.

The value of E for water is approximately 300,000 lb/in^2 or 2×10^9 N/m^2, varying slightly with temperature. This value is sufficiently high to permit the assumption in the solution of most hydraulic problems that liquids are incompressible.

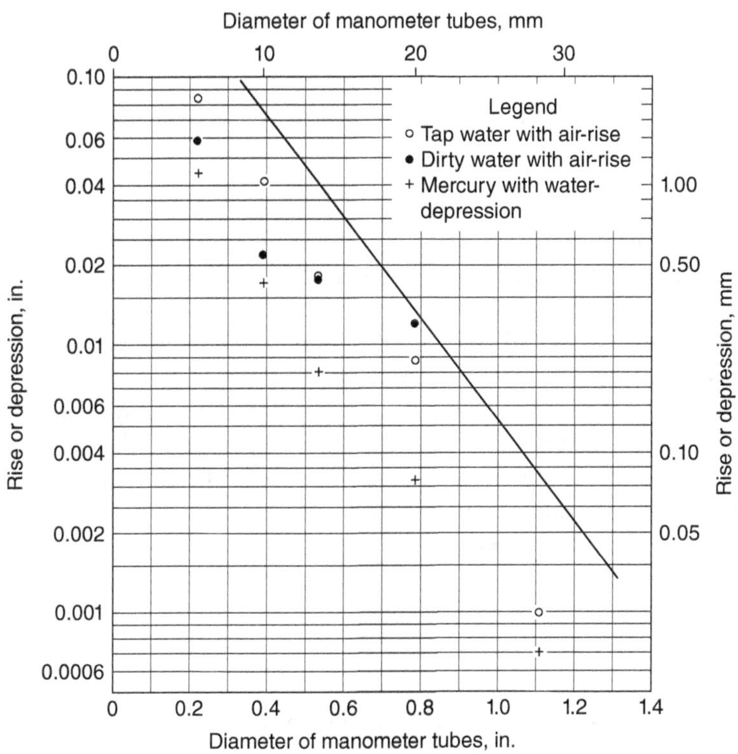

Figure 1.3 Capillary rise and fall in glass tubes.

Table 1.5 SI Prefixes

Prefix	Abbreviation	Meaning
Mega-	M	1,000,000
Kilo-	k	1,000
Hecto-	h	100
Deka-	da	10
Deci-	d	0.1
Centi-	c	0.01
Milli-	m	0.001
Micro-	μ	0.000001

The gaseous form of water is called *water vapor*, and the partial pressure due to the water vapor is called *vapor pressure*. Within a moderate range of temperatures and pressures and for conditions not too close to the point of condensation, water vapor obeys the gas laws reasonably well. For a given temperature, any given space can hold only a certain amount of water vapor in the presence of a solid or liquid surface. The pressure exerted by water vapor under saturation conditions is called *vapor pressure* p_v. Values of the vapor pressure of water for various temperatures are given in Table 1.2.

Hydraulic Units

The purpose of this subsection is to define the units used in hydraulic engineering. Most of the world currently uses the metric system in the form called International System (SI). The SI might also be called meter-newton-second system. The commonly used units of the U.S. customary system are given in Table 1.3 along with the factors for converting to the SI units and the abbreviations for both systems. The most used quantity in hydraulic engineering is the discharge, given in units of cubic meters per second. The prefixes commonly used in SI modify basic units by powers of 10. For example, milli means one thousandth and kilo means one thousand. The most used prefixes are shown in Table 1.5 together with their abbreviations and meanings.

Reference

1. R. G. Folsom, "Manometer Errors Due to Capillarity," *Instruments*, pp. 36–37, Feb. 1936.

SECTION 2

Hydrostatics

Hydrostatics is the branch of hydraulics that deals with the pressures and forces resulting from a fluid at rest. In this section, the principles of hydrostatics are presented and applications are made to hydraulic engineering problems. The use of these principles in cases where the fluid is in motion is discussed, and methods of determining the additional forces resulting from changes in velocity are presented.

Fluid Pressure

The force per unit area within a fluid is called *pressure, p,* and acts equally in all directions. The force on a boundary surface resulting from fluid pressure must be normal to the surface at all points because of the inability of fluids at rest to resist shear. The pressure variation with depth within a liquid may be evaluated by considering the forces acting on the vertical prism of height Δh and cross-sectional area Δa, shown in Fig. 2.1*a*. The summation of all the forces acting on this prism in the vertical direction as well as in all other directions must be equal to zero. The vertical forces consist of the weight and the forces due to both the pressure p_1 at the top and the pressure p_2 at the bottom. It is assumed here that the fluid is a liquid; that is, the specific weight does not vary with pressure. The summation of the vertical forces gives

$$p_2\,\Delta a = w\,\Delta h\,\Delta a + p_1\,\Delta a$$

Thus

$$p_2 = w\,\Delta h + p_1 \tag{2.1}$$

the downward direction of h being taken as positive. Equation (2.1) may be solved for Δh to give

$$\Delta h = \frac{p_2}{w} - \frac{p_1}{w} \tag{2.2}$$

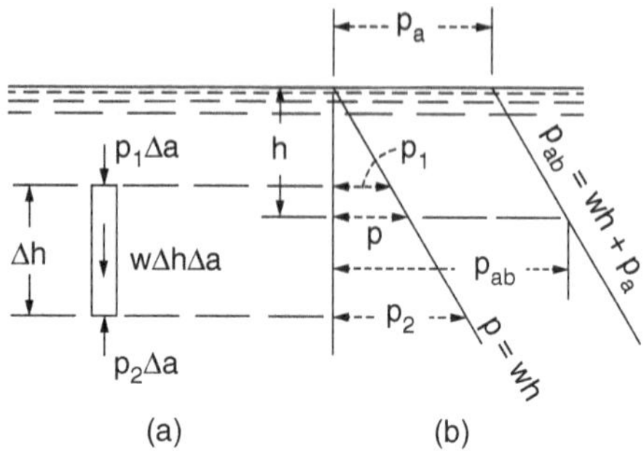

FIGURE 2.1 Fluid pressure.

If p_1 is taken as the pressure at the liquid surface, Δh becomes h, the vertical distance below the liquid surface to the point where the pressure is p. Furthermore, when the pressure at the liquid surface is the atmospheric pressure p_a, then p_2 is the absolute pressure at that point. The following expression for absolute pressure p_{ab} may now be obtained from Eq. (2.1):

$$p_{ab} = wh + p_a \tag{2.3}$$

More commonly used in engineering work is the gauge pressure p. The gauge-pressure scale is obtained by designating the atmospheric pressure as zero. The following expressions for gauge pressure may be obtained from Eq. (2.3) by letting P_a become zero:

$$p = wh \tag{2.4}$$

and

$$h = \frac{p}{w} \tag{2.5}$$

Equations (2.3) and (2.4) are illustrated graphically in Fig. 2.1*b*. The expression p/w is called *pressure head*. It expresses the depth in feet or meters of a liquid of specific weight w required to produce a pressure p. Values of absolute pressure are always positive, whereas the gauge pressure may be positive or negative, depending on whether the pressure is greater or less than the atmospheric pressure. Negative gauge pressures indicate that a partial vacuum exists.

Atmospheric Pressure

The variation of pressure with elevation in the atmosphere is shown in the following table. The straight-line variation derived for liquids does not apply to the atmosphere (or any other gas) because the density varies with the altitude.[1] The average value of the atmospheric pressure at sea level is usually taken as 101.3 kN/m^2 or 14.7 lb/in.^{2} (2116 lb/ft^2).

English units		Metric units	
Altitude, mi	**Pressure, lb/in.^{2}**	**Altitude, km**	**Pressure, kPa**
0	14.7	0	101.3
1	12.1	1	89.9
2	9.9	2	79.5
3	8.0	3	70.1
5	5.2	5	54.2
10	1.5	10	26.6
20	0.1	30	1.2

The pressure gauges used to measure atmospheric pressure are called *barometers*. A barometer of the type shown in Fig. 2.2 may be made by filling the tube with any liquid, temporarily sealing the open end, inverting the tube, and releasing the seal beneath the surface of the liquid in the pan. If the tube is sufficiently long, there will be a space at the top that is empty except for the vapor of the liquid. Evaporation will take place at the liquid surface in the tube until a saturated vapor and the corresponding vapor pressure p_v (see "Fluid Properties," Sec. 1) exist above the liquid surface. Example 2.1 illustrates that water is an impractical liquid for barometers. The liquid generally used is mercury. Mercury not only permits the use of a shorter tube but its vapor pressure is so low[2] that it may be neglected.

Example 2.1 Determine the value of h_b in Fig. 2.2 if the liquid is water at a temperature of 70°F and p_a is 14.7 lb/in.^{2} If Eq. (2.1) is written first from m to n and then from n to o, it will be seen that $p_o = p_m$ and

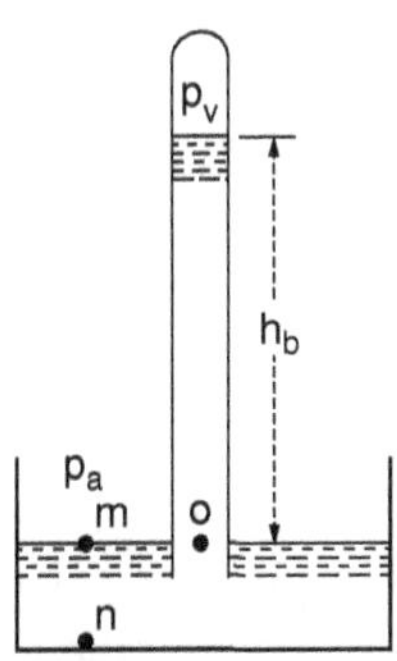

FIGURE 2.2 Barometer.

therefore that the pressure at o is atmospheric. Equation (2.2) may now be written as follows:

$$\Delta h = h_b = \frac{p_a}{w} - \frac{p_v}{w}$$

The value of p_a/w is $(14.7 \times 144)/62.3$, or 33.98 ft, the value of w being obtained from Table 1.2. The value of p_v is given in Table 1.2 as 52.3 lb/ft^2; therefore, $p_v/w = 52.3/62.3$, or 0.84 ft. Hence,

$$h_b = 33.98 - 0.84 = 33.14 \text{ ft} \quad \blacktriangle$$

Example 2.2 Determine the value of h_b in Fig. 2.2 if the liquid is mercury at a temperature of 21°C and $p_a = 101.3$ kN/m^2.

The vapor pressure of mercury is negligible, so that Eq. (2.2) becomes

$$\Delta h = h_b = \frac{p_a}{w} + 0$$

$$h_b = \frac{101.3}{13.55 \times 9.8} = 0.763 \text{ m} \quad \blacktriangle$$

Manometers

Manometers are tubes attached to reservoirs, pipes, or channels for the purpose of measuring the pressure. The equations of hydrostatics are used to determine pressures from manometer readings even though manometers are most frequently used to measure pressures in moving fluids. To ensure against including forces due to acceleration in the manometer readings, it is necessary to install the tube in a wall that is parallel to the flow lines in such a manner that the flow pattern is not disturbed by the opening. When the manometers contain only the fluid in the conduit, as illustrated by Fig. 2.3*a* and *b*, they are called *piezometers*. In Fig. 2.3*a* the pressure at l, p_l, is equal to the pressure at the center of the pipe, p_c. Consequently, the value of p_c may be obtained from any of the three tubes, using Eq. (2.4) arranged as follows:

$$p_c = wh_c \tag{2.6}$$

The left tube of Fig. 2.3*a* and the one shown in Fig. 2.3*b* may be used to measure negative as well as positive gauge pressures. Equation (2.6) may be applied to Fig. 2.3*b*, knowing the pressure at l to be zero and that values of h_c measured upward from the zero datum are negative.

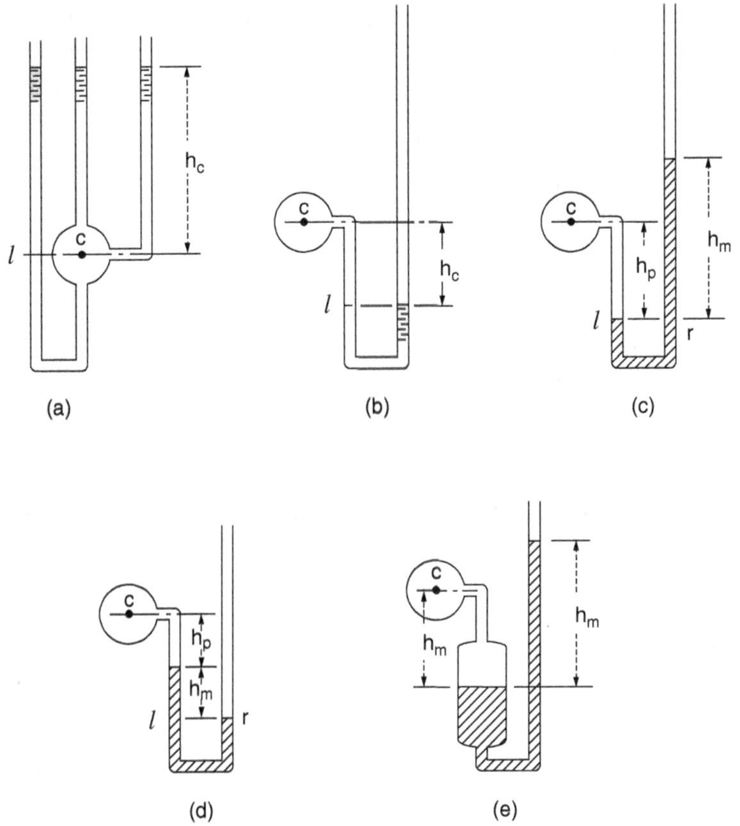

FIGURE 2.3 Manometers.

Example 2.3 Determine the value of h_c in Fig. 2.3*a* when the gauge pressure at *c* is 300 kN/m^2 if the pipe contains water.

Solving Eq. (2.6),[a]

$$h_b = \frac{p_c}{w} = \frac{300}{9.8} = 30.6\text{ m} \quad \blacktriangle$$

Example 2.4 Determine the pressure at *c* in Fig. 2.3*b* if h_c is 2.0 m and the pipe contains oil having a specific gravity of 0.95.

$$p_c = wh_c = -0.95 \times 9.8 \times 2.0 = -18.6\text{ kN/m}^2 \quad \blacktriangle$$

[a]Unless some particular temperature is stated, the specific weight of water will be taken as 62.4 lb/ft^3 or 9.8 kN/m^3 in numerical examples.

Piezometers are very sensitive pressure gauges but are impractical for the measurement of high pressures because of the excessive length of tube required, as illustrated by Examples 2.3 and 2.4. More satisfactory manometers for the measurement of high pressures are shown in Fig. 2.3*c* to *e*. The manometer liquid most commonly used is mercury. Any liquid of greater density than the pipe conduit fluid may be used provided that the two fluids are not miscible. Manometers of this type may be used whether the fluid is a gas or a liquid and for both positive and negative pressures.

Values of h_p and h_m in Fig. 2.3*c* are usually read from a scale placed between the two legs of the U tube. The location of the zero of this scale fluctuates greatly with variations in the pipe pressure, thus causing some inconvenience in making the readings. This objectionable feature may be partly eliminated by the arrangement shown in Fig. 2.3*e*. The amount of movement of the mercury surface in the left and right tubes will be inversely proportional to the square of the diameters of the tubes. If the difference in diameters is made large, the mercury surface in the left tube will be nearly stationary and the value of h_p will then be nearly constant. For many practical purposes the scale may be marked off to give the pressure at *c* directly from the values of h_m.

Example 2.5 Determine the value of h_m in Fig. 2.3*c* if $p_c = 300$ kN/m^2 gauge, $h_p = 2.0$ m, the pipe fluid is water, and the manometer fluid is mercury.

Let w_p be the specific weight of the pipe fluid and w_m the specific weight of the manometer fluid. From Eq. (2.1),

$$p_l = w_p h_p + p_c = 9.8 \times 2.0 + 300$$

and from Eq. (2.4),

$$p_r = w_m h_m = 13.6 \times 9.8\, h_m$$

Knowing p_l and p_r to be equal,

$$13.6 \times 9.8 h_m = 9.8 \times 2.0 + 300$$

from which h_m is found to be 2.4 m.[b] ▲

[b]Unless some particular temperature is stated, the specific gravity of mercury will be taken as 13.6 in numerical problems.

Example 2.6 Determine the value of p_c in Fig. 2.3*d* if $h_p = 0.20$ m and $h_m = 0.50$ m. The pipe fluid is water, and the manometer fluid is mercury.

Equation (2.4) gives

$$p_l = w_m h_m + w_p h_p + p_c = 0$$

or $$(13.6 \times 9.8 \times 0.50) + (9.8 \times 0.20) + p_c = 0$$

from which

$$p_c = -68.6 \text{ kN/m}^2 \text{ gauge} \quad \blacktriangle$$

Differential manometers of the type shown in Fig. 2.4 are employed to measure the difference in pressure between two points. Here it is desired to measure $p_1 - p_2$. In Fig. 2.4*a*, the specific weight of the manometer fluid w_m is greater than that of the pipe fluid w_p, whereas in Fig. 2.4*b* the reverse is true. In the arrangement shown in Fig. 2.4*a*, the fluid may be liquid or gas, whereas in Fig. 2.4*b* it must be a liquid. The manometer fluid in Fig. 2.4*b* may, however, be either liquid or gas. An application of the equations of hydrostatics will show that, for either U tube of Fig. 2.4,

$$p_l' = p_r' \tag{2.7}$$

and $$p_l - p_r = p_1 - p_2 \tag{2.8}$$

For Fig. 2.4*a*, Eq. (2.1) gives

$$p_l' = w_p \Delta h + p_l$$

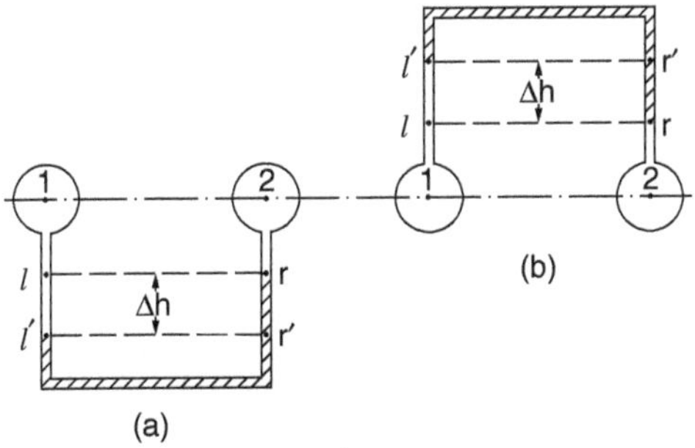

FIGURE 2.4 Differential manometers.

and $$p_r' = w_m \Delta h + p_r$$

From Eq. (2.7),

$$w_p \Delta h + p_l = w_m \Delta_h + p_r$$

or $$p_l - p_r = \Delta h(w_m - w_p)$$

and from Eq. (2.8),

$$p_1 - p_2 = \Delta h(w_m - w_p) \tag{2.9}$$

The corresponding equation for a U tube of the type shown in Fig. 2.4*b* will be

$$p_1 - p_2 = \Delta h(w_p - w_m) \tag{2.10}$$

Equations (2.9) and (2.10) show that the gauge becomes more sensitive as values of w_p and w_m become more nearly equal.

Pressure Forces on Plane Surfaces

In Fig. 2.5, *LM* represents any immersed plane surface which, if extended, would cut the liquid surface at *O* with an angle θ. Considering *O* the origin and *OM* the *y* axis, the pressure force *dP* on an elementary area *dA* is $wh\, dA = wy \sin\theta\, dA$. The total pressure force on the area *LM* is then

$$P = w \sin\theta \int y\,dA \tag{2.11}$$

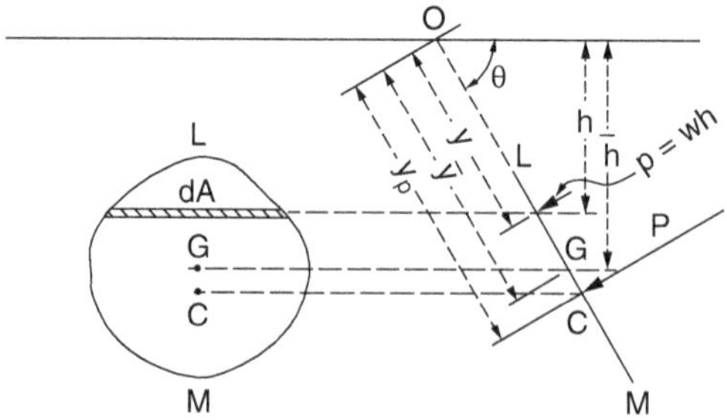

FIGURE 2.5 Pressure forces on plane surfaces.

If $\overline{y}$ is the distance of the center of gravity of the surface from O, then

$$\int y\, dA = A\overline{y}$$

and $$P = w\overline{y} \sin\theta\, A$$

or letting $\overline{h} = \overline{y} \sin\theta$,

$$p = w\overline{h}A \tag{2.12}$$

The *center of pressure* is the point on the immersed surface through which the resultant pressure acts. Because the intensity of pressure increases with the depth, the center of pressure is below the center of gravity. The distance y_p from an axis through O, formed by the intersection of the plane surface LM with the water surface, to the center of pressure may be found from a summation of the moments of the elementary forces about this axis as follows:

$$y_p = \frac{\int y\, dP}{P} \tag{2.13}$$

which from the preceding may be written

$$y_p = \frac{w \sin\theta}{w \sin\theta} \frac{\int y^2\, dA}{\int y dA} = \frac{fy^2\, da}{fy da} \tag{2.14}$$

The numerator of the fraction in Eq. (2.14) is the moment of inertia of the surface LM about the axis through O. This moment of inertia is equal to $Ak^2 + A\overline{y}^2$, where k is the radius of gyration of the surface. The denominator is equal to $A\overline{y}$. Substituting these values in Eq. (2.14) and reducing,

$$y_p = \overline{y} + \frac{k^2}{\overline{y}} \tag{2.15}$$

For surfaces such as are shown in Fig. 2.6, where the locus of the midpoints of horizontal lines is a straight line, the center of pressure lies on this straight line. For other surfaces, the horizontal position of the center of pressure can be obtained by taking moments as in the foregoing about an axis perpendicular to the one through O. The use of these equations is illustrated in the following examples. Problems involving fluid pressures on plane surfaces may also be solved more directly by considering the surface to be a slab with a nonuniform load

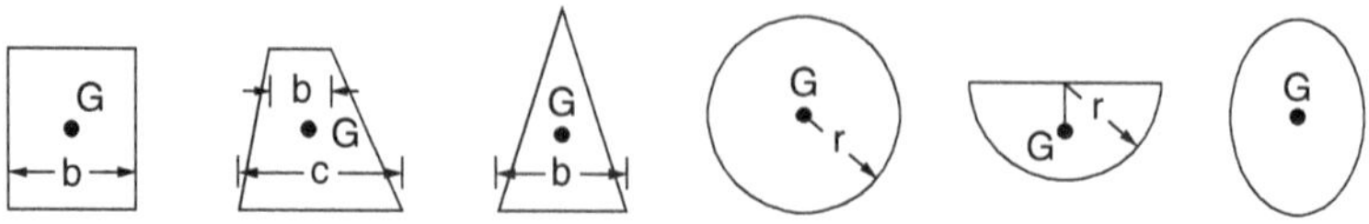

FIGURE 2.6 Center of pressure on plane surfaces.

equal in magnitude to the volume of the pressure-intensity diagram. Examples 2.7 and 2.8 give numerical applications of this method.

Example 2.7 Determine the magnitude and location of the resultant water-pressure force on a 1-m section of the gate *ab*, shown in Fig. 2.7, (1) by the use of Eqs. (2.12) and (2.15) and (2) by loading the gate with the pressure volumes.

1. $P_l = w\bar{h}A = 9.8 \times 6 \times 6 = 352.8$ kN

$$y_{pl} = \frac{9.8 \times 3 \times 6 \times 6 + 9.8 \times (6/2) \times 6 \times 7}{9.8 \times 6 \times 6}$$

$$= 3.0 + 7.0/2 = 6.5 \text{ m}$$

$$P_r = w\bar{h}A = 9.8 \times 3 \times 6 = 176.4 \text{ kN}$$

$$y_{pr} = \frac{9.8 \times 3 \times 6 \times 4}{9.8 \times 3 \times 6} = 4.0 \text{ m}$$

Note that the latter value might have been determined by inspection because the force on the right side of *ab* results from a triangular pressure distribution, the center of gravity of which would therefore be two-thirds of its altitude from the water surface, namely,

$$y_{pr} = 2/3 \times 6 = 4 \text{ m}$$

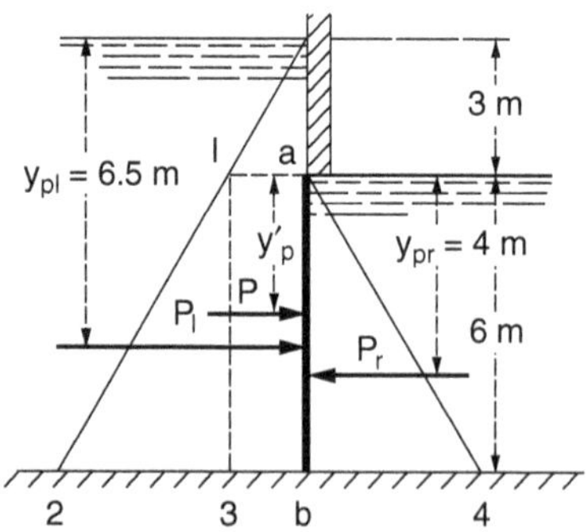

FIGURE 2.7 Pressure forces on plane surfaces.

The resultant force P is then obtained by subtracting P_r from P_l as follows:

$$P = 352.8 - 176.4 = 176.4 = 176.4 \text{ kN (to the right)}$$

The location of P may be found by taking moments about a,

$$y'_p\, 176.4 = 3.5 \times 352.8 - 4 \times 176.4$$

and $$y'_p = 3 \text{ m}$$

2. Inspection of Fig. 2.7 will show that the force resulting from the triangular pressure distribution 123 on the left is exactly balanced by the force on the right resulting from the pressure diagram *ab*4. Consequently the only unbalanced pressure is a uniform intensity of $3w$ N/m² on the left, as shown by the pressure diagram 13*ba*. Therefore,

$$P = 9.8 \times 3 \times 6 = 176.4 \text{ kN}$$

Because P is the resultant of a uniform pressure distribution, it must act on the center of gravity of the gate, that is, 3 m from a. ▲

Example 2.8 Determine the magnitude and location of the pressure force on the rectangular sloping gate shown in Fig. 2.8.

$$P = whA = 62.4 \times 8 \times (6 \times 10) = 30{,}000 \text{ lb}$$

$$y_{ab} = ((4 \times 5) + (4 \times 2/3 * 10))/8 = 5.85 \text{ ft} \quad ▲$$

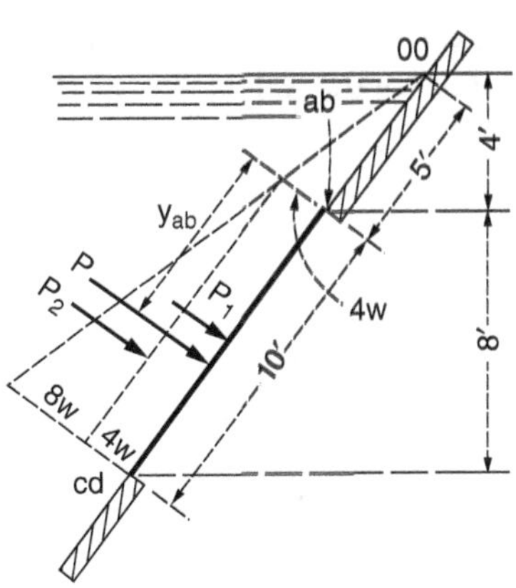

FIGURE 2.8 Pressure forces on plane surfaces.

Pressure Forces on Curved Surfaces

The resultant pressure force on a curved surface is made up of the sum of the small elements of force $p\,dA$, each acting perpendicular to the surface. The magnitude and the location of the resultant of these elementary forces are not easily determined by the methods used for plane surfaces. The horizontal and vertical components of the resultant, however, may be readily determined and their vectors combined.

Consider the forces acting on the prism of liquid shown in Fig. 2.9 bounded by the liquid surface $a0$, the vertical plane surface $0b$, and the curved surface ab. Acting downward on this volume is its weight W, and acting from right to left on $0b$ is the horizontal force $P_H = w\bar{h}A$, where A is the area of the imaginary vertical plane surface one edge of which is $0b$. These forces must be put into equilibrium by equal and opposite forces from the curved surface ab. It follows then that the horizontal component of the pressure force on a curved surface is equal to, and acts at the same point as, the force on a vertical plane surface formed by projecting the curved surface horizontally. Furthermore, the vertical component of the force on a curved surface is equal to the weight of the liquid above the surface and acts at its center of gravity. Similar reasoning will show that when the liquid is beneath the curved surface, the vertical component is equal to the weight of the imaginary volume of liquid above the surface and acts upward through its center of gravity. For example, the vertical component of the pressure force on the tainter gate of Fig. 2.10 is equal to the volume represented by LNM and acts upward through G as shown.

Example 2.9 Figure 2.11 shows a section through a water tank 20 m long. The wall of the tank abc is hinged at c and supported by a horizontal tie rod at a. The segment of the wall bc is a quarter circle having a 4-m radius.

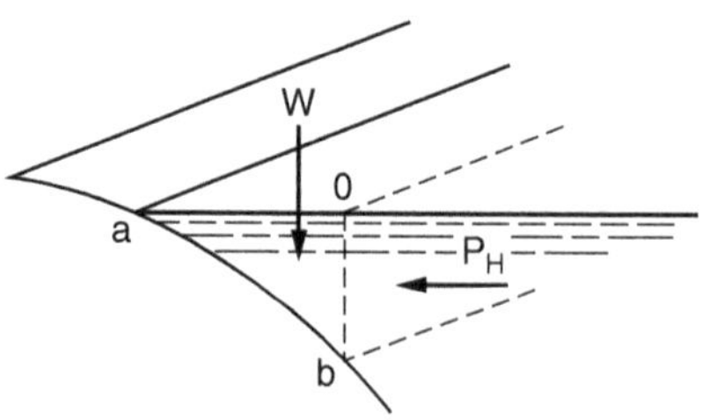

Figure 2.9 Pressure forces on curved surfaces.

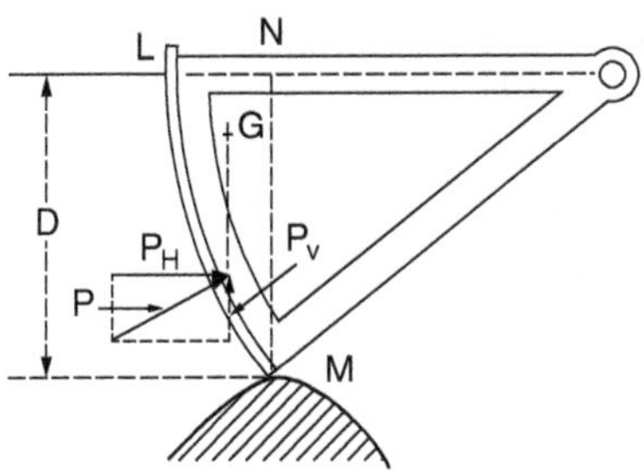

Figure 2.10 Pressure force on tainter gate.

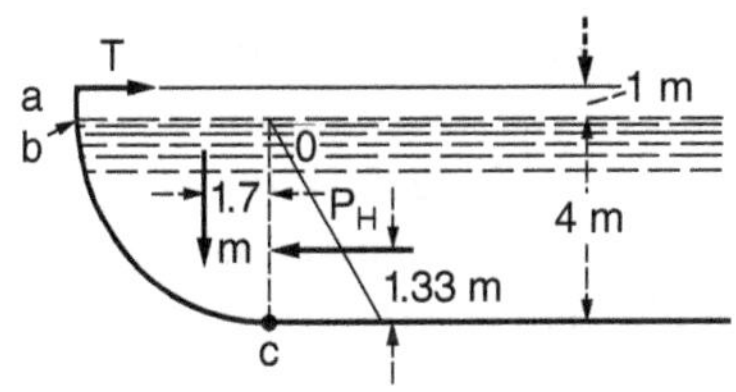

Figure 2.11 Pressure force on curved surfaces.

1. Determine the force T in the tie rod.

$$P_H = w\bar{h}A = 9.8 \times 2 \times 80 = 1568 \text{ kN}$$

The location of P is $1/3 \times 4 = 1.33$ m above c.

$$P_v = W = \frac{20 \times \pi r^2 \times w}{4} = \frac{20 \times \pi \times 4^2 \times 9.8}{4} = 2462 \text{ kN}$$

P_v acts at the center of gravity of the quarter circle which is to the left of $0c$ by the amount $4r/3\pi = 1.7$ m. T may then be obtained by taking moments about the hinge as follows:

$$5T = 4/3 \times 1568 + 1.7 \times 2462$$

and $$T = 1255 \text{ kN}$$

2. Determine the resultant water-pressure force on the wall.

$$P = \sqrt{1568^2 + 2462^2} = 2919 \text{ kN}$$

The direction and location of P may be found by combining vectors P_v and P_H at their intersection. Since all the elementary components of P are perpendicular to the gate and therefore pass through point O, it follows that P must also pass through O.

3. Determine the resultant force on the hinge, neglecting the weight of the wall.

   From $\Sigma H = 0$, the horizontal component is found to be 313 kN and from $\Sigma V = 0$, the vertical component is found to be 2462 kN. The resultant force on the hinge is therefore

$$\sqrt{2462^2 + 313^2} = 2482 \text{ kN} \quad \blacktriangle$$

Uniform Pressure on Cylindrical Surfaces

Figure 2.12 represents a cross section of a pipe or cylinder subjected to a uniform internal hydrostatic pressure, and Fig. 2.13 represents a similar cross section subjected to a uniform external pressure. The pressure at each point on the circumference is normal to the surface, as indicated by the arrows. The resultants of these normal pressures, on opposite sides of any diameter, are equal and in opposite directions and cause a tangential stress where

$$S = \frac{\gamma h d}{2} \tag{2.16}$$

S is tension for internal pressure and compression for external pressure.

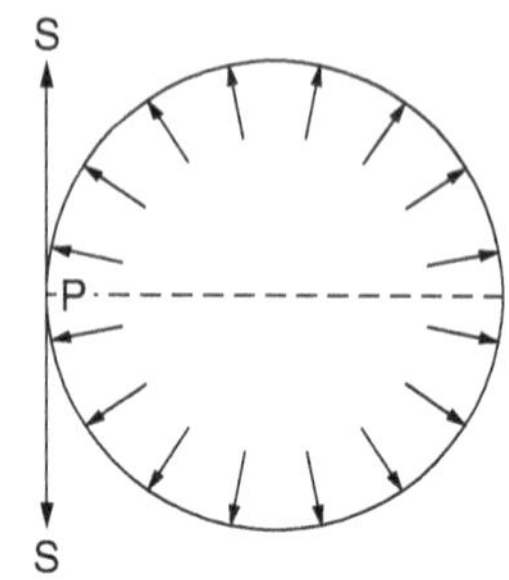

FIGURE 2.12 Pressure force on cylindrical surfaces.

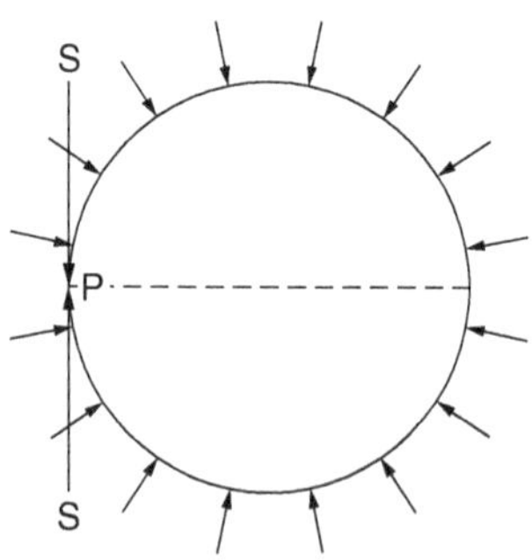

FIGURE 2.13 Pressure force on cylindrical surfaces.

If S is the stress in pounds per inch length, h the static head of water in feet, and d_i the diameter of the pipe in inches, Eq. (2.16) becomes

$$S = 0.217 h d_i$$

If S is the stress in newtons per meter length, h the static head of water in meters, and d_i the diameter of the pipe in meters, Eq. (2.16) becomes

$$S = 4900 h d_i$$

Equation (2.16) may be used for computing the tension in pipes, where h (the head to the center of the pipe) is large compared with the diameter. It may also be used for cylindrical tanks having a vertical axis and for thin circular arch dams. This formula applies to a segment of a cylinder, provided the edges are rigidly supported.

Uniform Pressure on Spherical Surfaces

The stress on the surface of a sphere of diameter d_i subjected to uniform hydrostatic pressure h is

$$S = \frac{\gamma h d_i}{4} \tag{2.17}$$

S will be tension when the pressure is applied to the inner surface and compression when applied to the outer surface.

If S is the stress in pounds per inch length on the surface of a sphere subjected to uniform hydrostatic pressure, h the static head in feet, and d_i the diameter of the sphere in inches, Eq. (2.17) becomes

$$S = 0.108 h d_i$$

If S is the stress in newtons per meter length on the surface of a sphere subjected to uniform hydrostatic pressure, h the static head in meters, and d_i the diameter of the sphere in meters, Eq. (2.17) becomes

$$S = 2450 h d_i$$

Pressures on Spillway Sections of Dams

If still water having a depth $H + D$ were acting on the upstream face of the overflow spillway shown in Fig. 2.14, the pressure distribution could be represented by the trapezoid 1234. The effect of the velocity of the water is to increase the pressure near the bottom and

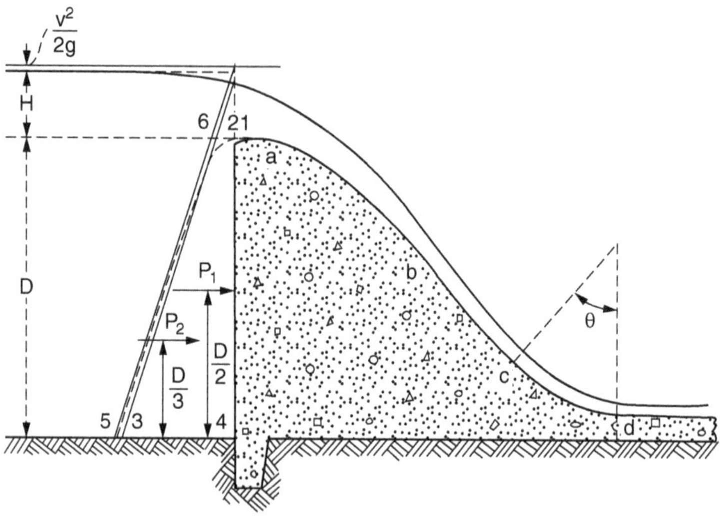

FIGURE 2.14 Pressures on spillways.

decrease it at the top, so that the actual pressure distribution is similar to that shown by the dashed curve 15. The increase in pressure near the bottom results from the impact of the approaching water, and the decrease near the top results from the fact that the pressure at the crest is approximately atmospheric. In computing the stability of such a section it is safe to assume that the trapezoidal pressure distribution 1234 exists. Based on these assumptions, the forces on the upstream face of the spillway are

$$P_1 = wDH \tag{2.18}$$

$$P_2 = \frac{wD^2}{2} \tag{2.19}$$

where P_1 and P_2 are in force per unit length. The locations of P_1 and P_2 are shown in Fig. 2.14.

In determining the stability of an overflow spillway section, it is also necessary to consider pressure forces acting on the downstream face. If the spillway slopes downward too rapidly, there is a tendency for pressures less than atmospheric to occur in the region *ab* in Fig. 2.14, thus adding to the overturning moment on the section. In region *cd*, a positive pressure is produced because of the impact of the jet as it is deflected through the angle θ. The downward component of this force occurs near the location of the maximum foundation stresses, which may make it an important factor in the design. The sketch shown in

Fig. 2.14 is based on the assumption that the hydraulic jump occurs downstream from the spillway. Where this is not the case, the tail-water pressure on the spillway must also be considered. Methods of determining the location of the hydraulic jump are described in Sec. 8.

Uplift Pressure on Dams

The water pressure under dams varies from a maximum at the upstream edge (b in Fig. 2.15) to a minimum at the downstream edge (f in Fig. 2.15). If there are no cutoff walls or impermeable layers in the foundation, the pressure at b is wD_u, and that at f is wD_d. For homogeneous isotropic foundation materials the pressure will vary from b to f approximately as the straight line de. This is because the energy loss is approximately the same for each meter of distance traveled by the water. The exact form of the pressure line even for this case varies slightly from a straight line and may be determined more precisely by constructing a flow net,[3,4] by the use of hydraulic models, or by means of an electrical analogy.[5] If the downstream edge of the dam were

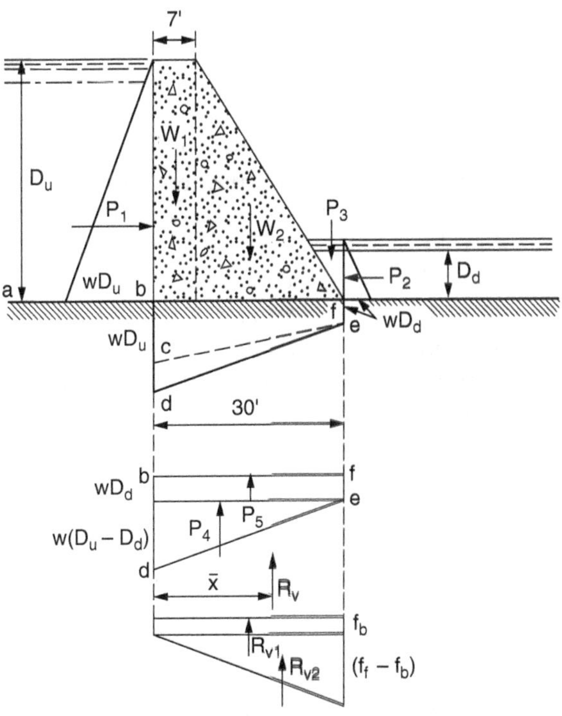

FIGURE 2.15 Forces on gravity dams.

extended beyond *f*, as by means of an impermeable apron, the uplift pressures would be increased because the drop in pressure from wD_u to wD_d would be extended over a longer distance. On the other hand, the construction of an impervious cutoff wall vertically downward at point *b* would reduce the uplift pressures, as illustrated schematically by the dashed line *ce*. This is because part of the total drop in pressure from point *b* to point *f* occurs as the water travels around the cutoff wall. A reduction in uplift pressure might also be accomplished by placing a layer of impermeable material in the region *ab*. This would cause most of the underflow to enter the ground upstream from point *a*, thus transferring some energy loss to upstream of point *b*. Measurements of uplift pressures under various conditions have been reported.[4,6,7] Much material of interest on this subject is also given in an article by Harza[8] and the accompanying discussions.

Stability of Dams

Gravity dams must be designed so that they are safe against overturning and sliding. If the materials in the dam and foundation were capable of withstanding unlimited stresses, the dam would be safe against overturning provided the resultant of all forces fell within the base of the dam. However, in order to limit the magnitude of the stresses and to ensure that the dam will maintain firm contact with the foundation over the entire base, it is required that the resultant fall within the middle third of the base. This requirement does not necessarily assure safety against overturning because the stresses at the downstream edge may still be sufficiently large to cause failure of the material in the dam or in the foundation. If, for example, the resultant falls at the downstream end of the middle third of the base, the stresses will vary from a maximum at the downstream edge to zero at the upstream edge, assuming a straight-line variation in stresses. The computations for a simple gravity dam are illustrated by the following example.

Example 2.10 In the gravity section shown in Fig. 2.15, $D_u = 38$ ft, $D_d = 9$ ft, and the other dimensions are as shown.

The magnitudes of the forces per foot of length are

$$P_1 = \frac{wD_u^2}{2} = 45{,}000 \text{ lb}$$

$$P_2 = \frac{wD_d^2}{2} = 2530 \text{ lb}$$

$$P_3 = wD_d\left(\frac{23}{38}\right)D_d\left(\frac{1}{2}\right) = 1530\text{ lb}$$

$$P_4 = w\frac{D_u - D_d}{2}\,30 = 27{,}100\text{ lb}$$

$$P_5 = wD_d \times 30 = 16{,}850\text{ lb}$$

$$W_1 = 150 \times 7 \times 38 = 39{,}900\text{ lb}$$

$$W_2 = 150 \times \frac{23 \times 38}{2} = 65{,}500\text{ lb}$$

The resultant vertical force is determined as follows:

$$R_v = W_1 + W_2 + P_3 - P_4 - P_5 = 63{,}000\text{ lb}$$

The summation of the moments about point b is

1. Clockwise:

$$P_1 \times \frac{38}{3} = 570{,}000$$

$$P_3\left(30 - \frac{1}{3} \times \frac{9}{38} \times 23\right) = 43{,}000$$

$$W_1 \times \frac{7}{2} = 140{,}000$$

$$W_2\left(7 + \frac{23}{3}\right) = 960{,}000$$

$$\text{Total} = 1{,}713{,}000$$

2. Counterclockwise:

$$P_2 \times \frac{9}{3} = 7600$$

$$P_4 \times \frac{30}{3} = 271{,}000$$

$$P_5 \times \frac{30}{2} = 253{,}000$$

$$\text{Total} = 531{,}600$$

Then $\bar{x} = \Sigma M/R_v = 18.8$ ft.

The approximate variation in foundation stresses may then be obtained as follows (see Fig. 2.15):

$$R_{v_1} = 30f_b \qquad \text{and} \qquad R_{v_2} = 30\frac{f_f - f_b}{2}$$

where f_b and f_f are the foundation stresses at b and f, respectively. Also,

$$R_{v1} + R_{v2} = R_v = 63{,}000 \text{ lb}$$

The summation of moments at point b yields the following equation:

$$15R_{v1} + 20R_{v2} = 18.8 \times 63{,}000$$

Solving these equations gives

$$f_b = 510 \text{ lb/ft}^2$$
$$f_f = 3700 \text{ lb/ft}^2$$

Safety against sliding requires that

$$R_H = C_f R_v \tag{2.20}$$

where R_H is the summation of the horizontal forces and C_f is the coefficient of friction between the dam and the foundation. Creager et al.[4] give C_f as 0.60 to 0.75 for masonry on masonry and masonry on good rock. For masonry on gravel, sand, and clay, they suggest 0.50, 0.40, and 0.30, respectively.

In this example, the required value of C_f from Eq. (2.20) would be

$$C_f = \frac{R_H}{R_v} = \frac{P_1 - P_2}{63{,}000} = \frac{42{,}470}{63{,}000} = 0.675$$

which would probably be satisfactory on rock but not on unconsolidated materials. However, this design might still be made safe against sliding by extending a concrete cutoff wall downward, as in Fig. 2.15. The principal effects of the cutoff wall are the establishment of the shear plane within the foundation material itself and the reduction of the hydrostatic uplift. For granular materials, the value of C_f is greater for a plane within the material itself than for a plane between the surface of the dam and the material. Furthermore, the reduction of the uplift pressures increases the value of R_v. Therefore, the use of a cutoff wall will increase R_H by increasing both C_f and R_v in Eq. (2.20). ▲

References

1. For a detailed discussion of pressure variations in gases see L. Prandtl and O. J. Tietjens, *Fundamentals of Hydro- and Aeromechanics*, McGraw-Hill, New York, pp. 29–35, 1934.
2. *Handbook of Physics and Chemistry*, 12th ed., p. 1269.
3. A. Casagrande, "Seepage through Dams," *J. New Engl. Water Works Assoc.*, vol. LI., no. 2, pp. 131–172, Jun. 1937.

4. W. P. Creager, J. D. Justin, and J. Hinds, *Engineering for Dams*, Wiley, New York, 1950.
5. E. W. Lane, "Flow Net and Electric Analogy," *Civil Eng. (N.Y.)*, Oct. 1934.
6. J. Hinds, "Upward Pressure under Dams," *Proc. ASCE*, Mar. 1938.
7. I. E. Hauck, "Uplift Pressure in Masonry Dams," *Civil Eng. (N.Y.)*, Sep. 1932.
8. L. F. Harza, "The Significance of Pore Pressure in Hydraulic Structures," *Trans. ASCE*, vol. 114, 1949.

SECTION 3

Fundamental Concepts of Fluid Flow

This section is included for the purpose of providing a review of the basic principles required to solve problems involving fluids in motion. The title of the section might well be *hydrokinetics* to distinguish it from the previous section, which dealt with hydrostatics.

Virtually no problems in hydraulics can be safely solved by means of tables or formulas without a firm understanding of continuity and energy. Other problems require the application of force and momentum concepts. These basic principles will be briefly set forth in this section along with brief discussions of flow classification, the Bernoulli equation, kinetic energy and momentum correction factors, and applications of the Bernoulli equation to siphons, venturi meters, and cavitation. Two other basic topics presented in this section are dimensional analysis and similitude.

Classification of Flow

In order to determine which basic principles are applicable to particular problems, it is necessary to classify flow. The various categories of flow lend themselves to presentation as a series of four pairs of alternative conditions.

The first pair of conditions to be discussed is *laminar flow* as distinguished from *turbulent flow*. In laminar flow, the fluid moves in parallel layers with no crosscurrents. In the case of a circular pipe, for example, the layers are circular cylindrical tubes. The fact that, in laminar flow, fluid particles move in parallel paths was demonstrated by Reynolds,[1] who introduced a thread of dye into flowing water.

Turbulent flow is characterized by pulsatory crosscurrent velocities. One result of the crosscurrent velocities of turbulent flow is the formation of a more uniform velocity distribution. It is shown in Sec. 6 that for laminar flow in a circular pipe the maximum velocity is twice the average velocity. For turbulent flow, the maximum velocity is more on the order of 1.25 times the average velocity. The more uniform velocity found in turbulent flow is brought about by the interchange of momentum between fast-moving particles near the center and slower ones nearer the walls. One of the most important practical differences between laminar and turbulent flow is the much greater energy loss in turbulent flow. The energy required to produce the crosscurrent velocities must be supplied by the pump, but it is of no help in transporting the fluid through the pipe.

The criterion that distinguishes laminar flow from turbulent was also developed by Reynolds[1] and has been named the Reynolds number R. For circular pipes flowing full, R is expressed as follows:

$$R = \frac{d\upsilon\rho}{\mu} = \frac{d\upsilon}{\nu} \tag{3.1}$$

where d is the diameter of the pipe, u the average velocity, p the density, μ the viscosity, and υ the kinematic viscosity. For pipes, laminar flow occurs when $R < 2000$ and flow is usually turbulent when $R > 4000$. For values of R between 2000 and 4000, flow is in a transition condition. The development of a Reynolds number for cases other than a circular pipe flowing full requires the use of different length terms. For example, for flow in open channels the length term may be the depth, the hydraulic radius, or one-fourth of the hydraulic radius. These variations in R are discussed in the various sections of this handbook where the situations arise.

The physical significance of R can be understood intuitively from an examination of the terms in Eq. (3.1). For example, it is easy to visualize that a decrease in diameter will inhibit crosscurrents and, therefore, that R should also decrease. However, perhaps the best method of understanding the physical significance of this parameter is to note that R can be expressed as the ratio of an inertial force to a viscous force as demonstrated later in this section. The inertial force represents the tendency of the fluid to develop turbulence at boundary irregularities, whereas the viscous force tends to damp out turbulence.

The numerical use of Eq. (3.1) will be demonstrated in the following example. The example illustrates the fact that fully laminar flow rarely exists in pipes larger than 2.5 cm when the fluid is water.

Example 3.1 Determine the value of the Reynolds number for a circular pipe having an inside diameter of 2.5 cm when the velocity is 8 cm/s and the water temperature is 21°C.

From Table 1.2,

$$w = 9787 \text{ N/m}^3$$
$$\mu = 0.00098 \text{ N·s/m}^2$$
$$\nu = 0.984 \times 10^{-6} \text{ m}^2\text{/s}$$

Then

$$\rho = \frac{w}{g} 997.7 \text{ kg/m}^3$$

and
$$R = \frac{dv\rho}{\mu} = \frac{0.025 \times 0.08 \times 997.7}{0.00098} = 2036$$

or
$$R = \frac{dv}{\nu} = \frac{0.025 \times 0.08}{0.984 \times 10^{-6}} = 2036 \quad ▲$$

Another pair of categories into which flow may be classified deals with steadiness of the discharge. Flow is classified as *steady flow* when the discharge is constant with respect to time. *Unsteady flow* occurs in pipes or open channels during and following changes in valve or gate openings. Typical unsteady flow problems involve such phenomena as water hammer in pipes, translatory waves in open channels, and the emptying of a tank. Another example of unsteady flow is the discharge of natural rivers, although during periods when there is no surface runoff[a] the change in discharge is so slow that many problems may be solved by assuming that flow is steady.

A special case of flow should be mentioned here in which the discharge varies with distance along the channel. This type of flow is called *spatially variable flow*, and it may be either steady or unsteady. Some examples of spatially variable flow are overland flow during rainstorms, flow in natural streams, and flow in side-channel spillways.

Flow may also be classified as *uniform flow* as differentiated from *nonuniform flow*. Flow is uniform if the cross-sectional area of the fluid remains constant. The term "varied flow" is frequently used as a synonym for nonuniform flow. The further distinction between *rapidly varied flow* and *gradually varied flow* are discussed in detail

---

[a]For a detailed discussion of flow in natural rivers see Wisler and Brater.[2]

in Sec. 8. Some examples of nonuniform flow are flow in venturi meters, flow in natural streams, and flow above and below gates in open channels.

Continuity

One of the fundamental concepts that must be satisfied in all flow problems is continuity of flow. Continuity states the fact that no fluid is lost or gained and no cavities are formed or destroyed as the fluid passes through a conduit. When the fluid is essentially incompressible, as for liquids, and in many cases also for gases, continuity may be expressed as follows:

$$Q = a_1V_1 = a_2V_2 = a_nV_n \tag{3.2}$$

where Q is the discharge, a the cross-sectional area, and V the average velocity. Equation (3.2) may be referred to the venturi meter in Fig. 3.4.

When the fluid is compressible, the densities at the various locations must be included, and the continuity equation becomes

$$\rho Q = \rho_1a_1V_1 = \rho_2a_2V_2 = \rho_na_nV_n \tag{3.3}$$

where ρQ represents the mass discharge in kilograms per second and the values of ρ are the densities at the various sections.

For spatially variable flow continuity may be expressed as follows for incompressible fluids:

$$a_1V_1 \pm q\,\Delta x = a_2V_2 \tag{3.4}$$

where q is the inflow or outflow per meter length of conduit, and Δx is the length of the conduit between sections 1 and 2.

Energy and the Bernoulli Equation

The basic principle most often used in hydraulics is the law of conservation of energy as expressed by the Bernoulli equation. All pipe and open-channel formulas have been developed to express the rate of energy dissipation based on the Bernoulli formula. Fundamental derivations of Bernoulli's equation are given in all standard texts.[3] A simplified derivation will be made here which illustrates the meaning of the terms.

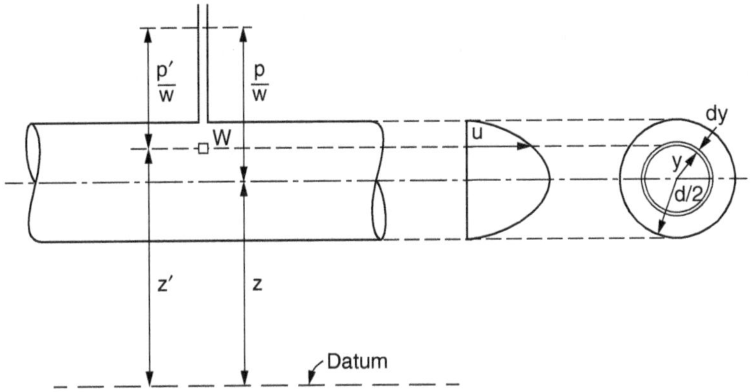

FIGURE 3.1 Energy of fluids in pipes.

If a particle of fluid of weight W is selected anywhere in the cross section of a pipe as in Fig. 3.1, its energy may be expressed as follows:

Potential energy due to elevation $= Wz'$

Potential energy due to pressure $= W\dfrac{p'}{w}$

Kinetic energy $= W\dfrac{u^2}{2g}$

Then the total energy of the particle W is

$$E = Wz' + W\frac{p'}{w} + W\frac{u^2}{2g} \tag{3.5}$$

It should be noted that the potential energy due to pressure is available only if the fluid is continuously supplied at the pressure p'. If Eq. (3.5) is divided by W, the following expression for energy at the location of W is obtained:

$$H' = z' + \frac{p'}{w} + \frac{u^2}{2g} \tag{3.5a}$$

This equation is not useful unless it can be generalized so that it applies to the entire cross section. The first two terms may be generalized by noting, from Fig. 3.1, that the potential energy can be readily converted between elevation and pressure,

$$z' + \frac{p'}{w} = z + \frac{p}{w} \tag{3.6}$$

in which the terms without superscripts refer to the center of the pipe.

The generalization of the kinetic-energy term is accomplished by introducing the average velocity V in place of u together with a kinetic-energy correction factor, which is discussed in detail later in this section.

The expression for the energy at any point in the cross section is then

$$H = z + \frac{p}{w} + \alpha\frac{V^2}{2g} \tag{3.7}$$

This expression is called the Bernoulli constant. Another useful form is obtained by multiplying Eq. (3.7) by w to obtain the energy as follows:

$$wH = wz + p + \alpha w\frac{V^2}{2g} \tag{3.8}$$

or, replacing w with ρg,

$$\rho gH = \rho gz + p + \alpha\frac{\rho V^2}{2} \tag{3.9}$$

A similar derivation can be made for an open channel. Reference to Fig. 3.2 will show that Eq. (3.5a) applies to open channels as well as to pipes. The first two terms can be generalized by noting that

$$z' + \frac{p'}{w} = z + D \tag{3.10}$$

Then, introducing V and a, as for closed conduits, the following expression for the Bernoulli constant is obtained:

$$H = z + D + \alpha\frac{V^2}{2g} \tag{3.11}$$

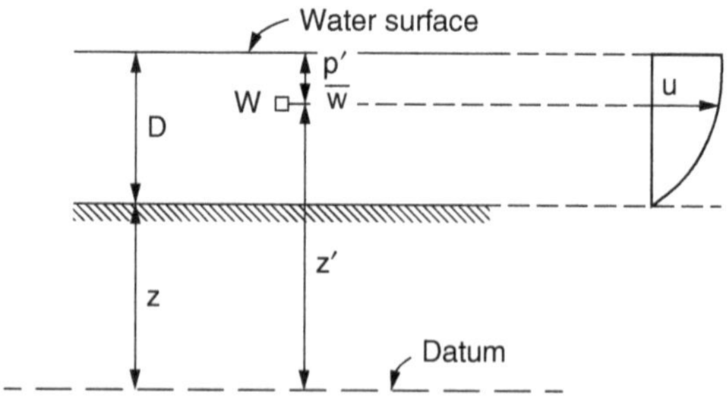

FIGURE 3.2 Energy of liquids in open channels.

The meaning of α, the *kinetic-energy correction factor*, can be demonstrated by writing an expression for the rate at which kinetic energy passes any section in a pipe as follows:

$$Qw\alpha \frac{V^2}{2g} = w \int_0^a u \, da \frac{u^2}{2g} \tag{3.12}$$

The term on the right is an exact expression for the total kinetic energy. In this term, da would be the area of an elemental ring in which the velocity is u, as shown in Fig. 3.1. The term on the left includes the total discharge Q and the average velocity V. If the variation of u over the area a can be expressed mathematically, the value of α can be determined exactly from Eq. (3.12). When the shape of the cross section and the velocity variation are too complicated for expression in mathematical terms, a graphic solution may be made.

One case for which a mathematical solution for α can be made is that of laminar flow in a circular pipe. In that case, the value of u is [see Eq. (6.13)]

$$u = \frac{p_1 - p_2}{4\mu l}\left(-y^2 + \frac{d^2}{4}\right) \tag{3.13}$$

where p_1 and p_2 are the pressures at two ends of a section of the pipe l meters long, d is the diameter, and y is defined in Fig. 3.1. For this case,

$$da = 2\pi y \, dy \tag{3.14}$$

The values of Q and V may be obtained from Eqs. (6.15) and (6.16). Inserting these values of Q, V, u, and da in Eq. (3.12) and integrating yields $\alpha = 2$. For turbulent flow in pipes, α usually is in the range from 1.01 to 1.10.[4] The value of α for laminar flow in open channels is 54/35. It may be determined in the same manner for pipes by making use of Eqs. (7.23), (7.25), and (7.26). For most practical problems α may be taken as unity without serious error. However, in some situations where high precision is desired, exact values of α should be used. This is especially true where α differs considerably from one point to another, as in a venturi meter.

A graphic method for deriving α in cases in which a direct mathematical solution cannot be obtained has been presented by O'Brien and Johnson.[5] The procedure consists in plotting lines of uniform velocity on the cross section, after which subdivisions of area between the lines Δa are determined and the right side of Eq. (3.12) is evaluated by adding values of $U_{av}^3 \Delta a$. The value of U_{av} is the average of the velocities bounding each subdivision of area.

Equation (3.7) can be written as follows between two points in along the pipe:

$$z_1 + \frac{p_1}{w} + \frac{p_1}{w} + \alpha_1 \frac{V_1^2}{2g} = z_2 + \frac{p_2}{w} + \alpha_2 \frac{V_2^2}{2g} + \Sigma h_l \tag{3.15}$$

where Σh_l is an energy loss between the two points. This equation states that the energy at one point in a fluid (point 1) is equal to the energy at any downstream location (point 2) plus the intervening losses Σh_l. The use of Bernoulli's equation will be demonstrated in various sections of this handbook. However, several applications are presented here to illustrate its use for basic devices such as siphons and venturi meters and for the study of cavitation.

Siphons

Figure 3.3 shows a siphon composed of three straight sections of pipe connected by elbows at points 3 and 4. If it were desired to determine the discharge through this siphon for given values of z and pipe sizes, the Bernoulli equation would be written from point 1 to point 5 as follows:

$$z_1 + \frac{p_1}{w} + \alpha_1 \frac{V_1^2}{2g} = z_5 + \frac{p_5}{w} + \alpha_5 \frac{V_5^2}{2g} + \Sigma h_l \tag{3.16}$$

Then the left side reduces to z_1 because p_1/w is zero on the gauge scale and V_1 is so small that it may be neglected. On the right side, z_5 and p_5/w are zero, so that only the kinetic energy term and the loss term remain. The loss term is defined as

$$\Sigma h_l = \begin{cases} \text{loss at entrance} \\ \text{loss at two elbows} \\ \text{loss in straight pipe sections} \end{cases}$$

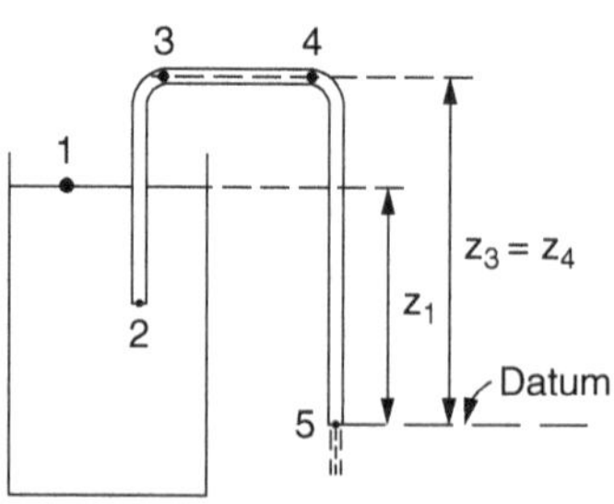

FIGURE 3.3 Application of Bernoulli equation to a siphon.

As will be shown in later sections of this handbook, all the energy losses can be expressed in terms of velocity. Upon inserting these loss terms in Eq. (3.16), the velocity, and hence the discharge, may be computed. Knowing the discharge, the pressure can be determined by writing the Bernoulli equation from point 1 to the point where it is desired to know the pressure, including only the loss between point 1 and the point in question.

Venturi Meters

A venturi meter is a constriction in a pipe that is used for the purpose of determining the discharge. The Bernoulli equation may be written from point 1 to point 2 in Fig. 3.4 as follows:

$$\frac{V_1^2}{2g} + \frac{p_1}{w} = \frac{V_2^2}{2g} + \frac{p_2}{w} + h_l \tag{3.17}$$

where V_1 is the average velocity in the pipe at 1, p_1 is the pressure at 1, V_2 and p_2 are the corresponding quantities at 2, and h_l is the energy loss between points 1 and 2. Using the continuity relationship

$$V_1 = \frac{d_2^2}{d_1^2} V_2$$

and rearranging terms in Eq. (3.17) the following expression is obtained:

$$V_2 = \sqrt{\frac{2g(p_1/w - p_2/w - h_l)}{1 - (d_2/d_1)^4}} \tag{3.18}$$

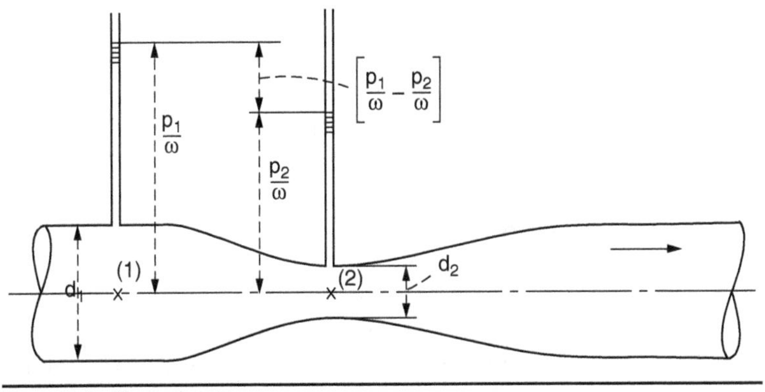

Figure 3.4 Venturi meter.

In the preceding expressions d_1 and d_2 are the diameters of the pipe and the throat of the venturi meter, respectively. The energy loss is usually taken care of by using a coefficient c having a value less than 1 and having a unique value for each meter, rather than by means of a direct subtraction, as in Eqs. (3.17) and (3.18). If this change is made and if both sides of Eq. (3.18) are multiplied by the throat area a_2, the following expression for discharge is obtained:

$$Q = ca_2\sqrt{\frac{2g(p_1/w - p_2/w)}{1-(d_2/d_1)^4}} \tag{3.19}$$

The value of the term $p_1/w - p_2/w$ must be obtained by means of a differential pressure gauge. Pressure gauges are discussed in Sec. 2 (see Fig. 2.4). Values of c, together with other information regarding the use of the venturi meters, are given in Sec. 12 under Venturi Meter.

Cavitation

Another illustration of the energy and continuity concepts will be made by an application to cavitation. If conditions are such that the pressure at some point in a conduit becomes very low because of a local increase in velocity or elevation, such as at the throat of a venturi meter or the highest point in a siphon, then cavitation may occur. Cavitation will begin when the pressure at any location reaches an absolute pressure equal to the saturated vapor pressure of the liquid (see Fluid Properties, Sec. 1). Usually, this condition is attained by increasing the discharge to the critical condition. If the discharge is increased beyond this critical condition, cavities filled with liquid vapor will alternately develop and collapse in the low-pressure region, often causing damage to the apparatus and always reducing the efficiency of flow greatly.

For the venturi meter, shown in Fig. 3.4, the discharge at which cavitation will begin for a given value of p_1 may be computed by using Eq. (3.17) in conjunction with the continuity equation. Absolute pressures (see Absolute Pressure, Sec. 2) must be used rather than gauge pressures, and p_2 is the vapor pressure corresponding to the temperature of the water. Values of vapor pressure for water are given in Table 1.2. The energy loss may be estimated as $0.05\ V_2^2/2g$ (see Minor Losses, Sec. 6).

A similar situation might occur at point 4 in the siphon, shown in Fig. 3.3. An increase in z_4 and/or z_1 would tend to reduce the pressure at point 4 to an absolute value near vapor pressure. The application of the Bernoulli equation would provide information about limiting conditions, although it is possible that the pipe from point 4 to point 5 may cease to flow full before this limiting condition is reached.

Momentum Concepts

Another fundamental principle involved in many hydraulics problems involves a change in momentum and is expressed by Newton's second law of motion. For the case of steady incompressible flow, the law may be stated in the following forms:

$$F_x = \frac{M_2 V_{2x} - M_1 V_{1x}}{\Delta t} \tag{3.20}$$

$$= \frac{M_1 (V_{2x} - V_{1x})}{\Delta t} + \frac{\Delta M V_{2x}}{\Delta t} \tag{3.21}$$

where F_x is the force in the x direction required to change the velocity of a mass M_1 from V_{1x} to V_{2x} in a time Δt and to change the velocity of any mass added ΔM between points 1 and 2 in the time Δt from zero to V_{2x}. Forces and velocities directed from left to right are considered positive. Then the force in Eq. (3.20) or (3.21) is the force acting on the fluid. For a flowing fluid Eq. (3.21) can be transformed as follows:

$$M_1 = \rho Q \, \Delta t \tag{3.22}$$

$$\Delta M = \rho i \, \Delta s \, \Delta t \tag{3.23}$$

Then

$$F_x = \rho Q (V_{2x} - V_{1x}) + \rho q \, \Delta s \, V_{2x} \tag{3.24}$$

In these equations q is the inflow per meter of length of conduit and Δs the distance along the conduit. The second term on the right side of Eq. (3.24) applies to spatially variable flow and will be given further consideration in Sec. 11. For the case of no addition or withdrawal of fluid, Eq. (3.24) becomes

$$F_x = \rho Q (V_{2x} - V_{1x}) \tag{3.25}$$

For precise work, correction factors are required in the momentum equation for the same reasons that the α factors are needed to correct for the use of the average velocity in the kinetic-energy term in the Bernoulli constant (see Energy and the Bernoulli Equation). In order to illustrate the nature of these factors, called β in this case, Eq. (3.25) may be rewritten as follows:

$$F_x = \rho Q V_{2x} - \rho Q V_{1x} \tag{3.26}$$

Noting from the continuity relation that

$$Q = a_1 V_1 = a_2 V_2$$

Equation (3.26) becomes

$$F_x = \rho a_2 V_{2x}^2 - \rho a_1 V_{1x}^2 \tag{3.27}$$

Then, knowing that Vx represents an average of velocities varying from zero to u_{max} (see Fig. 3.1), it may be seen that a correction factor greater than unity should be included. The value of β may be computed from the following equation, which expresses the rate at which momentum is passing a cross section of a conduit:

$$\rho a \beta V^2 = \rho \int_o^a u^2 \, da \tag{3.28}$$

For the special case of laminar flow in a circular pipe Eq. (3.28) can be integrated after inserting values of u and V from Eqs. (6.14) and (6.17), and the value of β is found to be 4/3. For more complicated velocity variations, a graphic integration is required. The method is identical with that previously described in this section for determining the kinetic-energy correction factor α. For turbulent flow in pipes, values of β vary from approximately 1.01 to 1.05.[4] In most practical problems β may be taken as unity, but where precise work is desirable and possible, it may be necessary to use correct values of β.

It should be reemphasized that Eqs. (3.20) to (3.25) give the force components for a particular direction based on momentum changes in that direction. The x direction was used simply as a convenient illustration of the type of equation that could be written for any direction. It should be further noted that the force F_x is the summation of all forces acting on the fluid, including fluid pressure, the force due to the pressure of the conduit boundaries, friction, and gravity. The use of the momentum principle will be illustrated by the following example. Additional illustrations of this principle are given in other sections of the handbook.

Example 3.2 In the reducing elbow shown in Fig. 3.5 the pressure at point 1 is 50 lb/in.^{2} and V_1 is 5 ft/s. The diameters at points 1 and 2 are 12 and 6 in., respectively. Point 2 is 1.5 ft above point 1, and ϕ is 45°. Determine the horizontal force on the reducing elbow.

The discharge is

$$Q = a_1 V_1 = (\pi/4) \times 1^2 \times 5 = 3.93 \text{ ft}^3/\text{s}$$

Then, from continuity,

$$V_2 = \frac{Q}{a_2} = \frac{3.93}{0.196} = 20 \text{ ft/s}$$

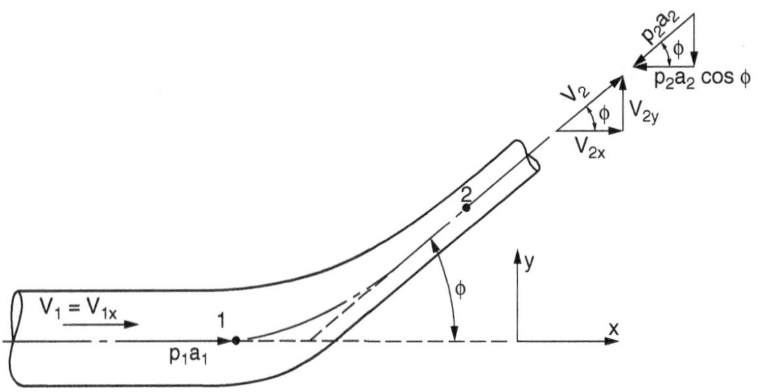

FIGURE 3.5 Momentum, Example 3.2.

The pressure at point 2 is obtained by writing the Bernoulli equation [Eq. (3.15)] from 1 to 2,

$$z_1 + \frac{p_1}{w} + \alpha_1 \frac{V_1^2}{2g} = z_2 + \frac{p_2}{w} + \alpha_2 \frac{V_2^2}{2g} + h_l \tag{3.29}$$

Neglecting losses and letting the values of α be unity, Eq. (3.29) becomes

$$0 + \frac{50x144}{62.4} + \frac{25}{64.4} = 1.5 + \frac{P_2}{62.4} + \frac{400}{64.4}$$

Then
$$P_2 = \frac{62.4 \times 108}{144} = 46.8 \text{ lb/in.}^2$$

Making use of Eq. (3.25),

$$\begin{aligned} F_x &= \rho Q(V_{2x} - V_{1x}) \\ &= \frac{62.4}{32.2} \times 3.93(0.707 \times 20 - 5) \\ &= 69.8 \text{ lb} \end{aligned}$$

Writing the summation of horizontal forces,

$$p_1 a_1 + E_x - p_2 a_2 \cos 45° = 69.8$$

where E_x is the force of the elbow on the water. Then, inserting numerical values,

$$50 \times 144 \times 0.785 + E_x - 6.8 \times 144 \times 0.196 \times 0.707 = 69.8$$

and $E_x = -4654$ lb

This means that the elbow is acting with a force of 4654 lb to the left against the water, or that the stresses in the pipe must resist a horizontal force to the right of this amount unless some external fastening is supplied. ▲

Dimensional Analysis and Similitude

This is a procedure for arranging a number of variables into one or more dimensionless groups.[b] For example, suppose the following variables are thought to be related to a particular flow phenomenon and it is desired to arrange them into a convenient dimensionless group π_1:

Typical length	d
Velocity	V
Density	ρ
Viscosity	μ

Assume now that π_1 is an exponential product of these variables, as shown by the following equation:

$$\pi_1 = d^a V^b \rho^c \mu^{-1} \tag{3.30}$$

The use of the negative exponent for μ is entirely arbitrary. A positive exponent or a symbolic one such as is used for the other three variables could also be employed. Then since π_1 is dimensionless, the following relationship must exist among the dimensions of the variables:

$$L^o M^o T^o = L^a \left(\frac{L}{T}\right)^b \left(\frac{M}{L^3}\right)^c \left(\frac{M}{LT}\right)^{-1} \tag{3.31}$$

The dimensions of the variables were obtained from Table 1.1. Because each of the basic dimensions must be eliminated from the dimensionless group π_1, the following relationships must hold for the exponents:

$$(L)\ 0 = a + b - 3c + 1$$

$$(M)\ 0 = c - 1$$

$$(T)\ 0 = -b + 1$$

Simultaneous solution of these equations yields the following results:

$$b = 1 \qquad c = 1 \qquad a = 1$$

---

[b]Only a single illustration of dimensional analysis is given here. For more elaborate and detailed discussions see Rouse[6] or Longhaor.[7]

Substitution of these values into Eq. (3.30) shows that π_1 must have the following form:

$$\pi_1 = \frac{dV\rho}{\mu} = R \tag{3.32}$$

This dimensionless group of variables is called the Reynolds number.

If a similar dimensionless analysis is made using in succession the gravitational acceleration g, the surface tension σ, and the modulus of elasticity E, the following additional dimensionless parameters are obtained:

$$\pi_2 = \frac{V^2}{gd} = F^2 \tag{3.33}$$

$$\pi_3 = \frac{dV^2\rho}{\sigma} = W \tag{3.34}$$

$$\pi_4 = \frac{\rho V^2}{E} = M^2 \tag{3.35}$$

In Eqs. (3.33) to (3.35) F is the Froude number, W the Weber number, and M the Mach number. The four dimensionless numbers presented in the foregoing equations have a particular significance, which will be further explained in the following paragraphs.

The *principles of similitude* form the basis for designing models so that the results can be converted to prototypes. In order to have similarity, it is first necessary that the model be geometrically similar to the prototype. Then the movement of the fluid in the model will be similar to that in the prototype if the ratios of the inertial force of the fluid to the various other forces acting on the fluid are the same in the model as in the prototype. There are four types of forces acting on the fluid, those due to viscosity, gravitation, surface tension, and elasticity.

The inertial force of a fluid can be expressed in terms of the momentum equation [Eq. (3.27)] as follows:

$$F_i = \rho QV = \rho aV^2 \tag{3.36}$$

A typical viscous force can be expressed as the product of shear stress and area,

$$F_\mu = \tau a \tag{3.37}$$

Noting from Eq. (1.3) that, for laminar flow,

$$\tau = \mu \frac{du}{dy} = \mu \frac{V}{y} \tag{3.38}$$

the friction force becomes

$$F_\mu = \mu \frac{V}{y} a \tag{3.39}$$

The ratio of the inertial force to the viscous force can now be written,

$$\frac{F_i}{F_\mu} = \frac{\rho a V^2 y}{\mu V a} = \frac{y V \rho}{\mu} = R \tag{3.40}$$

The preceding has shown that the ratio F_i/F_μ yields a particular value of the Reynolds number R. Consequently, for similitude insofar as viscous forces are concerned, the Reynolds number must be the same in the model and the prototype. A further implication is that R is a useful parameter for correlating coefficients that serve to take care of energy losses. Examples of such correlations are the friction factor f (see Sec. 6) in pipe flow and the discharge coefficients of pipe orifices (see Sec. 4).

The ratio of the inertial force to the gravitational force yields the Froude number, and similar ratios involving the surface tension and the elastic force result in the Weber and Mach numbers, respectively.

Conversion factors for computing prototype values from model values are obtained by equating the appropriate dimensionless parameters. For example, if studies are made of a flow phenomenon involving primarily the gravitational force, the conversion factors are obtained by equating the Froude number for the model F_m to that for the prototype F_p as follows, replacing d with L to represent any typical length:

$$F_m = F_p \tag{3.41}$$

or

$$\frac{V_m}{\sqrt{g_m L_m}} = \frac{V_p}{\sqrt{g_p L_p}} \tag{3.42}$$

Then the velocity ratio is

$$\frac{V_p}{V_m} = V_r = \sqrt{\frac{g_p L_p}{g_m d_n}} = \sqrt{g_r L_r} \tag{3.43}$$

Noting that the area ratio a_r is equal to L_r^2, the discharge ratio becomes

$$Q_r = \frac{a_p V_p}{a_m V_m} = g_r^{1/2} L_r^{5/2} \tag{3.44}$$

The time ratio is obtained by replacing V with L/T. Then

$$V_r = \frac{L_r}{T_r} = \sqrt{g_r L_r} \tag{3.45}$$

and

$$T_r = \sqrt{\frac{L_r}{g_r}} \tag{3.46}$$

Other conversion ratios can be determined in the same manner.

References

1. O. Reynolds, "An Experimental Investigation of the Circumstances Which Determine Whether the Motion of Water Shall Be Direct or Sinuous, and the Laws of Resistance in Parallel Channels," *Trans. R. Soc. (London)*, vol. 174, pp. 935–982, 1883.
2. C. O. Wisler and E. F. Brater, *Hydrology*, 2d ed., Wiley, New York, 1959.
3. See, for example, V. L. Streeter, *Fluid Mechanics*, 6th ed., McGraw-Hill, New York, pp. 134–144, 1975.
4. V. L. Streeter, "The Kinetic Energy and Momentum Correction Factors for Pipes and Open Channels of Great Width," *Civil Eng. (N.Y.)*, vol. 12, no. 4, pp. 212–213, 1942.
5. M. P. O'Brien and J. W. Johnson, "Velocity Head Correction for Hydraulic Flow," *Eng. News Rec.*, vol. 113, no. 7, pp. 214–216, 1934.
6. H. Rouse, *Fluid Mechanics for Engineers*, McGraw-Hill, New York, pp. 1–32, 1938.
7. H. L. Longhaor, *Dimensional Analysis and Theory of Models*, Wiley, New York, 1951.

SECTION 4

Orifices, Gates, and Tubes

An orifice is an opening with closed perimeter and of regular form through which water flows. If the perimeter is not closed or if the opening flows only partially full, the orifice becomes a weir (Sec. 5). An orifice with prolonged sides, such as a pipe 2 or 3 diameter in length or an opening in a thick wall, is called a *tube*. A *gate* is an opening in a hydraulic structure, usually including appurtenances for regulating the outflow. The same fundamental principles apply to each type of opening.

The stream of water that issues from an orifice is termed the "jet," and the depth of water producing discharge is the "head." An orifice with a sharp upstream edge is called a *sharp-edged* orifice. The *approach channel* is the channel leading up to an orifice, tube, or gate, and the mean velocity in this channel is the *approach velocity*. If the jet discharges into the air, the orifice is said to have *free discharge*. If the discharge is under water, it is called a *submerged orifice*. An orifice is spoken of as *vertical* or *horizontal*, depending on whether it lies in a vertical or a horizontal plane.

When the inner edge of the orifice consists of a sharp corner, the jet springs free, as illustrated in Fig. 4.1, and the orifice is a sharp-edged orifice. The jet issuing from a sharp-edged orifice contracts until it reaches the *vena contracta*. At the vena contracta the paths of all elements of the jet are parallel and the pressure in the jet can be assumed to be equal to that in the surrounding fluid. The vena contracta of a circular orifice is approximately ½ diameter downstream from the inner face of the orifice plate. The area at the vena contracta a_2 is related to the area of the orifice a as follows:

$$a_2 = C_c a \tag{4.1}$$

where C_c is the coefficient of contraction.

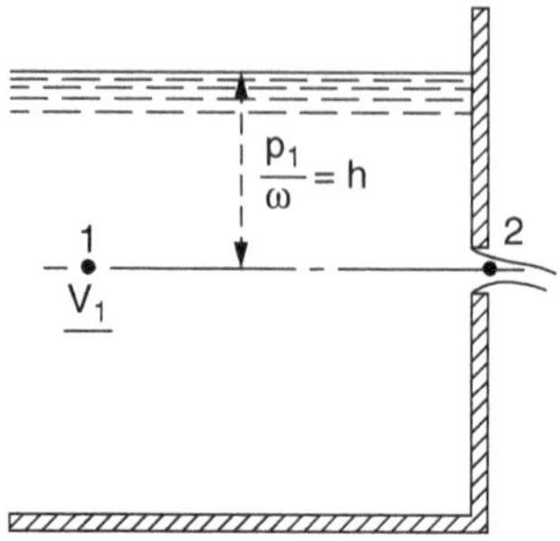

Figure 4.1 Sharp-edged orifice.

The contraction is caused by the fact that those elements of the fluid with approach from the side of the orifice have transverse velocity components directed toward the center of the orifice. Therefore the nearness of boundaries, such as for orifices in pipes, would tend to reduce the amount of contraction and increase C_c. If one edge of an orifice is flush with a wall, the contraction on that side will be entirely eliminated. However, the contraction from the other side will be increased by approximately the same amount. Increasing the roughness of the inner face of the orifice plate would also reduce the transverse velocity components slightly. Rounding the inner edge of the orifice reduces contraction, and the contraction can be eliminated completely by shaping the orifice to conform with the form of the contracting jet, as shown in Fig. 4.2.

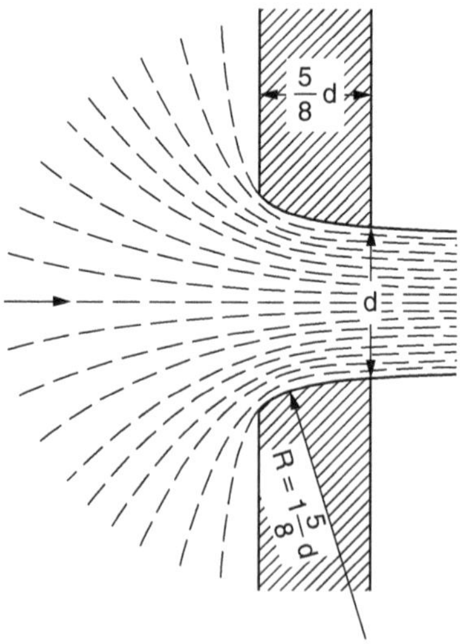

Figure 4.2 Bell-mouthed orifice with rounded upstream edge.

Fundamental Equations

The Bernoulli equation (Sec. 3), written from any point in the liquid, such as point 1 in Fig. 4.1, to the vena contracta (point 2), taking the datum plane through the center of the orifice, is

$$\frac{v_1^2}{2g}+\frac{p_1}{w}=\frac{v_2^2}{2g}+\frac{p_2}{w}+h_l \tag{4.2}$$

and

$$v_2=\sqrt{2g\left(\frac{p_1}{w}-\frac{p_2}{w}+\frac{v_1^2}{2g}-h_l\right)} \tag{4.3}$$

Point 2 being located where the jet has ceased to contract, its pressure is that of the surrounding fluid. For discharge into the atmosphere, p_2 is therefore zero on the gauge scale. For large tanks, v_1 is so small that it may be neglected. Replacing p_1/w with h and dropping the subscript of v_2, Eq. (4.3) may now be written

$$v_2=\sqrt{2g\,(h-h_l)} \tag{4.4}$$

Neglecting energy losses, the equation for the theoretical velocity becomes

$$v_t=\sqrt{2gh} \tag{4.5}$$

or

$$h=\frac{v_t^2}{2g} \tag{4.6}$$

The expression $v_t^2/2g$ is termed the "velocity head."

It is convenient to take care of the energy loss by introducing the coefficient of velocity C_v. Thus instead of subtracting the energy loss as in Eq. (4.4), the expression for the velocity at the vena contracta may be written

$$v_2=C_v\sqrt{2gh} \tag{4.7}$$

The discharge Q is the product of the velocity and the area at the vena contracta,

$$Q=a_2v_2 \tag{4.8}$$

Inserting the value of a_2 from Eq. (4.1) and the value of v_2 from Eq. (4.7),

$$Q=C_caC_v\sqrt{2gh} \tag{4.9}$$

The product of the coefficient of velocity C_v and the coefficient of contraction C_c is the coefficient of discharge C. Equation (4.9) may then be written

$$Q = Ca\sqrt{2gh} \tag{4.10}$$

The energy loss may be expressed in terms of C_v by equating the expressions for v given in Eqs. (4.4) and (4.7),

$$\sqrt{2g(h - h_t)} = C_v\sqrt{2gh}$$

Then

$$h_t = (1 - C_v^2)h_l \tag{4.11}$$

or introducing the value of h from Eq. (4.6),

$$h_t = \left(\frac{1}{C_v^2} - 1\right)\frac{v_t^2}{2g} \tag{4.12}$$

Path of Jet

Figure 4.3 illustrates a jet discharging from a vertical orifice under a head h. x and y are, respectively, the abscissa and the ordinate of any

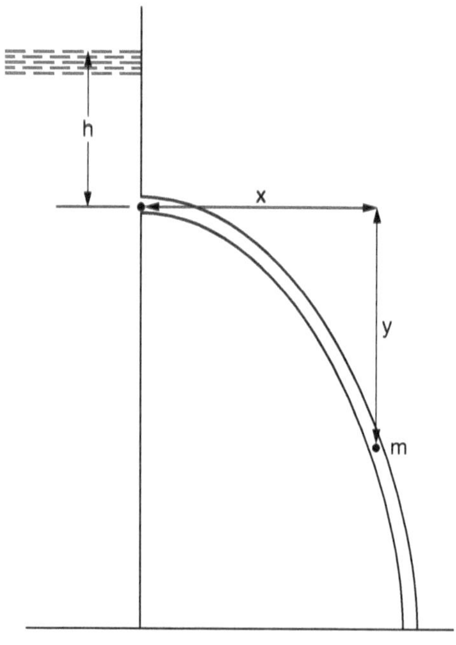

FIGURE 4.3 Path of jet.

point m in the path of the jet. If υ is the velocity in the vena contracta, at the end of time t,

$$x = \upsilon t \tag{4.13}$$

From the law of falling bodies,

$$y = \frac{1}{2} g t^2 \tag{4.14}$$

and eliminating t from Eqs. (4.13) and (4.14),

$$x^2 = \frac{2\upsilon^2}{g} y \tag{4.15a}$$

which is the equation of a parabola with its vertex at the orifice. Since

$$\upsilon = C_\upsilon \sqrt{2gh} \tag{4.7}$$

Equation (4.15*a*) can also be written

$$x^2 = 4C_\upsilon^2 h y \tag{4.15b}$$

Orifices under Low Heads

In deriving Eq. (4.10), the head producing discharge was assumed to be the head on the center of the orifice. When the head on a vertical orifice is small in comparison with the height of the orifice, there is an appreciable difference between the true theoretical discharge and the discharge given by Eq. (4.10).

Figure 4.4 shows a rectangular orifice of width L and height M. h_1 and h_2 are the respective heads on the upper and lower edges of the orifice. Neglecting the velocity of approach, the theoretical

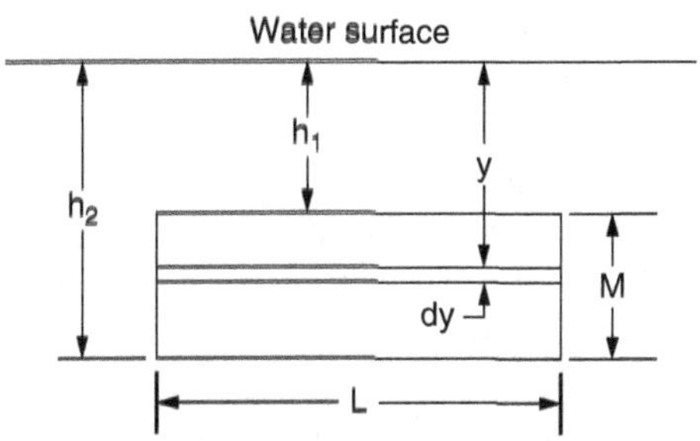

Figure 4.4 Rectangular orifice.

discharge through any elementary strip of area $L\,dy$, discharging under a head y, is

$$dQ_t = L\sqrt{2gy}\,dy$$

which, integrated between the limits h_2 and h_1, gives

$$Q_t = \frac{2}{3}L\sqrt{2g}\left(h_2^{3/2} - h_1^{3/2}\right) \tag{4.16}$$

When h_1 is zero,

$$Q_t = \frac{2}{3}L\sqrt{2g}\,h_2^{3/2} \tag{4.17}$$

which is the theoretical formula, without velocity-of-approach correction, for discharge over a weir.

Equation (4.16) gives the theoretical discharge for rectangular orifices. A similar but more complicated formula could be derived for circular orifices. For $h_1 = M$, Eq. (4.10) gives results about 1 percent greater than Eq. (4.16), and for $h_1 = 2M$, about 0.3 percent greater.

Equation (4.10) is generally employed for all orifices, including those discharging under low heads, the deviation from the theoretical form of the formula being corrected for in the coefficient.

Discharge under Falling Head

Figure 4.5 shows a vessel filled with water to a depth h_1. The time required to lower the water surface to a depth h_2 is to be found. a is the area of orifice, and A is the area of the water surface for a depth y. C is the coefficient of discharge. The increment of time dt required to lower the water the infinitesimal distance dy is

$$dt = \frac{A\,dy}{Ca\sqrt{2gy}} \tag{4.18}$$

From Eq. (4.18), if A can be expressed in terms of y, by integrating between limits h_1 and h_2 the time needed to lower the water surface the distance $h_1 - h_2$ can be obtained. Placing $h_2 = 0$ gives the time for emptying the vessel. Equation (4.18) applies to horizontal or inclined orifices, provided the water surface does not fall below the top of the orifice. For a cylinder or prism with vertical axis, A is constant, and Eq. (4.18) becomes, after integration,

$$t = \frac{2A}{Ca\sqrt{2g}}\left(\sqrt{h_1} - \sqrt{h_2}\right) \tag{4.19}$$

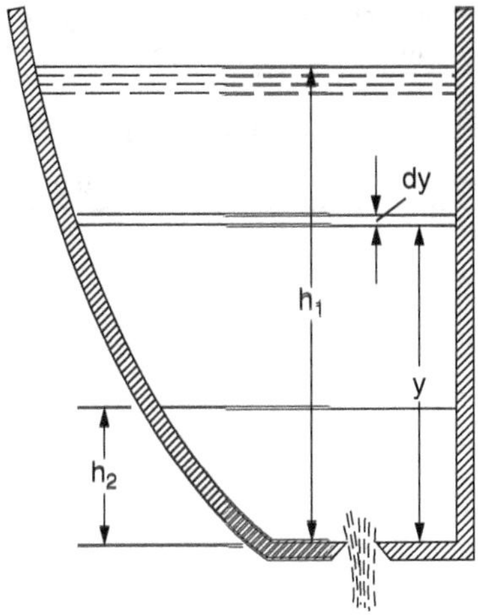

FIGURE 4.5 Discharge under falling head.

Orifice Coefficients

One of the earliest experimenters on sharp-edged orifices was Hamilton Smith, Jr.[1] His values of the coefficient of discharge for round and square orifices are given in Table 4.1. There have been many subsequent investigations of circular orifices, not all of which are in agreement. Investigations by Medaugh and Johnson[2] checked Smith's coefficients for orifices larger than 0.25 in. (6 mm) in diameter within 0.33 percent. Values of the coefficient of discharge for a 1-in. (2.5-cm) orifice as determined by various investigators and plotted by Medaugh and Johnson are shown in Fig. 4.6. The differences between the values are undoubtedly not entirely due to experimental errors. Many other factors may contribute, such as the ratio of the orifice diameter to the dimensions of the tank wall, the sharpness of the edge of the orifice, the roughness of the inner surface, the orifice plate, and the temperature of the water. The effect of having the tank wall approach the orifice is to suppress contraction and therefore to make C approach the value of C_v.

Smith and Walker[3] found values of C_v to vary from 0.954 to 0.991 for orifices varying in diameter from 0.75 to 2.5 in. (2 to 6 cm). They also found small variations with the head, these values being averages for heads varying from 1 to 60 ft (0.3 to 18 m).

Values of C_c for circular sharp-edged orifices were found to vary from approximately 0.67 for 0.75-in. (2-cm) orifices to 0.614 for 2.5-in. (6-cm) orifices when the head is 2 ft (60 cm) or more. Values are slightly larger for lower heads.

Table 4.1a Smith's Coefficients of Discharge for Circular and Square Orifices with Full Contraction, U.S. Customary System

	Diameter of circular orifices, ft						
Head, ft	**0.02**	**0.04**	**0.07**	**0.1**	**0.2**	**0.6**	**1.0**
0.4	....	0.637	0.624	0.618			
0.6	0.655	0.630	0.618	0.613	0.601	0.593	
0.8	0.648	0.626	0.615	0.610	0.601	0.594	0.590
1	0.644	0.623	0.612	0.608	0.600	0.595	0.591
1.5	0.637	0.618	0.608	0.605	0.600	0.596	0.593
2	0.632	0.614	0.606	0.604	0.599	0.597	0.595
2.5	0.629	0.612	0.605	0.603	0.599	0.598	0.596
3	0.627	0.611	0.604	0.603	0.599	0.598	0.597
4	0.623	0.609	0.603	0.602	0.599	0.597	0.596
6	0.618	0.607	0.602	0.600	0.598	0.597	0.596
8	0.614	0.605	0.601	0.600	0.598	0.596	0.596
10	0.611	0.603	0.599	0.598	0.597	0.596	0.595
20	0.601	0.599	0.597	0.596	0.596	0.596	0.594
50	0.596	0.595	0.594	0.594	0.594	0.594	0.593
100	0.593	0.592	0.592	0.592	0.592	0.592	0.592
	Side of square orifices, ft						
0.4	....	0.643	0.628	0.621			
0.6	0.660	0.636	0.623	0.617	0.605	0.598	
0.8	0.652	0.631	0.620	0.615	0.605	0.600	0.597
1	0.648	0.628	0.618	0.613	0.605	0.601	0.599
1.5	0.641	0.622	0.614	0.610	0.605	0.602	0.601
2	0.637	0.619	0.612	0.608	0.605	0.604	0.602
2.5	0.634	0.617	0.610	0.607	0.605	0.604	0.602
3	0.632	0.616	0.609	0.607	0.605	0.604	0.603
4	0.628	0.614	0.608	0.606	0.605	0.603	0.602
6	0.623	0.612	0.607	0.605	0.604	0.603	0.602
8	0.619	0.610	0.606	0.605	0.604	0.603	0.602
10	0.616	0.608	0.605	0.604	0.603	0.602	0.601
20	0.606	0.604	0.602	0.602	0.602	0.601	0.600
50	0.602	0.601	0.601	0.600	0.600	0.599	0.599
100	0.599	0.598	0.598	0.598	0.598	0.598	0.598

Table 4.1b Smith's Coefficients of Discharge for Circular and Square Orifices with Full Contraction, Metric System

	Diameter of circular orifices, cm						
Head, m	**0.5**	**1.0**	**2.0**	**5.0**	**10.0**	**15.0**	**30.0**
0.1		0.642	0.627	0.606			
0.2	0.657	0.637	0.618	0.604	0.598	0.595	
0.3	0.648	0.630	0.613	0.603	0.598	0.596	0.591
0.4	0.643	0.627	0.611	0.601	0.598	0.597	0.592
0.5	0.638	0.623	0.608	0.601	0.599	0.597	0.594
1	0.629	0.616	0.604	0.600	0.599	0.598	0.597
2	0.619	0.610	0.602	0.599	0.598	0.597	0.596
3	0.612	0.606	0.599	0.597	0.597	0.596	0.595
4	0.609	0.604	0.599	0.597	0.596	0.596	0.595
5	0.606	0.603	0.598	0.596	0.596	0.596	0.594
10	0.599	0.598	0.596	0.595	0.595	0.595	0.594
20	0.595	0.594	0.593	0.593	0.593	0.593	0.593
30	0.593	0.592	0.592	0.592	0.592	0.592	0.592
	Side of square orifices, cm						
0.1		0.649	0.631	0.606			
0.2	0.661	0.642	0.623	0.609	0.603	0.601	
0.3	0.651	0.635	0.619	0.608	0.604	0.602	0.599
0.4	0.647	0.631	0.617	0.606	0.604	0.603	0.600
0.5	0.642	0.627	0.614	0.606	0.604	0.603	0.602
1	0.634	0.621	0.609	0.605	0.605	0.604	0.603
2	0.624	0.615	0.607	0.604	0.604	0.603	0.602
3	0.617	0.611	0.605	0.603	0.603	0.602	0.601
4	0.614	0.609	0.604	0.603	0.602	0.602	0.601
5	0.611	0.608	0.603	0.602	0.602	0.602	0.600
10	0.605	0.603	0.602	0.601	0.601	0.600	0.600
20	0.601	0.600	0.600	0.599	0.599	0.599	0.599
30	0.599	0.598	0.598	0.598	0.598	0.598	0.598

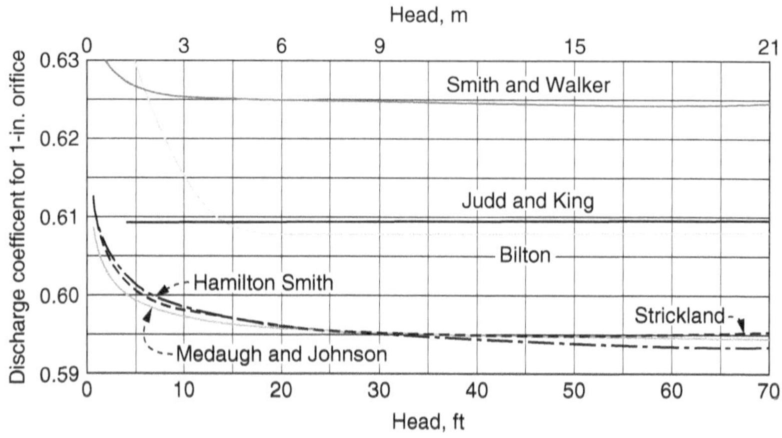

Figure 4.6 Orifice coefficients.

If there is suppression of contraction on one side and opportunity for complete contraction on the other sides, more water will approach with velocity components parallel to the face of the orifice on these sides, causing increased contraction. This will compensate to a large extent for loss of contraction on the other side. Williams[4] found, for rectangular orifices 30 in. (75 cm) wide and 2 to 4 in. (5 to 10 cm) high with full contraction at the top and completely suppressed contraction on the two sides and the bottom, that the average coefficient of discharge was 0.607. This value corresponds closely to the coefficient for orifices with complete contraction. For orifices having full contraction at the top, one side a sharp edge 6 in. (15 cm) from the side of the channel, and contraction suppressed at one side and the bottom, Williams obtained an average coefficient of 0.611. With these orifices, except that the top was beveled to an angle of 45°, he obtained values of C of 0.776 and 0.755, respectively. Table 4.2, from results compiled by Smith[1] (pp. 65–67), indicates the effect of suppression of contraction for small orifices. In this table, "suppressed contraction" means that the edge of the orifice coincides with the side of the channel, and "partly suppressed" means that the distance of the edge of the orifice from the side of the channel was 0.066 in. (2 cm). A special case of suppressed contraction is the pipe orifice, which will be discussed later.

When the inner edge of the orifice is rounded, as in Fig. 4.2, contraction is suppressed, C_c approaches 1, and C approaches the value of C_v. Values of C_v for such orifices are approximately the same as for sharp-edged orifices.

Roughening the inner surface of the orifice plate retards the velocity components along the wall, which cause contraction, thus

TABLE 4.2a Coefficients of Discharge for Rectangular Orifices 0.656 ft Wide with Partially Suppressed Contraction, U.S. Customary System

Description of contraction	Height, ft	Head, ft 1	3	5
Complete contraction	0.656	0.598	0.604	0.603
	0.328	0.616	0.615	0.611
	0.164	0.631	0.627	0.620
	0.098	0.632	0.628	0.623
	0.033	0.652	0.634	0.620
Suppressed at bottom only	0.656	0.620	0.624	0.625
	0.328	0.649	0.647	0.634
	0.164	0.671	0.668	0.666
	0.098	0.680	0.677	0.677
	0.033	0.710	0.705	0.696
Suppressed on both sides only	0.656	0.632	0.628	0.628
	0.328	0.637	0.630	0.630
	0.164	0.641	0.634	0.635
	0.098	0.653	0.643	0.639
	0.033	0.682	0.667	0.655
Suppressed at bottom and partly on one side	0.656	0.633	0.636	0.637
	0.328	0.658	0.656	0.654
	0.164	0.676	0.673	0.672
	0.098	0.682	0.683	0.681
	0.033	0.708	0.705	0.695
Suppressed at bottom and partly on two sides	0.656	0.678	0.664	0.663
	0.328	0.680	0.675	0.672
	0.164	0.687	0.680	0.673
	0.098	0.693	0.688	0.683
	0.033	0.708	0.705	0.698
Suppressed at bottom and two sides	0.656	0.690	0.677	0.672
Complete suppression	0.656	—	0.950	

Table 4.2b Coefficients of Discharge for 20-cm-Wide Rectangular Orifices with Partially Suppressed Contraction, Metric System

		Head, m		
Description of contraction	**Height, cm**	**0.5**	**1.0**	**1.5**
Complete contraction	20	0.600	0.604	0.603
	10	0.616	0.614	0.611
	5	0.630	0.626	0.620
	3	0.631	0.627	0.623
	1	0.646	0.632	0.620
Suppressed at bottom only	20	0.621	0.624	0.625
	10	0.648	0.645	0.634
	5	0.670	0.668	0.666
	3	0.679	0.677	0.677
	1	0.708	0.704	0.696
Suppressed on both sides only	20	0.631	0.628	0.628
	10	0.635	0.630	0.630
	5	0.639	0.634	0.635
	3	0.650	0.642	0.639
	1	0.677	0.665	0.655
Suppressed at bottom and partly on one side	20	0.634	0.636	0.637
	10	0.657	0.656	0.654
	5	0.675	0.673	0.672
	3	0.682	0.683	0.681
	1	0.707	0.704	0.695
Suppressed at bottom and partly on two sides	20	0.674	0.664	0.663
	10	0.678	0.675	0.672
	5	0.685	0.679	0.673
	3	0.692	0.687	0.683
	1	0.707	0.704	0.698
Suppressed at bottom and two sides	20	0.686	0.676	0.672
Complete suppression	20	—	0.950	

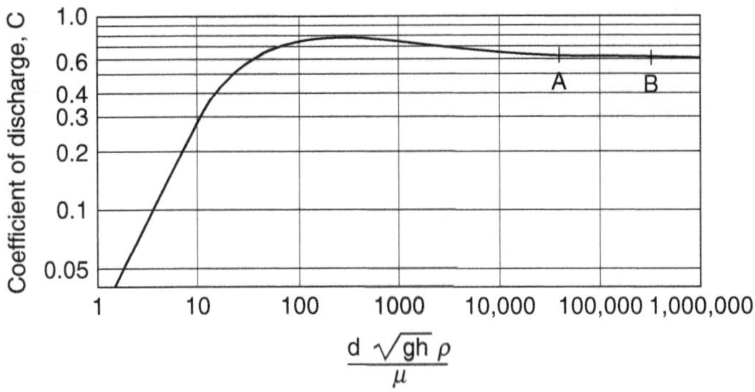

FIGURE 4.7 Orifice coefficients.

tending to increase the values of both C_c and C. Similar effects would result from having the orifice plate differ from a perfectly plane surface. If the plate were slightly warped outward, C_c would be increased, whereas an inward curvature would have the opposite effect (see "Standard Short Tubes," this section).

The effect of the water temperature is to change the viscosity and the density. Considering at the same time the variation of C with the velocity and the orifice diameter, as shown in Table 4.1, it becomes clear that here again the Reynolds number is likely to be the coordinating factor. This was shown to be the case by Lea,[5] who plotted more than 100 experimental values of C against R. The author's curve, derived from Lea's plotted points, is shown in Fig. 4.7. The fluids used in the tests were water, various mixtures of water and glycerin, and a number of oils. Flow is laminar for Reynolds numbers less than 12 and fully turbulent for R greater than 10,000 intervening values corresponding to a transition region. Except in the transition range, all points plotted by Lea show a spread of about 3 percent. A spread of approximately 15 percent occurs in the vicinity of $R = 1000$. The range of Reynolds numbers covered by the tests of Medaugh and Johnson is indicated on Fig. 4.7 by the limits A and B.

Investigations by Bilton[6] showed the coefficient of discharge of a circular orifice under any given head to be the same whether the jet is horizontal, vertical, or at any intermediate angle.

Table 4.3 presents extracts from tables by Fanning[7] and Bovey,[8] listing coefficients of discharge for various shapes and arrangements of orifices with complete contraction. Fanning's table, compiled from experiments from several sources, contains coefficients for orifices 1 ft (30 cm) wide, of various heights, and under a wide range of heads. Bovey's table, prepared from his own experiments on orifices of different shapes, each with the area of a circle 0.5 in. (1.25 cm) in diameter, indicates the effect of the shape of the opening on the coefficient.

Table 4.3a Coefficients of Discharge for Variously Shaped Orifices with Complete Contraction, U.S. Customary System

(a) Fanning's coefficients[7] for vertical rectangular orifices, 1 ft wide

	Height of orifice, ft					
Head, ft	**0.125**	**0.25**	**0.5**	**1**	**2**	**4**
1	0.622	0.616	0.611	0.605		
2	0.619	0.614	0.609	0.604	0.609	
4	0.614	0.610	0.607	0.603	0.606	0.608
6	0.610	0.608	0.604	0.601	0.604	0.605
8	0.608	0.606	0.603	0.601	0.603	0.604
10	0.606	0.604	0.602	0.601	0.602	0.603
15	0.607	0.603	0.601	0.601	0.602	0.603
20	0.607	0.604	0.602	0.601	0.602	0.603
30	0.609	0.604	0.603	0.601	0.603	0.605
50	0.614	0.607	0.605	0.602	0.606	0.609

(b) Bovey's coefficients[8] for variously shaped orifices, each 0.196 in.² area

				Rectangle, ratio of sides				
		Square		4:1		10:1		
Head, ft	**Circle**	**Sides vertical**	**Diagonal vertical**	**Long sides vertical**	**Long sides horizontal**	**Long sides vertical**	**Long sides horizontal**	**Triangle**
1	0.620	0.627	0.628	0.642	0.643	0.663	0.664	0.636
2	0.613	0.620	0.628	0.634	0.636	0.650	0.651	0.628
4	0.608	0.616	0.618	0.628	0.629	0.641	0.642	0.623
6	0.607	0.614	0.616	0.626	0.627	0.637	0.637	0.620
8	0.606	0.613	0.614	0.623	0.625	0.634	0.635	0.619
10	0.605	0.612	0.613	0.622	0.624	0.632	0.633	0.618
15	0.604	0.610	0.611	0.620	0.622	0.630	0.630	0.617
20	0.603	0.609	0.611	0.620	0.621	0.629	0.628	0.616

Submerged Orifices

Figure 4.8 shows an orifice in a tank with liquid on both sides. Application of the Bernoulli equation, taking the datum through the center of the orifice, yields Eq. (4.3) exactly in the same form as for free discharge,

$$v_2 = \sqrt{2g\left(\frac{p_1}{w} - \frac{p_2}{w} + \frac{v_1^2}{2g} - h_l\right)} \tag{4.3}$$

Table 4.3b Coefficients of Discharge for Variously Shaped Orifices with Complete Contraction, Metric System

(a) Fanning's coefficients[7] for vertical rectangular orifices, 30 cm wide

Head, m	Height of orifice, cm					
	4	8	15	30	60	120
0.3	0.622	0.616	0.611	0.605		
0.5	0.621	0.615	0.610	0.605	0.610	
1.0	0.616	0.611	0.608	0.603	0.607	0.609
2.0	0.609	0.607	0.604	0.601	0.604	0.605
2.5	0.608	0.606	0.603	0.601	0.603	0.604
3	0.606	0.604	0.602	0.601	0.602	0.603
4	0.606	0.603	0.601	0.601	0.602	0.603
5	0.607	0.603	0.601	0.601	0.602	0.603
10	0.609	0.604	0.603	0.601	0.603	0.606
15	0.613	0.607	0.605	0.602	0.606	0.609

(b) Bovey's coefficients[8] for variously shaped orifices, each 1.26 cm² in area

Head, m	Circle	Square		Rectangle, ratio of sides				Triangle
				4:1		10:1		
		Sides vertical	Diagonal vertical	Long sides vertical	Long sides horizontal	Long sides vertical	Long sides horizontal	
0.3	0.620	0.627	0.628	0.642	0.643	0.663	0.664	0.636
0.5	0.618	0.625	0.628	0.639	0.641	0.658	0.659	0.633
1.0	0.610	0.617	0.622	0.630	0.632	0.644	0.645	0.625
2.0	0.607	0.614	0.615	0.625	0.626	0.636	0.636	0.620
2.5	0.606	0.613	0.614	0.623	0.625	0.634	0.635	0.619
3	0.605	0.612	0.613	0.622	0.624	0.632	0.633	0.618
4	0.604	0.611	0.612	0.621	0.623	0.631	0.631	0.617
5	0.604	0.610	0.611	0.620	0.622	0.630	0.631	0.617

Noting that in this case

$$\frac{p_1}{w} - \frac{p_2}{w} = \Delta h \tag{4.20}$$

neglecting $v_1^2/2g$, and introducing C to take care of the contraction and energy loss, the equation for discharge is

$$Q = Ca\sqrt{2g\,\Delta h} \tag{4.21}$$

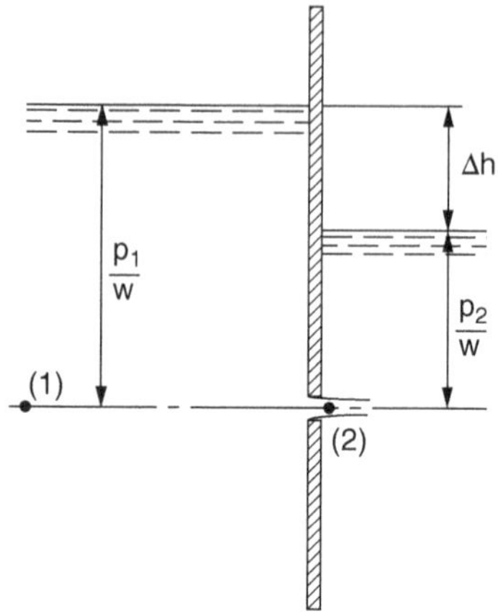

FIGURE 4.8 Submerged orifice.

The few experiments for determining C for submerged orifices that are available indicate that the value of the coefficient is not greatly affected by submergence. Table 4.4 gives values of C for sharp-edged submerged orifices, and Table 4.5 contains values of C for a submerged orifice 1 ft (30 cm) square with rounded corners.

Gates

Gates are openings in hydraulic structures to permit the passage of water, usually including means of regulating the outflow. Though gates have the hydraulic properties of orifices, there are no standards of design, and they therefore have varying degrees of contraction. The discharge may be either free or submerged. The formula for discharge is

$$Q = Ca\sqrt{2gh} \tag{4.10}$$

where h is the head on the center of the orifice for orifices with free discharge; for submerged orifices it is the difference in elevation of the water surfaces.

The results of experiments on the wooden models of gates shown in Figs. 4.9 and 4.10 were given by Unwin[9] and Bellasis.[10] Table 4.6 lists C values for each of these models for various depths of water above the top of the orifice.

The mean value of C for a sluice gate in the Argo dam at Ann Arbor, Michigan, discharging without submergence under a head of about 8 ft (2.4 m), was found by Ward[11] to be 0.545. The opening,

Table 4.4 Miscellaneous Coefficients of Discharge for Various Sharp-Edged Submerged Orifices*

Dimensions of orifice, ft (cm)	Authority	Head, ft/m							
		0.3	0.5	1.0	2.0	4.0	6.0	10.0	18.0
		0.1	0.15	0.3	0.6	1.2	1.8	3.0	5.5
Circle, $d = 0.05$ (1.5)	Smith[1]		0.599	0.597	0.595	0.595			
Circle, $d = 0.10$ (3)	Smith[1]	0.600	0.600	0.600	0.599	0.598			
Square, 0.05 by 0.05 (1.5 by 1.5)	Smith[1]		0.609	0.607	0.605	0.604			
Square, 0.10 by 0.10 (3 by 3)	Smith[1]	0.607	0.605	0.604	0.603	0.604			
Rectangle, $l = 3.0$ (90), $d = 0.05$ (1.5)	Smith[1]		0.621			0.620	0.620	0.618	
Circle, $d = 1.0$ (30)	Ellis				0.608	0.602	0.603	0.600	0.601
Square, 1.0 by 1.0 (30 by 30)	Ellis				0.601	0.601	0.603	0.605	0.606
Square, 4 by 4 (120 by 120)	Stewart	0.614							

* The two orifices experimented on by Ellis were horizontal; all other orifices were vertical.

TABLE 4.5a Coefficients of Discharge for Submerged Vertical Square Orifice with Rounded Corners,* U.S. Customary System

Dimensions of orifice, ft	Head, ft						
	3	5	6	10	12	14	18
Square, 1 by 1	0.952	0.946	0.945	0.943	0.943	0.944	0.944

*From experiments by Ellis.

TABLE 4.5b Coefficients of Discharge for Submerged Vertical Square Orifice with Rounded Corners,* Metric System

Dimensions of orifice, cm	Head, m						
	1.0	1.50	2.0	3.0	4.0	5.0	5.5
Square, 30 by 30	0.952	0.946	0.945	0.943	0.943	0.944	0.944

*From experiments by Ellis.

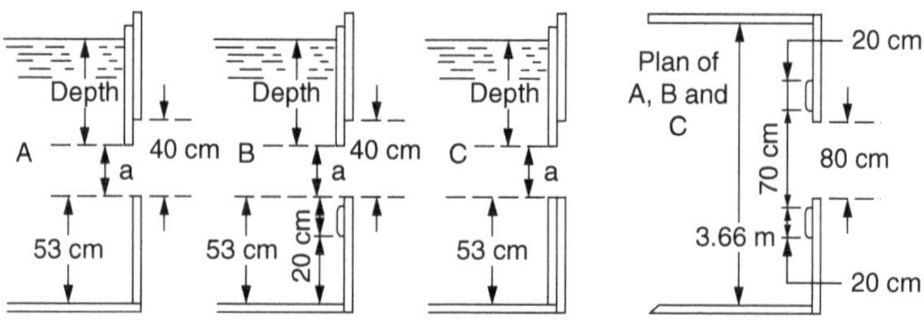

FIGURE 4.9 Gates discharging freely into air.

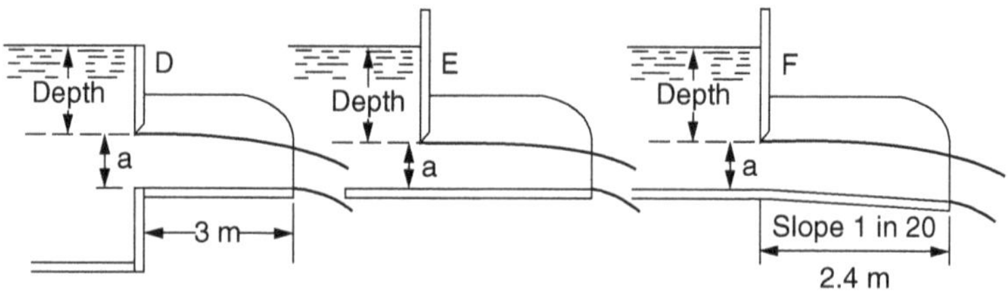

FIGURE 4.10 Gates with prolonged bottoms and sides.

Table 4.6 Coefficients of Discharge for Various Models, Figs. 4.9 and 4.10

Model	Depth of opening, ft (cm)	Depth of water above top of orifice, ft /m										
		0.07	**0.1**	**0.3**	**0.5**	**0.7**	**1.0**	**2.0**	**3.0**	**5.0**	**7.0**	**10.0**
		0.02	**0.03**	**0.10**	**0.15**	**0.20**	**0.30**	**0.60**	**1.00**	**1.50**	**2.00**	**3.00**
A	1.31 (40)			0.598	0.604	0.610	0.616	0.618	0.610	0.608	0.594	0.592
	0.66 (20)			0.633	0.638	0.640	0.641	0.640	0.638	0.637	0.636	0.634
	0.16 (5)			0.691	0.688	0.684	0.683	0.678	0.674	0.672	0.670	0.668
	10 (3)			0.709	0.700	0.695	0.692	0.688	0.681	0.677	0.675	0.672
B	1.31 (40)			0.644	0.650	0.654	0.656	0.649	0.634	0.620	0.615	0.611
	0.66 (20)			0.665	0.670	0.674	0.675	0.676	0.674	0.673	0.671	0.669
	0.16 (5)			0.665	0.681	0.688	0.693	0.695	0.694	0.692	0.691	0.689
	10 (3)			0.694	0.700	0.705	0.708	0.710	0.704	0.699	0.695	0.693
C	1.31 (40)			0.649	0.654	0.658	0.660	0.652	0.635	0.622	0.616	0.612
	0.66 (20)			0.668	0.673	0.676	0.678	0.679	0.677	0.674	0.672	0.670
	0.16 (5)			0.667	0.682	0.690	0.695	0.697	0.696	0.693	0.692	0.690
	10 (3)			0.696	0.702	0.707	0.710	0.712	0.705	0.699	0.695	0.693
D	0.66 (20)	0.487	0.495	0.543	0.562	0.577	0.588	0.601	0.601	0.601	0.601	0.601
	0.16 (5)	0.495	0.550	0.621	0.630	0.631	0.630	0.625	0.623	0.619	0.612	0.606
E	0.66 (20)	0.487	0.495	0.534	0.554	0.573	0.580	0.595	0.599	0.602	0.602	0.601
	0.16 (5)	0.495	0.544	0.602	0.612	0.618	0.623	0.627	0.628	0.627	0.622	0.617
F	0.66 (20)	0.530	0.535	0.571	0.584	0.595	0.600	0.608	0.610	0.610	0.609	0.608
	0.16 (5)	0.590	0.600	0.630	0.640	0.645	0.649	0.652	0.651	0.650	0.650	0.649

4 ft (1.2 m) wide by 5 ft (1.5 m) high, was between beveled piers. The base of the gate was 1 ft (30 cm) above the concrete floor, the bottom of the iron gate forming the top edge of the orifice.

Head gates and diversion gates usually have the general features illustrated in Fig. 4.11*a*. The bottom of the opening is nearly flush with the floor of the structure, the sides are flush with the piers, and the noses of the piers are beveled. There is full contraction at the top.

The values of C obtained by Newell[12] for the head gates of the Maxwell canal, Umatilla project, Oregon (Fig. 4.11*a*), are as follows:

h, ft (cm)	*v*, ft/s (m/s)	*C*
0.11 (3)	2.21 (0.67)	0.83
0.16 (5)	2.65 (0.81)	0.83
1.08 (33)	5.71 (1.74)	0.685
1.98 (60)	8.07 (2.46)	0.715

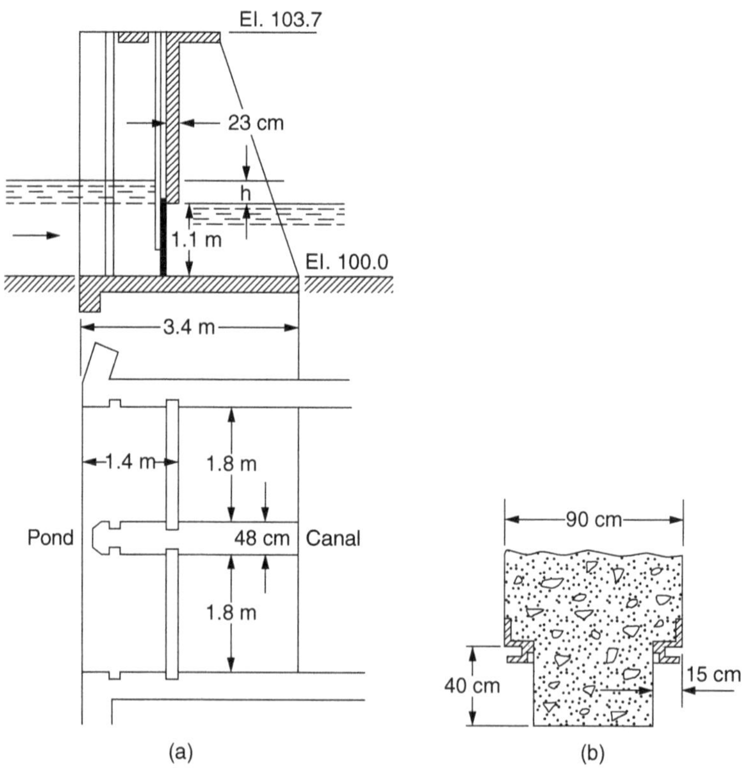

Figure 4.11 (*a*) Maxwell canal head gate. (*b*) Pier details, Boise canal head gate.

Only one gate was open for values in the last row. For the remaining values two gates were open.

The following values of C by Steward[13] are for the main head gates of the Boise canal, Idaho. There are eight gates, 3 ft (0.90 m) wide and 9 ft (2.75 m) high. G is the gate opening in feet (and meters). Except for the pier details shown in Fig. 4.11*b*, the structure is similar to that shown in Fig. 4.11*a*.

G, ft	G, m	h, ft	h, m	v, ft/s	v, m/s	C
0.67	0.20	7.6	2.3	18	5.5	0.81
0.72	0.22	8.1	2.5	16	4.9	0.70
0.76	0.23	7.3	2.2	17	5.2	0.77
0.80	0.24	7.6	2.3	17	5.2	0.78
1.10	0.34	5.0	1.5	16	4.9	0.91
1.40	0.43	7.0	2.1	18	5.5	0.86
1.45	0.44	9.1	2.7	19	5.8	0.77
1.57	0.48	12.2	3.7	21	6.4	0.75
1.85	0.56	7.1	2.1	17	5.2	0.81
2.13	0.65	7.4	2.3	18	5.5	0.81
2.22	0.68	9.0	2.7	19	5.8	0.78
2.85	0.87	9.6	2.9	19	5.8	0.78
3.41	1.04	7.0	2.1	16	4.9	0.74
3.75	1.14	5.3	1.6	15	4.6	0.81
4.46	1.36	5.1	1.9	14	4.3	0.68

G is the gate opening in meters. Only one gate was open for values in the seventh row. For the remaining values eight gates were open.

Parker[14] discussed experiments on head gates by Bornemann,[15] Chatterton,[16] and Benton[17] and published formulas by each. Chatterton's formula for C for values of h below 5 ft (1.5 m) is

$$C = 0.615 + 0.007 \times 2^{5-h} \qquad \text{(dimensions in ft)}$$

$$C = 0.615 + 0.007 \times 2^{5-3.28h} \qquad \text{(dimensions in meters)} \qquad (4.22)$$

Benton gives the following formula for heads below 5 ft (1.5 m) and widths of gate opening W_p to 10 ft (3 m):

$$C = 0.7201 + 0.0074W \qquad \text{(dimensions in ft)}$$

$$C = 0.7201 + 0.025W \qquad \text{(dimensions in meters)} \qquad (4.23)$$

Equations (4.22) and (4.23) are based on independent sets of observations. In Eq. (4.22) C varies only with h, and in Eq. (4.23) it varies only with W. Results of the two formulas agree quite closely for heads of 1 ft (0.3 m) or less, but differ by 10 to 25 percent for larger heads. Table 4.7 gives values of C computed by Eqs. (4.22) and (4.23).

The coefficient C for gates as given by these equations and tables includes the effect of the velocity of approach and is to be applied directly in Eq. (4.21). Knowledge of the coefficient is not sufficient to justify the use of a formula which contains a separate term to correct for velocity of approach. Experiments by Newell[12] show C to vary inversely with h at low heads and thus, in a general way, confirm Chatterton's results. Steward's results[13] do not indicate any tendency for C to vary with h at heads above 5 ft (1.5 m). It appears that for gates of this type (Fig. 4.11) conditions affecting side contractions have very little effect on C. If the gate sill is raised above the floor of the structure, a bottom contraction is produced which will have some effect in restricting discharge.

Discrepancies in experimental results (as indicated by the formulas of Chatterton and Benton) may result from differences in (1) structural details, which affect contraction; (2) channel dimensions, which affect the velocity of approach; and particularly, (3) the distance downstream from the gate at which the elevation of the water surface is measured. This may affect the value of h materially.

When a *sluice gate* discharges with the jet submerged, there is a characteristic dip at the downstream side of the gate, as shown in Fig. 4.12. In order to evaluate the head h on the gate, the magnitude of the depth D_s just downstream from the gate must be determined. An expression for D_s may be obtained by applying Newton's second law of motion to the water in the region from points 1 to 2 in Fig. 4.12. For the case of a rectangular channel having a gate extending across the entire width of the channel, the force applied to the quantity of water between points 1 and 2 may be taken as the difference between the hydrostatic forces at the two ends as follows:

$$F = \frac{wD_2^2}{2} - \frac{wD_s^2}{2} \tag{4.24}$$

The rate of change of momentum is

$$\frac{qw}{g}(V_1 - V_2)$$

where V_1 is the average velocity in the jet and V_2 the average velocity at point 2. By equating the preceding expressions and making use of the continuity equations, the following value of D_s may be obtained:

$$D_s = D_2\left[1 + \frac{2V_2^2}{gD_2}\left(1 - \frac{D_2}{D_1}\right)\right]^{1/2} \tag{4.25}$$

Table 4.7 Coefficients of Discharge for Submerged Gates from Chatterton's[16] and Benton's[17] Eqs. (4.22) and (4.23)

Head, ft	Head, m	Authority	Width of gate opening, ft/m					
			2	**4**	**6**	**8**	**10**	**12**
			0.5	**1.00**	**2.0**	**2.5**	**3.0**	**3.5**
0.02	0.01	Chatterton	0.83	0.83	0.83	0.83	0.83	0.83
		Benton	0.73	0.75	0.76	0.78	0.79	0.81
0.05	0.02	Chatterton	0.83	0.83	0.83	0.83	0.83	0.83
		Benton	0.73	0.75	0.76	0.78	0.79	0.81
0.10	0.03	Chatterton	0.82	0.82	0.82	0.82	0.82	0.82
		Benton	0.73	0.75	0.76	0.78	0.79	0.81
1.5	0.04	Chatterton	0.82	0.82	0.82	0.82	0.82	0.82
		Benton	0.73	0.75	0.76	0.78	0.79	0.81
0.2	0.06	Chatterton	0.81	0.81	0.81	0.81	0.81	0.81
		Benton	0.73	0.75	0.76	0.78	0.79	0.81
0.3	0.08	Chatterton	0.80	0.80	0.80	0.80	0.80	0.80
		Benton	0.73	0.75	0.76	0.78	0.79	0.81
0.4	0.10	Chatterton	0.78	0.78	0.78	0.78	0.78	0.78
		Benton	0.73	0.75	0.76	0.78	0.79	0.81
0.5	0.15	Chatterton	0.77	0.77	0.77	0.77	0.77	0.77
		Benton	0.73	0.75	0.76	0.78	0.79	0.81
0.75	0.20	Chatterton	0.75	0.75	0.75	0.75	0.75	0.75
		Benton	0.73	0.75	0.76	0.78	0.79	0.81
1.0	0.30	Chatterton	0.73	0.73	0.73	0.73	0.73	0.73
		Benton	0.73	0.75	0.76	0.78	0.79	0.81
1.5	0.45	Chatterton	0.69	0.69	0.69	0.69	0.69	0.69
		Benton	0.73	0.75	0.76	0.78	0.79	0.81
2.0	0.60	Chatterton	0.67	0.67	0.67	0.67	0.67	0.67
		Benton	0.73	0.75	0.76	0.78	0.79	0.81
2.5	0.75	Chatterton	0.65	0.65	0.65	0.65	0.65	0.65
		Benton	0.73	0.75	0.76	0.78	0.79	0.81
3.0	0.90	Chatterton	0.64	0.64	0.64	0.64	0.64	0.64
		Benton	0.73	0.75	0.76	0.78	0.79	0.81
3.5	1.00	Chatterton	0.64	0.64	0.64	0.64	0.64	0.64
		Benton	0.73	0.75	0.76	0.78	0.79	0.81
4.0	1.20	Chatterton	0.63	0.63	0.63	0.63	0.63	0.63
		Benton	0.73	0.75	0.76	0.78	0.79	0.81
4.5	1.35	Chatterton	0.63	0.63	0.63	0.63	0.63	0.63
		Benton	0.73	0.75	0.76	0.78	0.79	0.81
5.0	1.50	Chatterton	0.62	0.62	0.62	0.62	0.62	0.62
		Benton	0.73	0.75	0.76	0.78	0.79	0.81

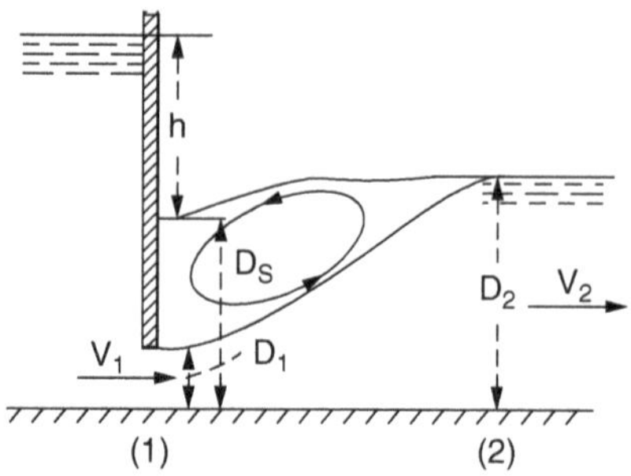

FIGURE 4.12 Flow through sluice gate.

Similar expressions may be obtained for channels other than rectangular or for cases where the gate width is less than the channel width. The principles involved in determining the nature of the water surface above and below the gate are presented in Sec. 8.

Pipe Orifices

A circular sharp-edged pipe orifice is shown in Fig. 4.13. The velocity at the vena contracta 2 is given by Eq. (4.3). Replacing the energy-loss term h_l with C_v, as was done for orifices in tanks, Eq. (4.3) becomes

$$v_2 = C_v \sqrt{2g}\left(\frac{p_1}{w} - \frac{p_2}{w} + \frac{v_1^2}{2g}\right) \tag{4.26}$$

In pipe orifices, the velocity of approach factor $v_1^2/2g$ cannot be omitted. Again letting $a_2 = C_c a$ and $C = C_c C_v$, the expression for the discharge becomes

$$Q = Ca\sqrt{2g\left(\frac{p_1}{w} - \frac{p_2}{w} + \frac{v_1^2}{2g}\right)} \tag{4.27}$$

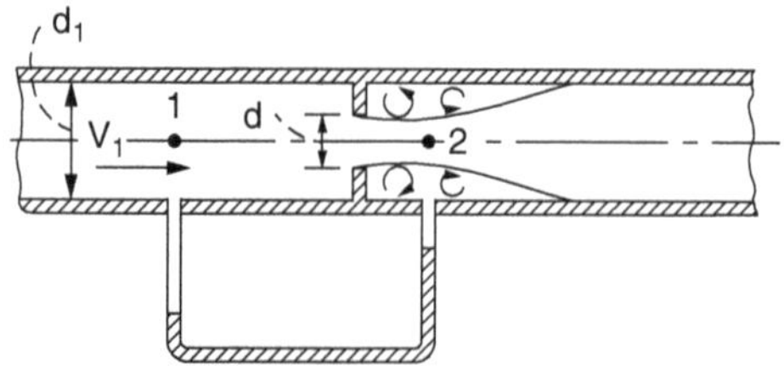

FIGURE 4.13 Pipe orifice.

In this form, the equation requires a cut-and-try solution. However, by replacing v_1^2 with its value obtained from the equation of continuity $(Q/a_1)^2$, a_1 being the area of the pipe, the following expression for Q is derived:

$$Q = Ca\sqrt{\frac{2g(p_1/w - p_2/w)}{1 - C^2(a/a_1)^2}} \tag{4.28}$$

The value of C varies not only with the factors that affect orifices in tanks, but also with the ratio of the orifice diameter d to the pipe diameter d_1 and with the location of the pressure taps. Tests indicate that the vena contracta is approximately ½ pipe diameter downstream from the orifice.[18] When the orifice is at the end of the pipe, p_2 in Eq. (4.28) is zero and only a single pressure tap is required. Curves showing the relation between C and the Reynolds number are plotted in Fig. 4.14. Portions of the curves are shown as broken lines because

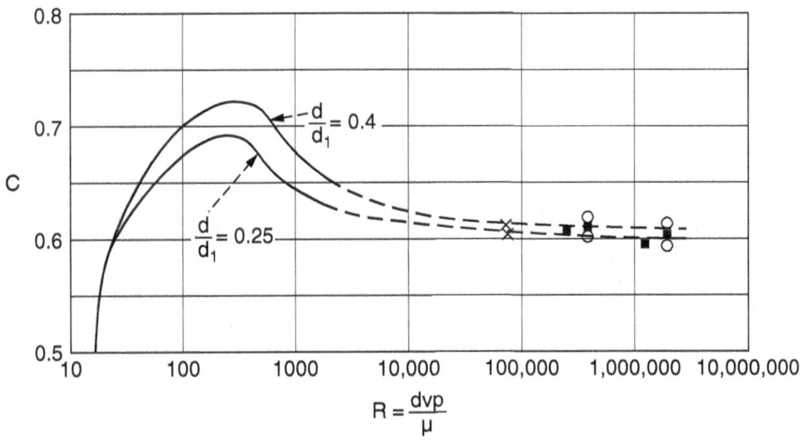

Symbol	Fluid	Experimenter	Source	Remarks
Solid lines	Castor oil	Kowalke, Bain, and Moss	King, Wisler, and Woodburn, "Hydraulics"*	
×	Air	Bean, Buckingham, and Murphy	King, Wisler, and Woodburn, "Hydraulics"*	Values of R were estimated
○	Water	Blackburn	King, Wisler, and Woodburn, "Hydraulics"*	Values of R were estimated
■	Water	Bailey	"Fluid Meters: Their Theory and Application," A.S.M.E. Research Pub.	Values of R were estimated

*H. W. King, C. O. Wisler, and J. G. Woodburn, *Hydraulics*, John Wiley and Sons/Toppan Company, Fifth Edition, 1948.

FIGURE 4.14 Coefficients for pipe orifices.

values of R are uncertain for the meager data available in this region. Equation (4.28) may be restated in the following form:

$$Q = Ka\sqrt{2g\left(\frac{p_1}{w} - \frac{p_2}{w}\right)} \tag{4.29}$$

where K is defined as

$$K = \frac{C}{\sqrt{1 - C^2(a/a_1)^2}} \tag{4.30}$$

In Eqs. (4.29) and (4.30) Q is the discharge, a and a_1 are the areas of the orifice and the pipe, respectively, p_1 and p_2 are the pressures at the upstream and downstream sides of the orifice, respectively, and w is the unit weight of the liquid. It will be noted from the preceding derivation that C takes care of the energy loss and the contraction of the jet. It follows from Eq. (4.30) that the coefficient K includes these factors as well as the velocity of approach.

Values of K for square-edged pipe orifices have been presented by the ASME[19] for three different types of pressure taps: flange taps, vena contracta taps, and radius taps. Flange taps are located with the center of the high-pressure tap 1 in. (2.5 cm) from the upstream face of the orifice plate and the center of the low-pressure tap 1 in. (2.5 cm) from the downstream side of the plate. For vena contracta taps, the high-pressure tap is located 1 pipe diameter d_1 upstream from the face of the orifice plate, and the low-pressure tap is located at the vena contracta. The mean distance from the orifice plate to the vena contracta, in terms of pipe diameters, for various values of the ratio of orifice diameter to pipe diameter d/d_1 is given in the following table:

d/d_1	Distance from orifice plate to vena contracta,* pipe diameters
0.8	0.32
0.7	0.45
0.6	0.57
0.5	0.66
0.4	0.73
0.3	0.80
0.2	0.87

*ASME, *Fluid Meters: Their Theory and Application*, pt. 1, 4th ed., 1937.

Radius taps are located 1 pipe diameter upstream and ½ pipe diameter downstream from the orifice plate.

Values of K are presented in terms of the Reynolds number,

$$R = \frac{dV}{\nu} \tag{4.31}$$

where d is the orifice diameter, V the velocity at the orifice, and ν the kinematic viscosity of the fluid. Values of ν for water may be obtained from Table 1.2. Values of K to be used with flange taps for 6- and 10-in. (15- and 25-cm) pipe are shown in Fig. 4.15. These were taken from the ASME Power Test Codes.[19] The curves for 6-in. (15-cm) pipe are quite similar to those for 3- and 4-in. (7- and 10-cm) pipe, whereas those for 10-in. (25-cm) pipe are similar to those for 8- and 14-in. (20- and 35-cm) pipe. Values of K for 2-in. (5-cm) pipe are somewhat lower than those for 6-in. (15-cm) pipe for $R < 10^5$ and slightly higher for $R > 2 \times 10^5$. If an installation of a pipe orifice is to be made without a calibration, it is recommended that reference be made to the original data for the particular pipe size, as well as to detailed installation requirements.[19]

Curves showing values of K for vena contracta taps are shown in Fig. 4.16. While these curves are for 6-in. (15-cm) pipe,[19] they are nearly identical to those for other pipe diameters.

Values of K for a pipe orifice in any particular location may be expected to differ somewhat from values given in Figs. 4.15 and 4.16 unless installation recommendations are carefully followed. Detailed instructions are available. Some of the more important restrictions

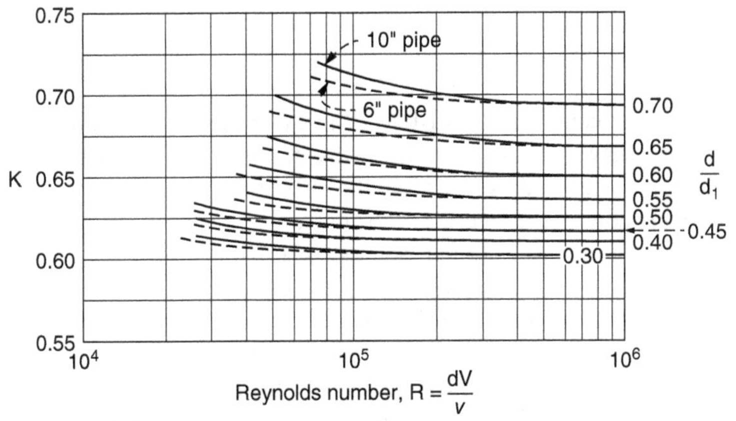

FIGURE 4.15 Pipe-orifice coefficients for flange taps.

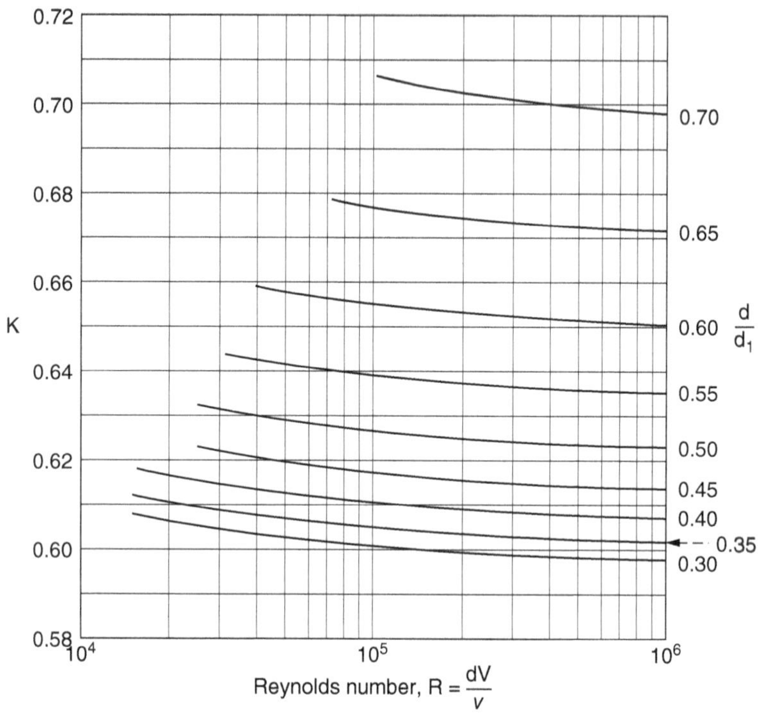

Figure 4.16 Pipe-orifice coefficients for vena contracta taps.

have to do with the location of pipe fittings in the vicinity of the orifice. For example, elbows or open valves must be located 5 pipe diameters upstream from the orifice for small values of d/d_1 and as many as 25 diameters upstream from the orifice for large values of d/d_1. No elbow should be placed nearer than 5 diameters from the downstream side of the orifice. Partially open valves should not be located upstream from the orifice. Where these restrictions cannot be followed, flow straighteners must be used, or the orifice must be calibrated in place. Studies regarding the effect of pipe fittings on orifice flow were conducted by the American Gas Association and ASME.[20]

The American National Standards Institute (ANSI) standard orifice is also a square-edged circular-pipe orifice. The pressure taps, however, are located immediately adjacent to the orifice plate. Coefficients for this type of taps have been published by the Verein Deutscher Ingenieure (VDI), translated into English by the American Society for Mechanical Engineers (ASME), and published by the National Advisory Committee for Aeronautics.[21]

Tubes

Borda's mouthpiece is a short cylindrical tube projecting inward. The edge of the tube must be relatively thin and sharp to ensure perfect contraction, and its length, about ½ diameter, such that the jet will not touch the sides of the tube. The following are average coefficients:

$$C = 0.51 \qquad C_v = 0.98 \qquad C_c = 0.52$$

Standard Short Tubes

A cylindrical tube having a length of 2 to 3 diameters, with the inner end flush with a flat wall so as to form a sharp-cornered entrance, is called a *standard short tube*. In such tubes the issuing jet first contracts and then expands, filling the tube. The coefficient of contraction is considered unity. The coefficient of discharge varies from 0.78 to 0.83. The mean value generally used is

$$C = 0.82$$

Short tubes projecting inward have coefficients of discharge varying from 0.72 to 0.80. The average value is

$$C = 0.75$$

Convergent short tubes are frustums of cones, as shown in Figs. 4.17 and 4.18. Figure 4.17 has the larger base flush with a flat wall forming a sharp-cornered entrance. Figure 4.18 has the entrance to the tube slightly rounded. The sides of the tube make an angle θ with the axis of the cone.

Experiments for these tubes give conflicting results. Fair average values are given by Unwin[22] as follows:

	C	
Angle θ, deg.	Fig. 4.17	Fig. 4.18
0	0.83	0.97
5¾	0.94	0.95
11¼	0.92	0.92
22½	0.85	0.88
45	—	0.75

A *converging bell-mouthed orifice* is illustrated in Fig. 4.19. If the surface of the opening is rounded to conform to the shape of the

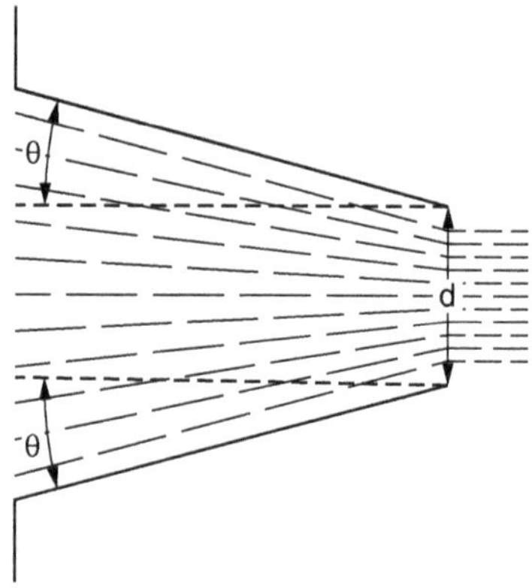

FIGURE 4.17 Convergent tube with sharp corner at entrance.

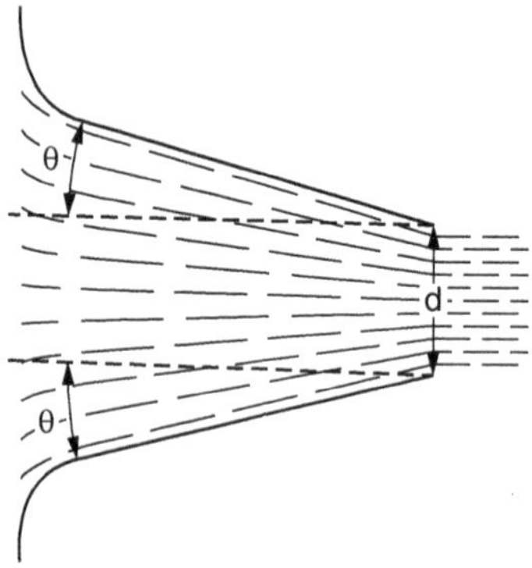

FIGURE 4.18 Convergent tube with rounded corner at entrance.

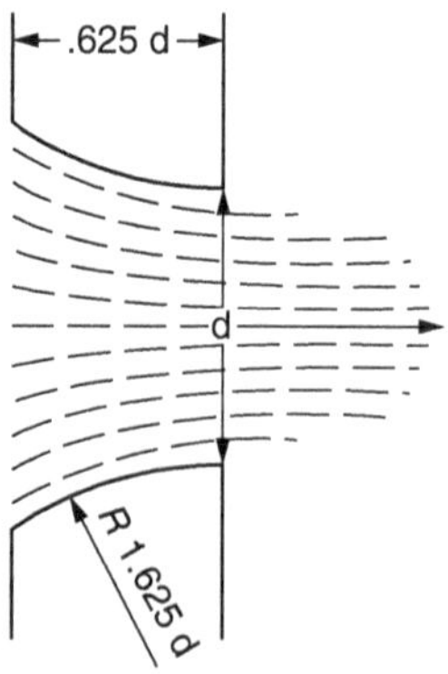

FIGURE 4.19 Converging bell-mouthed orifice.

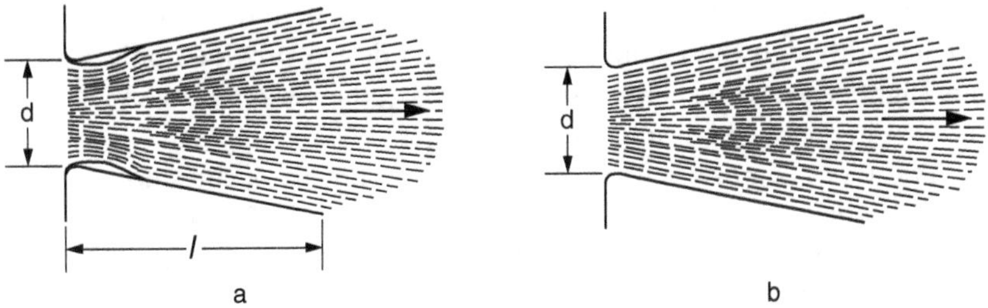

Figure 4.20 (*a*) Diverging tube with sharp corner at entrance. (*b*) Diverging tube with rounded corner at entrance.

contracted jet, C approaches unity. The following are coefficients by Weisbach[23] for $d = 0.033$ ft (1 cm). Other experiments indicate that these results hold approximately for larger orifices.

h, ft	*h*, m	*C*
0.066	0.02	0.959
1.640	0.5	0.967
11.480	3.5	0.975
55.770	17.0	0.994
337.930	103.0	0.994

Diverging conical tubes are shown in Fig. 4.20. The coefficient of discharge varies with the angle of divergence and the length of the tube. Experiments by Venturi showed the discharge to be a maximum with $l = 9d$ and an angle of divergence equal to 5°. If divergence is not too great, the tube will flow full. The coefficient of discharge is variable, but when so designed that the tube flows full, the following results may be obtained:

For Fig. 4.20*a* $C = 1.4$
For Fig. 4.20*b* $C = 2.0$

Nozzles

In 1888, Freeman[24] reported the results of tests made on various types of nozzles suitable for fire hoses. For many years no significant change in design had been developed. In 1952, results of tests carried on at the Iowa Institute of Hydraulic Research[25] were published. These tests were conducted with primary attention on the efficiency of the fire stream itself. This research resulted in improved designs of both monitors and nozzles. Although the recommended nozzle design provided only a very slight decrease in energy loss in the nozzle, there was a substantial decrease in the rate of disintegration of the jet.

Two types of nozzles were reported on by Freeman, the smooth type without jet contraction and the ring type shown in Fig. 4.21. The following are mean values of coefficients of discharge for smooth nozzles as determined by Freeman:

Diameter, in.	Diameter, cm	*C*
¾	2.0	0.983
7/8	2.2	0.982
1	2.5	0.972
1 1/8	2.8	0.976
1¼	3.1	0.971
1 3/8	3.4	0.959

The mean values of coefficients of discharge for ring nozzles as determined from Freeman's experiments are as follows. The ratio of diameter of opening to diameter just behind ring is given.

Ratio	*C*
0.50	0.63
0.60	0.65
0.70	0.68
0.80	0.71
0.85	0.73
0.90	0.77
0.95	0.87
1.00	0.975

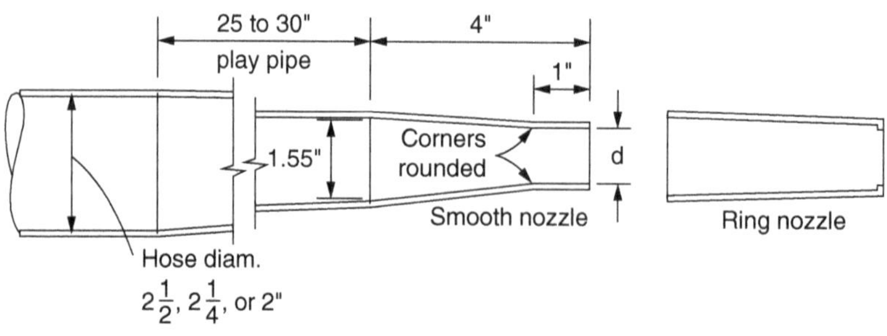

Figure 4.21 Smooth and ring nozzles.

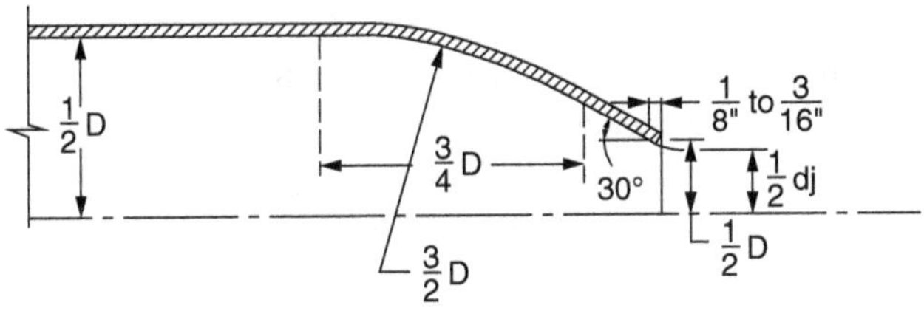

FIGURE 4.22 Nozzle developed from Iowa tests.

In Freeman's tests, as well as in the Iowa tests, the velocity of approach was neglected, and the coefficient of discharge was computed from the following equation:

$$Q = Ca\sqrt{2gh} \tag{4.10}$$

where a is the area of the nozzle and h is the pressure head just upstream from the nozzle.

The form of nozzle developed by the Iowa tests is shown in Fig. 4.22. Reported values of d/D, d_j/D, and C are given in the following table, with the terms d, d_j, and D defined in Fig. 4.22.

d/D	d_j/D	C	C_c^*	C_v^*
0.279	0.250	0.805	0.805	1.000
0.335	0.300	0.806	0.805	0.999
0.391	0.350	0.807	0.805	0.997
0.446	0.400	0.811	0.806	0.995
0.501	0.450	0.817	0.807	0.987
0.557	0.500	0.825	0.807	0.977

*Computed by the author.

Computations of C_c and C_v indicate that the energy losses are very small, as shown by the fact that values of C_v differ only slightly from unity. It may be seen from Eq. (4.11) that when $C_v = 0.977$, the energy loss is approximately 5 percent.

Information regarding standardized nozzles which are used to measure discharge in pipes is given in Sec. 12.

Submerged Tubes

Stewart[26] experimented on submerged tubes 4 ft (1.2 m) square and 0.31 to 14 ft (0.1 to 4.3 m) long under heads of 0.05 to 0.30 ft (1.5 to 9 cm). Entrance conditions included sharp edges and various degrees of suppressed contraction. Rogers and Smith[27] extended Stewart's experiments to include sharp-edged tubes 6, 8, and 10 in. (15, 20, and 25 cm) square under heads of 2.2 ft (67 cm) and less. They found C [Eq. (4.21)] to vary with L/D, L being the length of the tube and D the length of one side of the cross section, but independent of the head.

Table 4.8, obtained from the experiments described, expresses C as a function of L/p, p being the length of the perimeter of the tube cross section. For a square tube, $p = 4D$. Though the values of C

Table 4.8 Coefficients of Discharge for Submerged Tubes*

$\frac{L^\dagger}{p}$	All corners square	Condition of edges at entrance: Contractions suppressed on bottom only	Contractions suppressed on bottom and one side	Contractions suppressed on bottom and two sides	Contractions suppressed on bottom, two sides, and top
0.02	0.61	0.63	0.68	0.77	0.95
0.04	0.62	0.64	0.68	0.77	0.94
0.06	0.63	0.65	0.69	0.76	0.94
0.08	0.65	0.66	0.69	0.74	0.93
0.10	0.66	0.67	0.69	0.73	0.93
0.12	0.67	0.68	0.70	0.72	0.93
0.14	0.69	0.69	0.71	0.72	0.92
0.16	0.71	0.70	0.72	0.72	0.92
0.18	0.72	0.71	0.73	0.72	0.92
0.20	0.74	0.73	0.74	0.73	0.92
0.22	0.75	0.74	0.75	0.75	0.91
0.24	0.77	0.75	0.76	0.78	0.91
0.26	0.78	0.76	0.77	0.81	0.91
0.28	0.78	0.76	0.78	0.82	0.91
0.30	0.79	0.77	0.79	0.83	0.91
0.35	0.79	0.78	0.80	0.84	0.90
0.40	0.80	0.79	0.80	0.84	0.90
0.60	0.80	0.80	0.81	0.84	0.90
0.80	0.80	0.80	0.81	0.85	0.90
1.00	0.80	0.81	0.82	0.85	0.90

*Compiled from experiments by Stewart[26] and Rogers and Smith.[27]

†L = length of tube, p = perimeter of tube cross section.

contained in Table 4.8 are derived from experiments on square tubes, they will probably apply quite accurately to other tubes which do not depart far from this form of section.

Culverts

The total loss of head in a culvert is the sum of the entrance and outlet losses, which vary with the design of the ends, plus the loss of head due to friction, which varies directly with the length. It is convenient to treat culverts flowing full, with both ends submerged, as submerged tubes and to include all losses of head in the coefficient of discharge C [Eq. (4.21)]. To obtain the maximum discharge, the intake must be submerged to a depth equal to at least the sum of the velocity head and the loss of head at the entrance. The discharge is entirely independent of the slope of the tube. It depends only on h, the difference in elevation of the water surfaces at the intake and the outlet.

The results of 1480 experiments on concrete, vitrified clay, and corrugated-metal-pipe culverts and 1821 experiments on concrete box culverts were published by Yarnell et al.[28] Experiments were conducted on pipes 12, 18, 24, and 30 in. (30, 45, 60, and 75 cm) in diameter and on box culverts 2 by 2, 3 by 3, 4 by 4, 4 by 3, 4 by 2¼, 4 by 2, 4 by 1, and 4 by ½ all in feet (0.6 by 0.6, 0.9 by 0.9, 1.2 by 1.2, 1.2 by 0.9, 1.2 by 0.7, 1.2 by 0.6, 1.2 by 0.3, and 1.2 by 0.15, all in meters). The lengths of both pipes and box culverts experimented upon were 24, 30, and 36 ft (7.3, 9.1, and 11.0 m). If d is the diameter of pipe culverts, r the hydraulic radius of box culverts, and L the length of the culvert, the following are expressions for determining C in Eq. (4.21).

For concrete pipe, beveled lip entrance,

$$C = \left(1.1 + \frac{0.026L}{d^{1.2}}\right)^{-1/2} \qquad \text{(dimensions in feet)}$$

$$C = \left(1.1 + \frac{0.0205L}{d^{1.2}}\right)^{-1/2} \qquad \text{(dimensions in meters)} \qquad (4.32)$$

For concrete pipe, beveled lip entrance,

$$C = \left(1 + 0.31d^{0.5} + \frac{0.026L}{d^{1.2}}\right)^{-1/2} \qquad \text{(dimensions in feet)}$$

$$C = \left(1 + 0.56d^{0.5} + \frac{0.0205L}{d^{1.2}}\right)^{-1/2} \qquad \text{(dimensions in meters)} \qquad (4.33)$$

For vitrified clay pipe, bell-end upstream,

$$C=\left(1+0.023d^{1.9}+\frac{0.022L}{d}\right)^{-1/2} \qquad \text{(dimensions in feet)}$$

$$C=\left(1+0.22d^{1.9}+\frac{0.022L}{d}\right)^{-1/2} \qquad \text{(dimensions in meters)} \qquad (4.34)$$

For corrugated metal pipe,

$$C=\left(1+0.1+d^{0.6}+\frac{0.106L}{d^{1.2}}\right)^{-1/2} \qquad \text{(dimensions in feet)}$$

$$C=\left(1+0.33d^{0.6}+\frac{0.084L}{d^{1.2}}\right)^{-1/2} \qquad \text{(dimensions in meters)} \qquad (4.35)$$

For concrete box culverts with rounded lip entrance,

$$C=\left(1.05+\frac{0.0045L}{r^{1.25}}\right)^{-1/2} \qquad \text{(dimensions in feet)}$$

$$C=\left(1.05+\frac{0.0033L}{r^{1.25}}\right)^{-1/2} \qquad \text{(dimensions in meters)} \qquad (4.36)$$

For concrete box culverts with square-cornered entrance,

$$C=\left(1+0.57r^{0.3}+\frac{0.0045L}{r^{1.25}}\right)^{-1/2} \qquad \text{(dimensions in feet)}$$

$$C=\left(1+0.57r^{0.3}+\frac{0.0033L}{r^{1.25}}\right)^{-1/2} \qquad \text{(dimensions in meters)} \qquad (4.37)$$

Inasmuch as the experiments from which these formulas were derived include only lengths of 24 to 36 ft (7 to 11 m), it would appear best, for long culverts, to apply the formulas only to a portion of the length say 25 to 50 ft (10 to 15 m) and compute the head lost in the remainder of the conduit by an open-channel or pipe formula. The h to be used in Eq. (4.21) is then the difference in elevation of the water surfaces at the entrance and the outlet minus this computed loss of head.

The experiments indicate that 45° wing walls used in connection with pipe culverts increase the discharge by 1 to 10 percent over that obtained with a straight end wall and that wing walls are more efficient when set flush with the edge of the pipe than when set 6 in. (15 cm) back from the edge. U-type wing walls used in connection with vitrified clay pipe produced a slightly smaller discharge than that obtained with a straight end wall. The beveled lip at the entrance to concrete pipes greatly reduced the entrance loss. It was found that a 24-in. (60-cm) clay pipe 38 ft (11.5 m) long with a straight end wall and the bell-end upstream carried about 10 percent more water than the same culvert with a square-cornered entrance. Rounding the entrance of a 24-in. (60-cm) vitrified-clay-pipe culvert increased the capacity by 13 percent. It was found also that the discharge is decreased slightly by projecting the pipe through the head wall. In a square-cornered entrance there is little difference whether the pipe projects 3 in. or 2 or 4 ft (7.5 cm or 60 or 120 cm). An 18-in. (45-cm) corrugated metal pipe with a 3-in. (7.5-cm) projection carries slightly more water than the same pipe with a 2- or 4-ft (60- or 120-cm) projection. Doubling the end area of the submerged outlet of an 18-in. (45-cm) vitrified-clay-pipe culvert by attaching a 10° diverging end was found to increase the discharge by 40 percent over that obtained through the same culvert without the diverging end.

The culvert discharge may also be computed by considering the culvert as a short pipe. Bernoulli's equation written from point 1 to point 2 in Fig. 4.23, taking the datum plane at the elevation of the water surface at point 2, gives

$$H = h_e + h_f + h_o \tag{4.38}$$

In this equation h_e is the entrance loss, h_f the friction loss, and h_o the loss at the outlet. The entrance and outlet losses are usually expressed in terms of the velocity head,

$$h_e = K_e \frac{V^2}{2g} \tag{4.39}$$

and

$$h_o = K_o \frac{V^2}{2g} \tag{4.40}$$

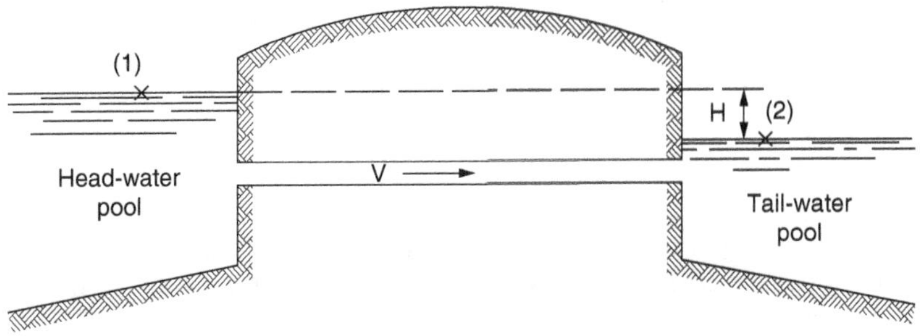

FIGURE 4.23 Pipe culvert.

where V is the average velocity in the pipe. The friction loss may be determined from the Manning formula or the Darcy-Weisbach equation, as described in Secs. 6 and 7.

Straub and Morris[29] have reported values of K_e and K_o as well as values of the Manning coefficient n and the Darcy-Weisbach coefficient f, based on a series of laboratory tests on concrete and corrugated-metal-pipe culverts. Full-scale tests were made on culverts 193 ft (59 m) long for pipes flowing full and for the case of uniform flow with a free surface. Round pipes of 18, 24, and 36 in. (45, 60, and 90 cm) diameters were used, as well as arched corrugated metal pipes having the same perimeters as the round pipes. The dimensions of the corrugated pipe refer to the inside diameter, and the height of the corrugations was ½ in. (1.25 cm). The entrance to the concrete culverts consisted of the groove end of a length of tongue-and-groove pipe. Tests were made with the inlet end flush with the wall of the headwater pool and also with a projecting inlet.

For the concrete pipe, design values of the entrance-loss coefficient K_e were found to be 0.10 for the flush inlet and 0.15 for the projecting inlet. These values apply to full flow and to part full flow conditions. For corrugated metal pipes flowing full, practical design values of K_e were found to be 0.5 for a flush entrance and 0.9 for the projecting entrance. For the case of flow with a free surface in corrugated pipes, the corresponding values of K_e were 0.4 and 0.7, respectively.

The outlet-loss coefficient K_o was found to be nearly 1.0 in all cases.

Values of the Manning n varied somewhat with the pipe size, the shape of the pipe, and the Reynolds number. Practical design values taken near the upper end of the range are 0.011 for concrete pipe and 0.025 for corrugated metal pipe. Similar values of the Darcy-Weisbach f for concrete pipe are 0.018 for the 18-in. (45-cm) pipe and 0.016 for the 24- and 36-in. (60- and 90-cm) pipe. The corresponding values of f for corrugated metal pipe are 0.105 for the 18 in. (45 cm) size, 0.090 for the 24 in. (60 cm) size, and 0.075 for the 36 in. (90 cm) size. The results obtained from Eq. (4.38), using the coefficients given here, are in close agreement with values computed from Eq. (4.21), using appropriate values of C from Table 4.8.

References

1. H. Smith, Jr., *Hydraulics*, Wiley, New York 1886.
2. F. W. Medaugh and G. D. Johnson, "Investigation of the Discharge and Coefficient of Small Circular Orifices," *Civil Eng.* (*N.Y.*), pp. 422–424, Jul. 1940.

3. D. Smith and W. J. Walker, "Orifice Flow," *Proc. Inst. Mech. Eng. (London)*, pp. 23–36, 1923.
4. Williams, experiments performed at University of Michigan, 1928, unpublished.
5. F. C. Lea, *Hydraulics*, 6th ed., Longmans, Green, New York, 1942.
6. H. J. I. Bilton, "Coefficients of Discharge through Circular Orifices," paper read before Victorian Institute of Engineers, Apr. 1908; *Eng. News*, vol. IX Jul. 9, 1908.
7. J. T. Fanning, *Water-Supply Engineering*, Van Nostrand, Princeton, NJ, pp. 205–206, 1906.
8. H. T. Bovey, *Hydraulics*, Wiley, New York, p. 40, 1909.
9. W. C. Unwin, "Hydraulics," in *Encyclopaedia Britannica*, 11th ed., vol. 14, p. 41, 1912.
10. E. L. Bellasis, *Hydraulics*, 2d ed., Spon & Chamberlain, New York, p. 68, 1911.
11. C. N. Ward, thesis, University of Michigan, 1915, unpublished.
12. H. D. Newell, U.S. Bureau of Reclamation, Experimental Data 9, 1916.
13. W. G. Steward, U.S. Bureau of Reclamation, Experimental Data 35, 1923.
14. Philip à Morley Parker, *The Control of Water*, Routledge & Kegan Paul, London, 1925, pp. 164–168.
15. Bornemann, *Civilingenieur*, vol. 26, p. 297, 1880.
16. Chatterton, Hydraulic experiments in the Kistna delta.
17. Benton, Punjab Irrigation Branch, Paper 8.
18. ASME, *Flow Meters: Their Theory and Application*, 3d ed., 1931.
19. ASME, "Flow Measurement by Means of Standardized Nozzles and Orifice Plates," Power Test Codes, Supplement to "Instruments and Apparatus," pt. 5, Chap. 4, 1949, prepared and published by ASME under the auspices of the Power Test Codes Committee.
20. "Investigation of Orifice Meter Installation Requirements," Interim Research Report, American Gas Association and ASME, Mar. 1951.
21. "Standards for Discharge Measurement with Standardized Nozzles and Orifices," German Industrial Standard, 4th ed., 1952; NACA Tech. Mem. 952.
22. W. C. Unwin, *A Treatise on Hydraulics*, A & C Black, London, p. 89, 1907.
23. J. Weisbach, *Ingenieur und Machinen-Mechanik*, F. Vieweg und sohn, Braunschweig, p. 969, 1875.
24. J. R. Freeman, "Experiments Relating to Hydraulics of Fire Streams," *Trans. ASCE*, vol. 21, pp. 303–482, 1888.
25. H. Rouse, J. W. Howe, and D. E. Metzler, "Experimental Investigation of Fire Monitors on Nozzles," *Trans. ASCE*, vol. 117, pp. 1147–1188, 1952.
26. C. B. Stewart, "Investigation of Flow through Large Submerged Orifices and Tubes," *Univ. Wis. Bull.*, no. 216, vol. 4, no, 4, pp. 249–332, 1908.
27. T. C. Rogers and T. L. Smith, "Experiments with Submerged Orifices and Tubes," *Eng. News*, vol. 76, no. 18, pp. 825–827, Nov. 2, 1916.
28. D. L. Yarnell, F. A. Nagler, and S. M. Woodward, "Flow of Water through Culverts," *Univ. Iowa Studies in Eng., Bulletin 1*, 1926.
29. L. G. Straub and H. M. Morris, "Hydraulic Tests on Concrete Culvert Pipes," Tech. Paper 4, ser. B; "Hydraulic Tests on Corrugated Metal Culvert Pipes," Tech. Paper 5, ser. B, St. Anthony Falls Hydraulic Laboratory, Minneapolis, MN, 1950.

SECTION 5
Weirs

A weir is a notch of regular form through which water flows. The term is also applied to the structure containing such a notch. Thus a weir may be a depression in the side of a tank, reservoir, or channel, or it may be an overflow dam or other similar structure. Classified in accordance with the shape of the notch, there are rectangular weirs, triangular or V-notch weirs, trapezoidal weirs, parabolic weirs, and weirs of irregular shape.

The edge or surface over which the water flows is called the *crest* of the weir. The overflowing sheet of water is termed the *nappe*. The depth of water producing the discharge, H in Fig. 5.1, is the *head*. A weir with a sharp upstream corner, or edge, so formed that the water springs clear of the crest, is called a *sharp-crested* weir. Sharp-crested weirs are usually made by machining a 90° angle on the inner edge of the crest. All other weirs, for want of a better term, are classed as weirs *not* sharp-crested, or *round-crested* weirs. A *broad-crested* weir may be defined as one having a horizontal or nearly horizontal crest sufficiently long in the direction of flow so that the nappe will be supported and hydrostatic pressures will be fully developed for at least a short distance. The approach channel is the channel leading up to the weir, and the mean velocity in this channel is the approach velocity. If the nappe discharges into the air (Fig. 5.1), the weir has free discharge. If the discharge is partially under water (Fig. 5.5), the weir is said to be *submerged*, or *drowned*.

Sharp-crested weirs are useful as a means of measuring flowing water. Weirs not sharp-crested are commonly incorporated in hydraulic structures, and while they are sometimes employed to measure flow, this is usually a secondary function.

Fundamental Principles

The equation for discharge over a weir cannot be derived exactly because not only does the flow pattern of one weir type differ from that of another, but the flow pattern for a given weir varies with the

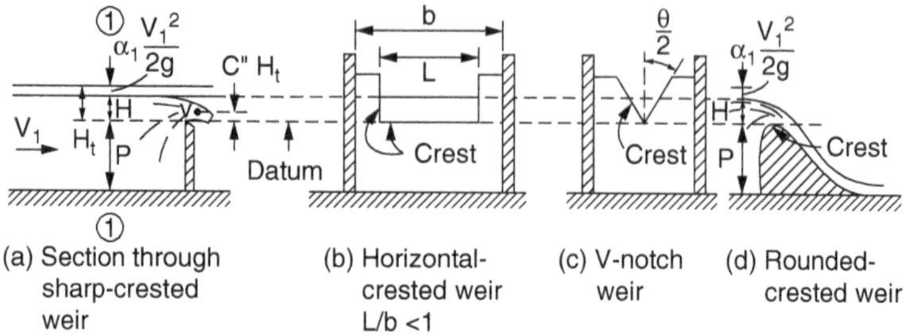

FIGURE 5.1 Weirs, definition sketch.

discharge. Furthermore, the number of variables involved is so great as to defy a rigorous analytical approach. Approximate derivations are presented in most texts.[1] These derivations show effects of gravitational forces in an approximate manner, but do not include the effects of viscosity, surface tension, the ratios of the dimensions of the weir to the dimensions of the approach channel, the nature of the weir crest, the velocity distribution in the approach channel, and the roughness of the inside surface of the weir and approach channel. A simplified derivation will be made here to show the general nature of the relationship between the discharge and the most important variables and to demonstrate the nature of the effect of some of the variables. The derivation will be made for sharp-crested weirs, but as will be shown later, a similar derivation would apply to weirs that are not sharp-crested.

The Bernoulli equation,[a] written from any point in section 1 of Fig. 5.1 to υ, which is the center of gravity of the vena contracta of the nappe located $C'H_t$ above the lowest point on the crest, is

$$H + \alpha_1 \frac{V_1^2}{2g} = H_t = C'H_t + \alpha_\upsilon \frac{V_\upsilon^2}{2g} + h_t \tag{5.1}$$

It may be seen from Fig. 5.1 and Eq. (5.1) that H_t is the total energy head above the datum on the upstream side of the weir, which includes both the potential energy H and the kinetic energy, or velocity of approach, term $\alpha_1(V_1^2/2g)$. Solving Eq. (5.1) for V_υ,

$$V_\upsilon = \sqrt{2g}\sqrt{\frac{1}{\alpha_\upsilon}\left[H_t(1-C') - h_l\right]} \tag{5.2}$$

[a]See "Energy and the Bernoulli Equation," Sec. 3.

Introducing an arbitrary coefficient C'' to serve the combined functions of h_l, C', and α_v permits the rearrangement of Eq. (5.2) into the following simpler form:

$$V_v = C''\sqrt{2gH_t} \tag{5.3}$$

The discharge Q may be expressed as

$$Q = A_v V_v \tag{5.4}$$

The area at the vena contracta A_v may be expressed in terms of the weir opening and a coefficient of contraction,

$$A_v = C_c A \tag{5.5}$$

where A is the area of the weir opening bounded at the top by the plane of the water surface in the approach channel. Then

$$Q = C_c A C''\sqrt{2gH_t} \tag{5.6}$$

Letting

$$C_1 = C_c C''\sqrt{2g} \tag{5.7}$$

Equation (5.6) becomes

$$Q = C_1 A\sqrt{H_t} \tag{5.8}$$

If the weir has a *horizontal* crest of length L, as shown in Fig. 5.1*b*,

$$A = HL$$

and Eq. (5.8) becomes

$$Q = C_1 LHH_t^{1/2} \tag{5.9}$$

If now the coefficient is made to include the effect of the velocity of approach, H_t can be replaced by H, and

$$Q = CLH^{3/2} \tag{5.10}$$

This is the general form of the equation used for rectangular weirs.

If the weir is a V-notch weir, reference to Fig. 5.1*c* will show that

$$A = H^2 \tan\frac{\theta}{2} \tag{5.11}$$

and

$$Q = C_1 \tan\frac{\theta}{2} H^2 H_t^{1/2} \tag{5.12}$$

Again, including the velocity of approach in the coefficient,

$$Q = C_1 \tan\frac{\theta}{2} H^{5/2} \tag{5.13}$$

Letting

$$C = C_1 \tan\frac{\theta}{2} \tag{5.14}$$

the expression for discharge becomes

$$Q = CH^{5/2} \tag{5.15}$$

This is the general form of the equation for V-notched weirs.

The same approximate derivation would apply to round-crested weirs, as illustrated in Fig. 5.1*d*, the only difference being that the section corresponding to the vena contracta would occur at or near the crest and any contraction of the nappe would occur only at the top of the nappe.

As stated previously, the coefficient C must include the effect of all variables except those included in Eqs. (5.10) and (5.15).

The most convenient method of including these variables can be established by dimensional analysis. In general, it can be shown that

$$C = f\left(R, W, \frac{x_1}{y_1}, \frac{x_2}{y_2}, \ldots, \frac{x_n}{y_n}\right) \tag{5.16}$$

where R is the Reynolds number, W is the Weber number (see Sec. 3), and the x/y ratios indicate significant length ratios such as, for example, the head divided by the height of the weir (H/P), a roughness length divided by the head, or the side slope of a V notch. The particular parameters that are of importance for each type of weir, indicated generally by Eq. (5.16), will be discussed in detail in the appropriate subsections that follow.

Sharp-Crested Weirs

All the tests on weirs of this type were made with the nappe fully aerated. When the crest length L of a horizontal weir (Fig. 5.1) is shorter than the width of the channel b, as well as in the case of V-notch weirs, aeration is automatic. However, for horizontal weirs extending over the full width of the channel, $L/b = 1$, air at atmospheric pressure must be provided by vents. Otherwise the air beneath the nappe will be exhausted, causing a reduction of pressure beneath the nappe, with a corresponding increase in the discharge for a given head but also possible pressure fluctuations that will lead to unstable flow and fatigue loading on the structure.

Horizontal Sharp-Crested Weirs with $L/b = 1$

Such weirs are also described as weirs with suppressed end contraction because the length of the nappe at the vena contracta is the same as the length of the weir. This does not mean, however, that the discharge is the same over each unit of length of the weir, because of a reduced velocity in the boundary layer which causes a slight reduction in discharge near the walls of the channel.

A consideration of the terms on the right side of Eq. (5.16) for horizontal weirs will show that a convenient method of including the variation in the velocity of approach, as well as the contraction of the jet, is to relate C to H/P. When H/P is small, the area of the approach channel is relatively large compared with the area of the nappe at the vena contracta, and the kinetic-energy term is of relatively little importance, whereas, as H/P becomes larger, the velocity in the approach channel becomes larger in relation to the velocity in the nappe.

The contraction of the nappe from the top is caused by the conversion of potential energy to kinetic energy as the flow falls over the weir. However, at the bottom of the nappe the contraction is caused primarily by the vertical component of the velocity along the inner face of the weir. When H/P is small, the contraction is relatively large, but for large values of H/P the influence of the vertical components is smaller and there is less contraction. Thus in regard to the velocity of approach and the contraction of the nappe, there will be a larger discharge for a given H when H/P is large, and this increase in discharge is reflected in the variation of C with H/P. Even some of the very earliest weir equations included H/P in the expression for C.

The effect of viscosity is represented in Eq. (5.16) by the Reynolds number R. This factor is related to the energy loss as well as to the velocity variation near the boundaries. The length term in the Reynolds number may be taken as either the head or the length, so that alternative forms of R are

$$R = \frac{VH}{\nu} \tag{5.17}$$

and

$$R = \frac{VL}{\nu} \tag{5.18}$$

If it is noted from Eq. (5.3) that $V \propto H^{1/2}$, it may be seen that, for constant values of kinematic viscosity,

$$R \propto H^{3/2} \text{ or } R \propto H^{1/2} L \tag{5.19}$$

Thus it is demonstrated that R itself need not appear in the equation for C to reflect the effect of viscosity, but that a term containing H or L, or both, can accomplish this purpose. This explains to a large extent the fact that even some of the earliest equations for C in a horizontal sharp-crested weir without end contractions contained a term involving H. For this type of weir the effect of L is apparently very small.

The surface tension affects weir discharge, particularly at low heads, because of differences both in the manner in which the nappe adheres to the weir and in the form of the nappe. The effect of surface tension is represented in Eq. (5.16) by the presence of the Weber number, which can be written for horizontal weirs with either the head or the width as the length parameter,

$$W = \frac{V\sqrt{H}}{\sqrt{\sigma/\rho}} \tag{5.20}$$

or

$$W = \frac{V\sqrt{L}}{\sqrt{\sigma/\rho}} \tag{5.21}$$

However, again as in the case of the Reynolds number, $V \propto H^{1/2}$, so that the effect of surface tension can be related to H [Eq. (5.20)] or to H and L [Eq. (5.21)], this being an additional reason for having H or L, or both, in the expressions for C.

Some values of the weir coefficients C for horizontal sharp-crested weirs with $L/b = 1$ will be presented next. It would be impractical to attempt to present all the data available in the literature. Only representative data will be given to illustrate the range of values of C.

Experiments on Sharp-Crested Rectangular Weirs

Horton[2] described early experiments in France.[b] In all the earlier experiments on weirs only small quantities of water were available. In most cases results are given in the form of C in Eq. (5.10). The Francis[3] experiments were performed in Lowell, Massachusetts, in 1852. Francis had facilities for using larger quantities of water than had been available for the earlier experiments. Fteley and Stearns,[4] in 1877 and 1879, near Boston, experimented with two sharp-crested suppressed

[b]Horton[2] gives a very complete description of most of the weir experiments performed prior to 1907, including the experiments of Francis,[3] Fteley and Stearns,[4] and Bazin.[5]

weirs. They also experimented on a weir with end contractions. The greatest velocity of approach was 1.0 ft/s. The Bazin[5] experiments on suppressed weirs were performed near Dijon, France, in 1886. There were in all 381 experiments. The Frese[6] experiments were performed in Hanover, Germany, prior to 1890. Comparatively large volumes of water were used in testing weirs under a wide range of conditions. King[7] conducted experiments at the University of Michigan. The Rehbock[8] experiments were performed at the Karlsruhe Hydraulic Laboratory in Karlsruhe, Germany. The quantities of water used were not large, but conditions were favorable for unusual accuracy and for conducting experiments under a wide range of conditions. Rehbock also presented the results of extensive experiments by the Swiss Society of Engineers and Architects. The experiments of Schoder and Turner[9] were performed at Cornell University in New York, between 1904 and 1920. With the published results of these experiments 1162 experiments by others were included. In all, 2438 separate volumetric measurements for 152 different heads were made.

Tests on weirs of this type were also conducted by Kindsvater and Carter.[10] Their tests covered a range of values of H/P from approximately 0.10 to 2.5, a range of heads from 0.10 to 0.72 ft, and weir heights from 0.30 to 1.44 ft. They also varied the weir length and channel width from 0.10 to 2.68 ft. In presenting their data they adopted the method used by Rehbock of including the effect of H in the main body of the equation. Kindsvater and Carter also introduced a method that includes the effect of the weir length L in the main body of the equation. Their method is shown in the following three equations, in which the notation has been changed to correspond to that used in this book:

$$Q = C_e L_e H_e^{3/2} \tag{5.22}$$

$$L_e = L + k_L \tag{5.23}$$

$$H_e = H + k_H \tag{5.24}$$

In these equations k_L and k_H are factors representing the effects of viscosity and surface tension, and the subscript e indicates effective values, that is, H_e is the effective head. By treating the variables in this manner the authors were able to obtain a single linear relationship between C_e and H/P for all values of H. The values of k_L and k_H were obtained by trying successive values of k_L and k_H until the values of C_e were obtained that were the most independent of H and L.

$$C_e = 3.22 + 0.40\frac{H}{P} \tag{5.25}$$

$$k_H = 0.003 \qquad k_L = -0.003$$

For metric units use the following equation.

$$C_e = 1.78 + 0.22\frac{H}{P} \tag{5.26}$$

$$k_H = 0.001 \qquad k_L = -0.001$$

For practical purposes as well as for a comparison of results it is of interest to develop the relation between C in Eq. (5.10) and C_e in Eq. (5.22),

$$Q = CLH^{3/2} = C_e L_e H_e^{3/2}$$

Inserting values of L_e and H_e from Eqs. (5.23) and (5.24),

$$CLH^{3/2} = C_e(L + k_L)(H + k_H)^{3/2} \tag{5.27}$$

Then

$$C = C_e\left(1 + \frac{k_L}{L}\right)\left(1 + \frac{k_H}{H}\right)^{3/2} \tag{5.28}$$

By using this equation, values of C were computed by the author from the Kindsvater and Carter tests and the USBR tests for comparison with those given from earlier experiments. The comparisons are shown graphically in Fig. 5.2 for H = 0.2 ft and for H = 1.0 ft.

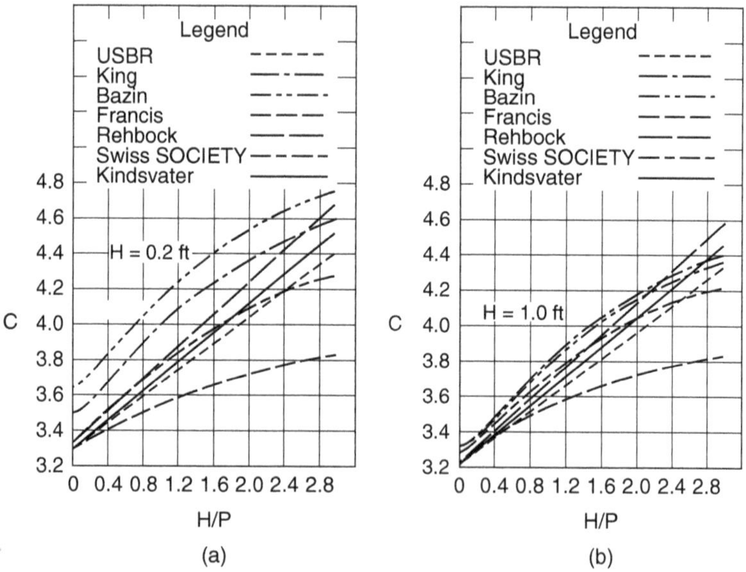

FIGURE 5.2 Coefficients, horizontal sharp-crested weirs.

To prevent confusion, the results of the tests of Fteley and Stearns, Frese, and Schoder and Turner are not shown. The Fteley and Stearns test results are similar to those of the Swiss Society; the Frese values of C are similar to those given by the King and Bazin equations; and the Schoder and Turner tests are nearly identical with the Rehbock values. In studying the variations in test results shown by these figures, it should be realized that each of the curves represents the best fit through points which may scatter over a range of 1.5 to 2.5 percent on either side of the curves.

It may be seen from Fig. 5.2 that the Francis tests give smaller values of C than the others for all values of H. For H = 0.2 ft (Fig. 5.2*a*) the Bazin and King values are considerably higher than the others. Neglecting the Francis results, there is a spread of approximately 5 percent in values of C for $H = 1.0$ ft and $H/P = 2.0$. In general it must be concluded that even among tests for which conditions appear to be quite similar, there are rather great differences in discharges for the same head, and that although the weir is a very useful measuring device, its limitations should be recognized and understood (see "Selection and Use of Weirs," in this section).

The computations involved in dealing with horizontal sharp-crested weirs without end contractions will be illustrated by a numerical example.

Example 5.1 Estimate the discharge over a sharp-crested weir extending over the full width of a rectangular channel 5 ft wide if the weir blade is 3 ft high and the head is 0.84 ft.

$$\frac{H}{P} = 0.84/3.0 = 0.28$$

From Kindsvater and Carter tests, using Eqs. (5.22) to (5.25) and noting that $k_L = 0.003$ and $k_L = -0.003$, we have, from Eq. (5.23),

$$L_e = 5 - 0.003 = 4.997 \text{ ft}$$

from Eq. (5.24),

$$H_e = 0.84 + 0.003 = 0.843 \text{ ft}$$

and from Eq. (5.25),

$$C_e = 3.22 + 0.40\frac{H}{P} = 3.22 + 0.11 = 3.33$$

Then, from Eq. (5.22),

$$Q = C_e L_e H_e^{3/2}$$
$$= 3.33 \times 4.997 \times (0.843)^{3/2} = 12.9\ \text{ft}^3/\text{s} \quad \blacktriangle$$

Horizontal Sharp-Crested Weirs with End Contractions

When the weir length L is less than the channel width b, the weir opening is in the form of a notch, with the ends of the notch having sharp edges and causing contractions in the same manner as the horizontal portion of the crest. A much smaller number of tests have been made on weirs of this type than on weirs without end contractions.

Tests made by Francis[3] indicated that the effect of the end contractions could be taken care of by reducing L in Eq. (5.10) by $0.2H$. The later work of Kindsvater and Carter[10] indicates that this procedure is not accurate for all conditions. Efforts to compensate for the end contractions led to the development of the *Cipolletti weir.*

This type of weir has ends that are not vertical but have a slope of 1:4, horizontal to vertical. This slope was chosen to provide sufficient additional discharge to compensate for the end contractions. However, experiments by Flinn and Dyer[11] and others indicate that the side slope of the notch should be greater than 1:4.

The experimentation by Kindsvater and Carter reported previously included tests on horizontal weirs with end contractions. Their procedure provides what is probably the best available method for estimating the discharge through rectangular notched sharp-edged weirs. The results are given in the form of two graphs, which are reproduced here in Fig. 5.3. Figure 5.3*a* shows a curve relating k_L to L/b, and Fig. 5.3*b* shows lines of C_e versus H/P for various values of

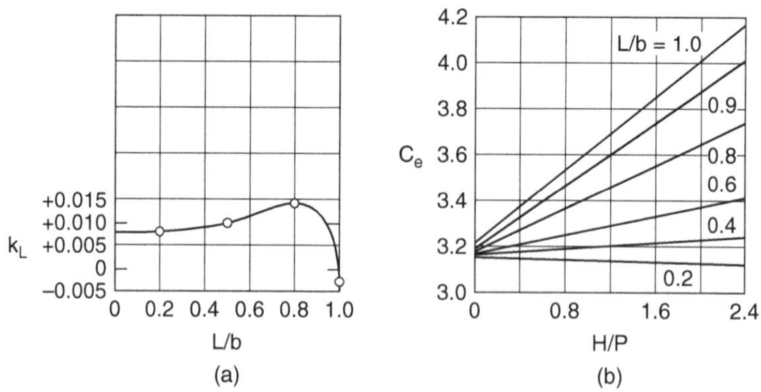

FIGURE 5.3 Coefficients, horizontal sharp-crested weirs with end contractions.

L/b, the top line being the line representing Eq. (5.25), which applies to the case of $L/b = 1$. The use of these curves will be illustrated by means of an example.

Example 5.2 A rectangular notched sharp-crested weir having a horizontal crest 1 m long located 60 cm above the bottom is centered in a rectangular channel 1.5 m wide. Determine the discharge when the head is 12 cm.

$$\frac{L}{b} = \frac{1}{1.5} = 0.667 \qquad \frac{H}{P} = \frac{0.12}{0.6} = 0.20$$

Then, from Fig. 5.3*b*,

$$C_e = 1.77$$

From Fig. 5.3*a*, $k_L = +0.004$, and taking $k_H = 0.001$,

$$L_e = 1.0 + 0.004 = 1.004 \text{ m}$$

and

$$H_e = 0.12 + 0.001 = 0.121 \text{ m}$$

Then

$$Q = C_e L_e H_e^{3/2}$$
$$= 1.77 \times 1.004 \times 0.121^{3/2} = 0.075 \text{ m}^3/\text{s} \quad \blacktriangle$$

V-Notch Weirs

Triangular, or V-notch, weirs permit the accurate measurement of much lower discharges than do horizontal-crested weirs. The discharge of a notched weir increases more rapidly with the head than in the case of a horizontal crested weir. For example, at a head of 2 ft the discharge through a 90° V-notch weir is approximately the same as that of a horizontal weir 1.5 ft long, whereas at a head of 0.1 ft the discharge of the horizontal weir is 20 times that of the 90° V-notch weir. The V-notch weir therefore presents distinct advantages, where low discharges are included in the range to be measured.

The basic expression for discharge through a V-notch weir is given by Eqs. (5.13) to (5.15),

$$Q = C_1 \tan\frac{\theta}{2} H^{5/2} \tag{5.13}$$

Letting

$$C = C_1 \tan\frac{\theta}{2} \tag{5.14}$$

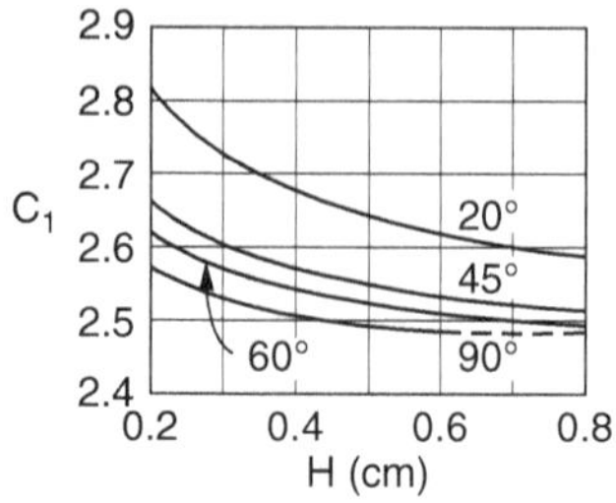

FIGURE 5.4 Coefficients, V-notch sharp-crested weirs.

then

$$Q = CH^{5/2} \tag{5.15}$$

The meanings of H and θ are given in Fig. 5.1*a* and *c*. In these equations C_1 and C must include the effects of contraction, velocity of approach, viscosity, and surface tension. Usually the velocity of approach is negligible for V-notch weirs unless H becomes very large, in which case it would be better to use Eq. (5.12).

Lenz[12] has developed a procedure for including the effects of viscosity and surface tension. His paper shows applications to his own experimental data for water and several oils, as well as to data of other experimenters. For fluids other than water, the reader is referred to Lenz's paper. For water Lenz's results are summarized by the curves shown in Fig. 5.4. Here again, the effects of viscosity and surface tension are reflected by the variation of C_1 or C with H, as discussed earlier.

Another method of taking into account the effect of H for a particular weir and constant temperature is to hold C constant and let the exponent of H be such as to fit the experimental data. In that case Eq. (5.15) would be modified as follows:

$$Q = CH^n \tag{5.15a}$$

Extensive tests conducted at the University of Michigan on a 90° V-notch weir for water at heads varying from 0.2 to 1.8 ft resulted in the following equation:

$$Q = 2.52H^{2.47} \tag{5.29}$$

From experiments by Barr,[13]

$$Q = 2.48H^{2.48} \tag{5.30}$$

The University of Michigan experiments were performed on a weir cut from unpolished commercial steel plate, whereas the Barr tests were made with a weir cut in a polished brass plate. Barr found that the discharge from his weir could be increased about 2 percent by roughening the upstream face with an application of varnish and emery dust.

Because tan ($\theta/2$) is unity for a 90° V notch, Eqs. (5.32) and (5.33) indicate that the value of C_1 in Eq. (5.13) is approximately 2.5. Using the value, the following general expression for discharge over notched weirs is derived:

$$Q = 2.5 \tan\frac{\theta}{2} H^{2.5} \tag{5.31}$$

Experiments conducted at the University of Michigan on a 22½° V-notch weir yielded the equation

$$Q = 0.5H^{2.43} \tag{5.32a}$$

Equation (5.31) gives the following equation for this weir:

$$Q = 0.5H^{2.5} \tag{5.32b}$$

Tests made at the University of Michigan on a 60° V-notch weir gave the result

$$Q = 1.47H^{2.51} \tag{5.33a}$$

as compared with the following expression derived from Eq. (5.31):

$$Q = 1.44H^{2.5} \tag{5.33b}$$

Hertzler[14] gives equations derived from tests on six different 120° V-notch weirs. The average values of C and n for these weirs yield the expression

$$Q = 4.36H^{2.48} \tag{5.34a}$$

whereas Eq. (5.31) gives the following result for 120° weirs:

$$Q = 4.33H^{2.5} \tag{5.34b}$$

It may be seen from the foregoing examples that Eq. (5.31) provides a means for deriving the discharge equation for a V-notch weir of any angle θ with a fair degree of dependability. It is especially useful in designing a weir to measure a particular range of discharges. The use of the equations for V-notch weirs is illustrated by the following example.

Example 5.3 Estimate the discharge over a sharp-crested 90° V-notch weir when $H = 15$ cm and $P = 30$ cm. The fluid is water, and the channel above the weir is 1.2 m wide.

Assuming that the velocity of approach is negligible, the discharge can be estimated from Eqs. (5.14) and (5.15) together with Fig. 5.4. From Fig. 5.4. $C = C_1 = 1.38$ and $Q = 1.38 \times 0.15^{2.5}$. Then $Q = 1.38 \times 0.0087 = 0.012$ m^3/s.

The velocity of approach is

$$V_1 = \frac{0.012}{0.45 \times 1.2} = 0.022 \text{ m/s}$$

and the velocity head is

$$\frac{V_1^2}{2g} = 0.000025 \text{ m}$$

This quantity is negligible compared with an H of 0.15 m and can be safely ignored. ▲

Submerged Sharp-Crested Weirs

A profile through a submerged sharp-crested weir is illustrated in Fig. 5.5. The discharge Q over such a weir is related not only to the head H_1 on the upstream side of the weir, but also to the head H_2 on the downstream side and, to a lesser extent to the height P of the weir crest above the floor of the channel. Early experiments by Francis (1848, 1883), Fteley and Stearns (1882), Bazin (1894), Cone (1916), and Cox (1928) have been summarized by Vennard and Weston.[15] They showed that the various test data could be presented in an orderly manner by selecting as variables Q/Q_1 and H_2/H_1, where Q_1 is the discharge at head H_1, computed from the equation for free discharge (unsubmerged), which is expressed in general terms as

$$Q_1 = CH_1^n \tag{5.35}$$

By plotting Q/Q_1 against H_2/H_1 they found that the various data tended to fall on a single curve, except for small values of P/H_1.

In 1947 Villemonte[16] presented the results of a series of tests on submerged sharp-crested weirs. He conducted tests on rectangular, triangular, parabolic, cusped, and proportional weirs. He showed that the results for all types could be represented by the single equation:

$$\frac{Q}{Q_1} = \left[1 - \left(\frac{H_2}{H_1}\right)^n\right]^{0.385} \tag{5.36}$$

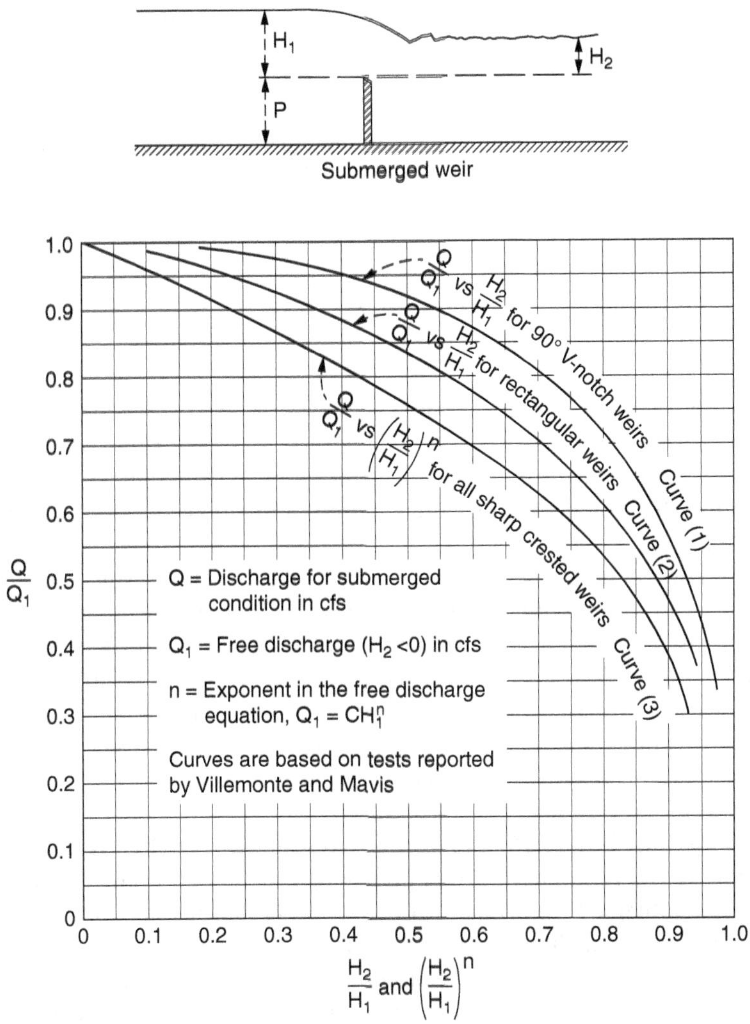

FIGURE 5.5 Discharge for submerged weirs.

where n is the exponent in the free-discharge equation [Eq. (5.35)], and the other terms are as defined previously. This equation was found to satisfy all test results, with a maximum deviation of 5 percent for some of the individual results.

In 1949, Mavis[17] presented results of tests on rectangular, triangular, parabolic, circular, sutro, and cusped weirs. He found that a single equation could be used to express the results for all the tests.

His equation, with subscripts changed to conform to the usage in this book, is

$$\frac{Q}{Q_1}=1-\left(0.45S+\frac{0.40}{2^{10-10S}}\right) \tag{5.37}$$

where S is defined as

$$S=\frac{a_2\sqrt{H_2}}{a_1\sqrt{H_1}} \tag{5.38}$$

In Eq. (5.38), a_2 is the weir area corresponding to H_2 and a_1 is the weir area corresponding to H_1. Mavis also presented some interesting data which resulted from tests done in 1717 by Poleni. It was found that Poleni's results agreed within 2 to 4 percent with Mavis's data.

The curves shown in Fig. 5.5 are based on the results of the works of Villemonte and Mavis. It may be noted that the channel widths differed for the two sets of tests, that P was different for all cases, and that rectangular weirs of four different widths were tested. Curves 1 and 2 are composite curves based on the results of the two investigators for the 90° V-notch weirs and the rectangular weirs, respectively. Curves 1 and 2 differ by no more than 1 percent from the test results.

Because Eqs. (5.36) and (5.37) both indicate that Q/Q_1 is a function of $(H_2/H_1)^n$, curve 3 was prepared, which is an average of the results obtained from Eqs. (5.36) and (5.37). Results obtained from either equation differ by less than 1 percent from curve 3. Curve 3 may be used to compute the discharge of a submerged sharp-crested weir of any shape. This curve is also in reasonable agreement with the results of the investigations summarized by Vennard and Weston, as well as with data presented by Stevens.[18] It should be noted, however, that for some of the weirs tested, the results could be represented more closely by an equation differing slightly from Eqs. (5.36) and (5.37) and by a curve differing slightly from curves 1 to 3. Therefore, if great accuracy is essential, it is recommended that the particular weir, or a similar one, be tested in a laboratory under conditions comparable to field conditions. When using the curves, shown in Fig. 5.5, it is recommended that H_1 be measured at least $2.5H_1$ upstream from the weir and that H_2 be measured beyond the turbulence caused by the nappe.

Example 5.4 Determine the discharge of a 90° V-notch weir if $H_1 = 0.9$ ft, $H_2 = 0.3$ ft, and $Q_1 = 2.50H_1^{2.5}$.

1. Use curve 1 of Fig. 5.5.

$$Q_1 = 2.50 \times 0.9^{2.5} = 1.92 \text{ ft}^3/\text{s}$$

$$\frac{H_2}{H_1} = \frac{0.3}{0.9} = 0.333$$

$$\frac{Q}{Q_1} = 0.972 \text{ (from curve 1)}$$

$$Q = 0.972 \times 1.92 = 1.86 \text{ ft}^3/\text{s}$$

2. Use curve 3 of Fig. 5.5.

$$\left(\frac{H_2}{H_1}\right)^n = 0.333^{2.5} = 0.064$$

$$\frac{Q}{Q_1} = 0.972 \text{ (from curve 3)}$$

$$Q = 0.972 \times 1.92 = 1.86 \text{ ft}^3/\text{s} \quad \blacktriangle$$

Example 5.5 Determine the discharge of a parabolic weir if $H_1 = 24$ cm, $H_2 = 12$ cm, and $Q_1 = 0.61H_1^{2.0}$.

$$Q_1 = 0.61 \times (0.24)^{2.0} = 0.035 \text{ m}^3/\text{s}$$

$$\left(\frac{H_2}{H_1}\right)^n = \left(\frac{12}{24}\right)^{2.0} = 0.25$$

$$\frac{Q}{Q_1} = 0.89 \text{ (from curve 3)}$$

$$Q = 0.89 \times 0.035 = 0.031 \text{ m}^3/\text{s} \quad \blacktriangle$$

Weirs Not Sharp-Crested

Sharp-crested weirs, if used to obtain discharge records for comparatively long periods, are difficult to maintain. The crest is likely to become dulled or rusted, or it may be damaged by floating ice and debris. Under such conditions it may be advisable to use a weir with a thicker crest. It is often convenient to use an existing weir or overflow dam for measuring discharges. Weirs of various dimensions and shapes are used in hydraulic structures. When designing such structures it is important to be able to estimate approximately the discharges over these weirs (see Sec. 2).

The amount of water that will pass over a weir, not sharp-crested, depends to a large extent on its sectional form and the shape of its crest, and it is necessary to resort to experiment to determine the discharge over any particular shape. Inasmuch as the number of shapes of weirs is unlimited, it is not to be expected that experimental data are or ever will be available for them all. There are available, however, the results of several series of experiments on weirs of different cross sections, which furnish much valuable information for determining discharges over weirs of the same or similar shapes.

The available experiments are not extensive enough for a comprehensive study of the effect of the velocity of approach on weirs not sharp-crested. The coefficients given in this chapter probably apply more accurately where the velocity of approach is not high. From a consideration of sharp-crested weirs, it appears that discharges for high velocities of approach will be somewhat greater than given by Eq. (5.10).

Since experimental conditions will seldom be duplicated in practice, it is probable that errors may result from the general use of the coefficients given in this chapter. Extreme accuracy, however, is not always necessary in a design where uncertainty as to the exact quantity of water to be provided for may exist.

The problem of establishing a fixed relation between head and discharge for weirs not sharp-crested is complicated by the fact that the nappe may assume a variety of forms in passing over the weir. For each modification of nappe form there is a corresponding change in the relation between head and discharge. The effect of this condition is more noticeable for low heads.

The nappe may undergo several of these modifications in succession as the head is varied. The successive forms that appear with an increasing stage may differ from those pertaining to similar stages with a decreasing head. The head at which the changes of nappe form occur varies with the rate of change of the head, whether increasing or decreasing, and with other conditions.

Among weirs of irregular section, there is a large class for which, from the nature of their section, the nappe can assume only one form unless drowned. Such weirs, it is suggested, may, if properly calibrated, equal or exceed the usefulness of the thin-edged weir for purposes of stream gauging because of their stability of section and because the thin-edged weir is not free from modifications of the nappe form for low heads.

Broad-Crested Weirs

A weir approximately rectangular in cross section is termed a broad-crested weir. Unless otherwise noted, it will be assumed to have vertical faces, a plane level crest, and sharp right-angled corners. Figure 5.6 represents a broad-crested weir of breadth b. The head H should be measured at least $2.5H$ upstream from the weir. Because of the sharp upstream edge, contraction of the nappe occurs. Surface contraction begins at a point slightly upstream from the weir.

The discharge over broad-crested weirs is usually expressed by the equation

$$Q = CLH^{3/2} \tag{5.10}$$

Experiments on broad-crested weirs have been performed by Blackwell, Bazin, Woodburn, the U.S. Deep Waterways Board, and the U.S. Geological Survey. These experiments cover a wide range of conditions as to the head, breadth, and height of the weir. Considerable discrepancy exists in the results of the different experiments, especially for heads below 0.5 ft. For heads from 0.5 to about 1.5 ft the coefficient becomes more uniform, and for heads from 1.5 ft to the point at which the nappe becomes detached from the crest, the coefficient as given by the different experiments is nearly constant and equals approximately 2.63. When the head reaches one to two times the breadth of the weir, the nappe becomes detached and the weir becomes essentially sharp-crested. The effect on discharge of the roughness of the crest can be computed by applying the principles of flow in open channels.

In order to put the results of the various experiments in a form convenient for use, Table 5.1 has been prepared by graphically interpolating the results of all experiments, giving more weight to those of the U.S. Geological Survey. This table should give values of C within the limits of accuracy of the original experiments.

The effect of rounding the upstream corner of a broad-crested weir is to increase the discharge for a given head. Table 5.2 lists experiments

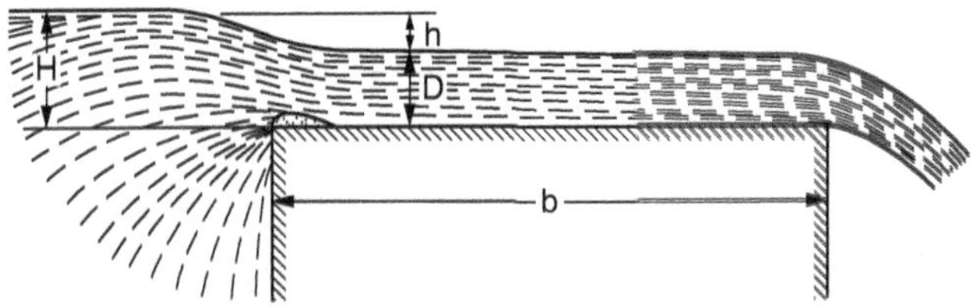

Figure 5.6 Broad-crested weir.

TABLE 5.1a Values of C in Formula $Q = CLH^{3/2}$ for Broad-Crested Weirs, U.S. Customary System

Measured head, ft	Breadth of crest of weir, ft										
	0.50	**0.75**	**1.00**	**1.50**	**2.00**	**2.50**	**3.00**	**4.00**	**5.00**	**10.00**	**15.00**
0.2	2.80	2.75	2.69	2.62	2.54	2.48	2.44	2.38	2.37	2.49	2.68
0.4	2.92	2.80	2.72	2.64	2.61	2.60	2.58	2.54	2.50	2.56	2.70
0.6	3.08	2.89	2.75	2.64	2.61	2.60	2.68	2.69	2.70	2.70	2.70
0.8	3.30	3.04	2.85	2.68	2.60	2.60	2.67	2.68	2.68	2.69	2.64
1.0	3.32	3.14	2.98	2.75	2.66	2.64	2.65	2.67	2.68	2.68	2.63
1.2	3.32	3.20	3.08	2.86	2.70	2.65	2.64	2.67	2.66	2.69	2.64
1.4	3.32	3.26	3.20	2.92	2.77	2.68	2.64	2.65	2.65	2.67	2.64
1.6	3.32	3.29	3.28	3.07	2.89	2.75	2.68	2.66	2.65	2.64	2.63
1.8	3.32	3.32	3.31	3.07	2.88	2.74	2.68	2.66	2.65	2.64	2.63
2.0	3.32	3.31	3.30	3.03	2.85	2.76	2.72	2.68	2.65	2.64	2.63
2.5	3.32	3.32	3.31	3.28	3.07	2.89	2.81	2.72	2.67	2.64	2.63
3.0	3.32	3.32	3.32	3.32	3.20	3.05	2.92	2.73	2.66	2.64	2.63
3.5	3.32	3.32	3.32	3.32	3.32	3.19	2.97	2.76	2.68	2.64	2.63
4.0	3.32	3.32	3.32	3.32	3.32	3.32	3.07	2.79	2.70	2.64	2.63
4.5	3.32	3.32	3.32	3.32	3.32	3.32	3.32	2.88	2.74	2.64	2.63
5.0	3.32	3.32	3.32	3.32	3.32	3.32	3.32	3.07	2.79	2.64	2.63
5.5	3.32	3.32	3.32	3.32	3.32	3.32	3.32	3.32	2.88	2.64	2.63

Table 5.1b Values of C in Formula $Q = CLH^{3/2}$ for Broad-Crested Weirs, Metric System

Measured head, m	Breadth of crest of weir, m										
	0.15	0.20	0.30	0.45	0.60	0.75	0.90	1.20	1.50	3.00	4.50
0.10	1.61	1.55	1.50	1.46	1.44	1.44	1.43	1.40	1.38	1.41	1.49
0.20	1.70	1.60	1.52	1.46	1.44	1.44	1.48	1.49	1.49	1.49	1.49
0.30	1.83	1.73	1.65	1.52	1.47	1.46	1.46	1.48	1.48	1.48	1.45
0.40	1.83	1.80	1.77	1.61	1.53	1.48	1.46	1.46	1.46	1.48	1.46
0.50	1.83	1.82	1.81	1.70	1.60	1.52	1.48	1.47	1.46	1.46	1.45
0.60	1.83	1.83	1.83	1.67	1.57	1.52	1.50	1.48	1.46	1.46	1.45
0.80	1.83	1.83	1.83	1.81	1.70	1.60	1.55	1.50	1.48	1.46	1.45
0.90	1.83	1.83	1.83	1.83	1.77	1.69	1.61	1.51	1.47	1.46	1.45
1.0	1.83	1.83	1.83	1.83	1.83	1.76	1.64	1.52	1.48	1.46	1.45
1.2	1.83	1.83	1.83	1.83	1.83	1.83	1.70	1.54	1.49	1.46	1.45
1.4	1.83	1.83	1.83	1.83	1.83	1.83	1.83	1.59	1.51	1.46	1.45
1.5	1.83	1.83	1.83	1.83	1.83	1.83	1.83	1.70	1.54	1.46	1.45
1.7	1.83	1.83	1.83	1.83	1.83	1.83	1.83	1.83	1.59	1.46	1.45

Table 5.2a Values of C in Formula $Q = CLH^{3/2}$ for Models of Broad-Crested Weirs with Rounded Upstream Corner, U.S. Customary System

Experimenter	Radius of curve, ft	Breadth of weir, ft	Height of weir, ft	Head H, ft									
				0.4	0.6	0.8	1.0	1.5	2.0	2.5	3.0	4.0	5.0
Bazin	0.33	2.62	2.46	2.93	2.97	2.98	3.01	3.04					
Bazin	0.33	6.56	2.46	2.70	2.82	2.87	2.89	2.92					
U.S. Deep Waterways	0.33	2.62	4.57		2.77	2.80	2.83	2.92	3.00	3.08	3.17	3.34	3.50
U.S. Deep Waterways	0.33	6.56	4.56			2.83	2.83	2.83	2.82	2.82	2.82	2.82	2.81

Table 5.2b Values of C in Formula $Q = CLH^{3/2}$ for Models of Broad-Crested Weirs with Rounded Upstream Corner, Metric System

Experimenter	Radius of curve, m	Breadth of weir, m	Height of weir, m	Head H, m								
				0.10	0.20	0.30	0.45	0.60	0.75	0.90	1.20	1.50
Bazin	0.10	0.80	0.75	1.62	1.64	1.66	1.68					
Bazin	0.10	2.00	0.75	1.49	1.56	1.60	1.61					
U.S. Deep Waterways	0.10	0.80	1.40		1.53	1.56	1.61	1.66	1.70	1.75	1.85	1.93
U.S. Deep Waterways	0.10	2.00	1.40			1.56	1.56	1.56	1.56	1.56	1.56	1.55

on this type of weir. The effect of rounding the upstream corner on a radius of 10 cm is to increase the coefficient C by approximately 9 percent. Coefficients by Woodburn[19] for flat weirs with rounded upstream corners and gently sloping crests are given in Table 5.3.

Blackwell experimented with three weirs 3 ft broad, having a slightly inclined crest. Inclining the crest appears to increase the coefficient of discharge slightly. The results of these experiments are rather inconsistent, especially for low heads. Table 5.3 has been compiled from Blackwell's experiments. Sloping the top of a broad-crested weir makes it similar to a triangular weir with the upstream face vertical. The coefficients given in Tables 5.4 and 5.5 will therefore be helpful in selecting coefficients for broad-crested weirs with sloping crests.

If the upstream corner of a weir is so rounded as to prevent contraction entirely, and if the slope of the crest is as great as the loss of head due to friction, flow occurs at critical depth, and discharge is given by the rational formula

$$Q = 3.087LH^{3/2} \tag{5.39}$$

where H and L are in feet and Q is in ft^3/s.

For further discussion of flow at critical depth, see Sec. 8. It should be noted that $C = 3.087$ is the maximum value of the coefficient that is obtainable for broad-crested weirs under any conditions.

Weirs of Triangular Section

Figure 5.7 represents the cross section of a weir having the upper face vertical and the lower face inclined downward, the two faces meeting in a sharp angle, which forms the crest.

Bazin has experimented with weirs of this type 2.46 ft high, having various slopes of the downstream face. The coefficients resulting from those experiments are given in Table 5.4.

It will be observed that the coefficient for a given slope, in each case shown by the experiments, is nearly constant for heads above 0.7 ft. It seems fair to assume, therefore, that these values could be extended to higher heads with reasonable assurance. The average values of the coefficients given in Table 5.4 for heads above 0.7 ft were plotted logarithmically and found to fall very accurately on a straight line. This line was then extended to include slopes of 20:1, horizontal to vertical, from which the values given in Table 5.5 were taken. Table 5.5 may be used for computing discharges over weirs of the types shown in Fig. 5.7 for heads above 0.7 ft. These coefficients are to be used for broad-crested weirs with inclined tops only when the breadth is sufficient to prevent the nappe from springing clear. In the latter case the weir becomes in principle a thin-edged weir.

Table 5.3a Values of C in Formula $Q = CLH^{3/2}$ for Broad-Crested Weirs with Crests Inclined Slightly Downward, U.S. Customary System

	(1) Energy head H_e, ft					
Crest	**0.5**	**0.6**	**0.8**	**1.0**	**1.2**	**1.4**
Level	2.78	2.79	2.81	2.83	2.85	2.85
Slope 0.004	2.95	2.94	2.92	2.90	2.88	2.87
Slope 0.026	3.07	3.06	3.04	3.02	3.00	2.99

		(2) Head H, ft						
Slope of crest	**Breadth of weir, ft**	**0.1**	**0.2**	**0.3**	**0.4**	**0.5**	**0.6**	**0.7**
12:1	3.0	2.58	2.87	2.57	2.60	2.84	2.81	2.70
18:1	3.0	2.91	2.92	2.53	2.60	2.80	2.74	2.62
18:1	10.0	2.52	2.68	2.73	2.60	2.90	2.80	2.68

Table 5.3b Values of C in Formula $Q = CLH^{3/2}$ for Broad-Crested Weirs with Crests Inclined Slightly Downward, Metric System

	(1) Energy head H_e, cm					
Crest	**15**	**20**	**25**	**30**	**40**	**45**
Level	1.54	1.54	1.55	1.56	1.57	1.57
Slope 0.004	1.63	1.62	1.61	1.60	1.59	1.59
Slope 0.026	1.70	1.69	1.68	1.67	1.65	

		(2) Head H, cm						
Slope of crest	**Breadth of weir, m**	**3**	**6**	**9**	**12**	**15**	**18**	**21**
12:1	0.9	1.43	1.59	1.42	1.44	1.57	1.55	1.49
18:1	0.9	1.61	1.61	1.40	1.44	1.55	1.51	1.45
18:1	3.0	1.39	1.48	1.51	1.55	1.60	1.55	1.48

Bazin also experimented with weirs of triangular cross sections 1.64 ft high, having both faces inclined (Fig. 5.8). Coefficients covering the range of these experiments are given in Table 5.6.

Weirs of Trapezoidal Section

Figure 5.9 represents a weir of trapezoidal section with both upstream and downstream faces inclined. Experiments on this type of weir were

TABLE 5.4a Values of C in Formula $Q = CLH^{3/2}$ for Weirs of Triangular Cross Section with Vertical Upstream Face and Sloping Downstream Face, U.S. Customary System

Slope of downstream face, horizontal: vertical	Height of weir, ft	Head H, ft										
		0.2	0.3	0.4	0.5	0.6	0.7	0.8	0.9	1.0	1.2	1.5
1:1	2.46	3.88	3.85	3.85	3.85	3.85	3.85	3.85	3.85	3.85	3.85	3.85
2:1	2.46	3.48	3.48	3.49	3.49	3.5	3.5	3.5	3.50	3.50	3.51	3.51
2:1	1.64	3.56	3.47	3.47	3.51	3.54	3.57	3.58	3.58	3.58	3.59	3.57
3:1	1.64		2.90	3.11	3.22	3.26	3.33	3.37	3.40	3.40	3.41	3.41
5:1	2.46		3.08	3.06	3.05	3.05	3.07	3.09	3.12	3.13	3.13	3.13
10:1	2.46		2.82	2.83	2.84	2.86	2.89	2.90	2.91	2.91	2.92	2.93

TABLE 5.4b Values of C in Formula $Q = CLH^{3/2}$ for Weirs of Triangular Cross Section with Vertical Upstream Face and Sloping Downstream Face, Metric System

Slope of downstream face, horizontal: vertical	Height of weir, cm	Head H, cm							
		5	10	15	20	25	30	35	45
1:1	75	2.14	2.13	2.13	2.13	2.13	2.13	2.13	2.13
2:1	75	1.92	1.93	1.93	1.93	1.93	1.93	1.94	1.94
2:1	50	1.97	1.92	1.94	1.96	1.98	1.98	1.98	1.97
3:1	50		1.72	1.78	1.82	1.86	1.88	1.88	1.88
5:1	75		1.69	1.69	1.70	1.71	1.73	1.73	1.73
10:1	75		1.56	1.57	1.59	1.60	1.61	1.61	1.62

Table 5.5 Values of C in Formula $Q = CLH^{3/2}$, Being the Mean and Extension of Experimental Results on Weirs of Triangular Cross Section with Vertical Upstream Face and Sloping Downstream Face*

Slope of downstream face, horizontal: vertical	C (English)	C (metric)
1:1	3.85	2.13
2:1	3.54	1.96
3:1	3.35	1.85
4:1	3.21	1.77
5:1	3.13	1.73
6:1	3.07	1.70
7:1	3.02	1.67
8:1	2.98	1.65
9:1	2.94	1.62
10:1	2.92	1.61
12:1	2.86	1.58
14:1	2.80	1.55
16:1	2.76	1.52
18:1	2.72	1.50
20:1	2.69	1.49

*Table should only be used for heads above 20 cm.

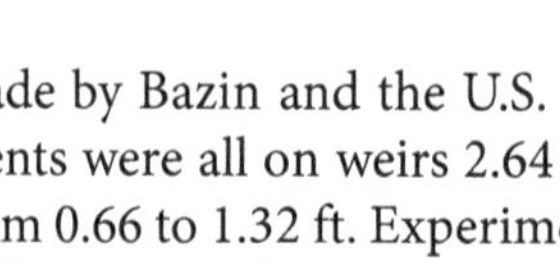

Figure 5.7 Triangular weir.

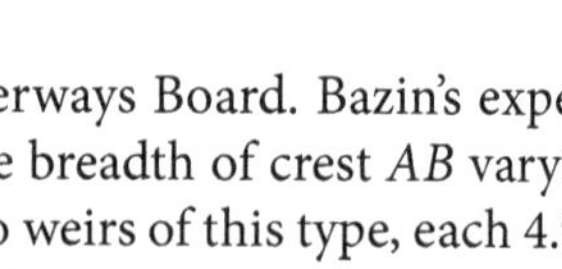

Figure 5.8 Triangular weir.

made by Bazin and the U.S. Deep Waterways Board. Bazin's experiments were all on weirs 2.64 ft high, the breadth of crest *AB* varying from 0.66 to 1.32 ft. Experiments on two weirs of this type, each 4.9 ft high, were performed by the U.S. Deep Waterways Board.

Coefficients covering the range of Bazin's experiments are given in Table 5.7. Table 5.8 gives coefficients resulting from the experiments by the U.S. Deep Waterways Board.

For weirs of trapezoidal cross section with sloping upstream face and vertical downstream face (Fig. 5.10) there are five series of experiments by the U.S. Deep Waterways Board. All the models for

Table 5.6a Values of C in Formula $Q = CLH^{3/2}$ for Weirs of Triangular Cross Section with Both Faces Inclined,* U.S. Customary System

Slope of upstream face, horizontal: vertical	Slope of downstream face, horizontal: vertical	Head H, ft										
		0.2	**0.3**	**0.4**	**0.5**	**0.6**	**0.7**	**0.8**	**0.9**	**1.0**	**1.2**	**1.5**
1:1	1:1		4.26	4.20	4.14	4.11	4.11	4.11	4.10	4.08	3.93	3.75
1:1	2:1	3.82	3.80	3.77	3.77	3.79	3.82	3.84	3.85	3.85	3.85	3.84
1:1	3:1		3.55	3.52	3.48	3.46	3.45	3.46	3.47	3.48	3.47	3.46
2:1	2:1	3.88	3.85	3.83	3.81	3.81	3.83	3.86	3.87	3.87	3.87	3.87
1:2	2:1	3.74	3.71	3.68	3.69	3.72	3.73	3.73	3.74	3.74	3.73	3.71
1:3	2:1	3.65	3.64	3.64	3.67	3.68	3.69	3.69	3.69	3.69	3.68	3.66
Vertical	2:1	3.56	3.47	3.47	3.51	3.54	3.57	3.58	3.58	3.58	3.59	3.57

*For heads above 1.5 ft use value of C given for head of 1.5 ft.

Table 5.6b Values of C in Formula $Q = CLH^{3/2}$ for Weirs of Triangular Cross Section with Both Faces Inclined,* Metric System

Slope of upstream face, horizontal: vertical	Slope of downstream face, horizontal: vertical	Head H, cm							
		5	10	15	20	25	30	40	50
1:1	1:1		2.35	2.29	2.27	2.27	2.25	2.17	2.07
1:1	2:1	2.11	2.10	2.08	2.10	2.12	2.13	2.13	2.12
1:1	3:1		1.96	1.92	1.91	1.91	1.92	1.92	1.91
2:1	2:1	2.14	2.13	2.10	2.11	2.13	2.14	2.14	2.14
1:2	2:1	2.07	2.05	2.04	2.06	2.06	2.07	2.06	2.05
1:3	2:1	2.02	2.01	2.03	2.03	2.04	2.04	2.03	2.02
Vertical	2:1	1.97	1.92	1.94	1.96	1.98	1.98	1.98	1.97

*For heads above 50 cm use value of C given for head of 50 cm.

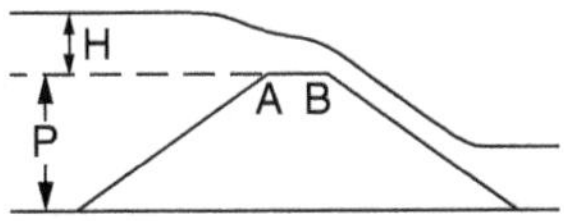

FIGURE 5.9 Trapezoidal weir.

these experiments were approximately 4.9 ft high, and the breadth of crest *AB* was either 0.33 or 0.66 ft. The length of all weirs was 6.58 ft. Table 5.9 gives coefficients derived from these experiments.

Weirs of Irregular Section

Irregular weir sections are shown in Figs. 5.11 to 5.23. Figures 5.11 to 5.15 represent models of weirs experimented on by the U.S. Deep Waterways Board, under the direction of G. W. Rafter, at the hydraulic laboratory of Cornell University. From four to seven experiments were run on each model, the head varying approximately from 1.0 to 5.5 ft. Values of *C* obtained from these experiments are tabulated in Table 5.10.

In 1903 experiments were performed at the hydraulic laboratory of Cornell University for the U.S. Geological Survey, under the direction of Robert E. Horton, to determine the coefficients of discharge of weirs modeled after various types of dams. Figures 5.16 to 5.25 show the forms of crests of some models in the experiments. The weirs were all 11.25 ft high and either 8 ft or 15 ft long. The purpose of the experiments was to enable the Geological Survey to determine more accurately discharges over weirs at gauging stations. Coefficients obtained from these experiments are listed in Table 5.11.

Figure 5.24 is a cross section of the old dam at Austin, Texas. Five series of gaugings of flow over this dam were made with a current meter by Taylor[20] in 1900. The head ranged from 0.42 to 1.44 ft.

Figure 5.25 is a cross section of the Blackstone River dam at Albion, Massachusetts. Five current meter measurements of the water passing over this dam were made by Dwight Porter. The head in each case was about 1 ft, and the resulting values of *C* vary from 3.41 to 3.94.

The last two lines in Table 5.11 give mean values of *C* as measured from the flow over these dams.

Spillway Sections

The typical overflow spillway section of a dam, as illustrated in Fig. 5.26, is a weir with rounded crest. The coefficient varies from 3.0 to more than 4.0, depending primarily on the shape of the crest, the extent of end contraction, and the head. Because of

Table 5.7a Values of C in Formula $Q = CLH^{3/2}$ for Weirs of Trapezoidal Cross Section with Both Faces Inclined,* U.S. Customary System

Slope of upstream face, horizontal: vertical	Slope of downstream face, horizontal: vertical	Width of crest, ft	Head H, ft										
			0.2	0.3	0.4	0.5	0.6	0.7	0.8	0.9	1.0	1.2	1.5
1:2	1:1	0.66	2.70	2.82	2.89	3.02	3.13	3.24	3.34	3.44	3.52	3.66	3.82
1:2	2:1	0.66	2.71	2.79	2.83	2.92	3.03	3.14	3.27	3.32	3.38	3.50	3.61
1:2	3:1	0.66	2.70	2.76	2.80	2.91	3.00	3.07	3.14	3.21	3.27	3.37	3.45
1:2	4:1	0.66	2.71	2.74	2.84	2.88	2.98	3.06	3.12	3.17	3.21	3.28	3.35
1:2	5:1	0.66	2.71	2.80	2.86	2.88	2.93	3.02	3.08	3.12	3.17	3.23	3.26
1:2	2:1	1.32		2.71	2.77	2.80	2.80	2.84	2.88	2.93	2.98	3.08	3.22
1:2	4:1	1.32		2.76	2.80	2.82	2.82	2.85	2.88	2.91	2.94	3.01	3.10
1:2	6:1	1.32			2.79	2.80	2.82	2.85	2.87	2.90	2.93	2.98	3.08
2:1	2:1	0.67	2.82	2.94	3.04	3.13	3.20	3.26	3.32	3.38	3.43	3.51	3.61
1:1	2:1	0.67	2.73	2.86	2.92	3.02	3.12	3.21	3.29	3.36	3.42	3.53	3.65
1:3	2:1	0.67	2.50	2.62	2.75	2.87	2.99	3.09	3.18	3.27	3.34	3.46	3.55
Vertical	2:1	0.67	2.55	2.58	2.66	2.77	2.90	2.99	3.09	3.18	3.26	3.39	3.51

*Table indicates that values of C increase slightly for heads above 1.5 ft. See also Table 5.8.

TABLE 5.7b Values of C in Formula $Q = CLH^{3/2}$ for Weirs of Trapezoidal Cross Section with Both Faces Inclined,* Metric System

Slope of upstream face, horizontal: vertical	Slope of downstream face, horizontal: vertical	Width of crest, cm	Head H, cm							
			5	10	15	20	25	30	40	50
1:2	1:1	20	1.49	1.56	1.67	1.76	1.85	1.94	2.02	2.11
1:2	2:1	20	1.50	1.54	1.61	1.71	1.81	1.87	1.93	1.99
1:2	3:1	20	1.49	1.52	1.61	1.68	1.73	1.81	1.86	1.91
1:2	4:1	20	1.50	1.51	1.59	1.67	1.72	1.77	1.81	1.85
1:2	5:1	20	1.50	1.55	1.59	1.64	1.70	1.75	1.78	1.80
1:2	2:1	40		1.50	1.55	1.56	1.59	1.65	1.70	1.78
1:2	4:1	40		1.52	1.56	1.57	1.59	1.62	1.66	1.71
1:2	6:1	40			1.55	1.57	1.59	1.62	1.65	1.70
2:1	2:1	20	1.56	1.62	1.73	1.78	1.83	1.90	1.94	1.99
1:1	2:1	20	1.51	1.58	1.67	1.74	1.82	1.89	1.95	2.02
1:3	2:1	20	1.38	1.45	1.59	1.68	1.76	1.85	1.91	1.96
Vertical	2:1	20	1.41	1.43	1.53	1.62	1.71	1.80	1.87	1.94

*Table indicates that values of C increase slightly for heads above 50 cm. See also Table 5.8.

Table 5.8a Values of C in Formula $Q = CLH^{3/2}$ for Weirs of Trapezoidal Cross Section with Both Faces Inclined,* U.S. Customary System

Slope of upstream face, horizontal: vertical	Slope of downstream face, horizontal: vertical	Width of crest, ft	Head H, ft									
			1.6	1.8	2.0	2.5	3.0	3.5	4.0	4.5	5.0	5.5
2:1	2:1	0.67	3.57	3.56	3.56	3.57	3.58	3.60	3.62	3.65	3.68	3.70
2:1	5:1	0.33	3.58	3.56	3.53	3.48	3.44	3.43	3.48	3.54	3.57	3.58

*See also Table 5.7.

Table 5.8b Values of C in Formula $Q = CLH^{3/2}$ for Weirs of Trapezoidal Cross Section with Both Faces Inclined,* Metric System

Slope of upstream face, horizontal: vertical	Slope of downstream face, horizontal: vertical	Width of crest, cm	Head H, m									
			0.50	0.55	0.60	0.75	0.90	1.10	1.20	1.40	1.50	1.70
2:1	2:1	20	1.97	1.97	1.97	1.97	1.98	1.99	2.00	2.02	2.03	2.04
2:1	5:1	10	1.98	1.97	1.95	1.92	1.90	1.90	1.92	1.96	1.97	1.98

*See also Table 5.7.

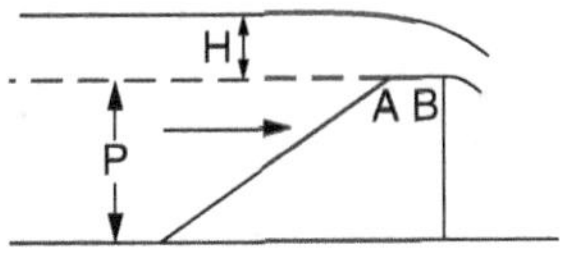

Figure 5.10 Trapezoidal weir.

the variation in C for different weirs, it is necessary to calibrate each weir if a high degree of accuracy is desired. However, the shape of the crest is often designed as a "standard crest," which was developed[21] to fit the shape of the underside of the nappe of a sharp-crested weir. For this type of crest it is possible to coordinate the coefficients of discharge for various overflow sections to some extent. The shape of this crest may be obtained from tables[22] or from the following equation[23] for the portion of the spillway downstream from the crest:

$$Y = 0.47X^{1.80} \tag{5.40}$$

where X and Y are in feet.

Combined with a circular curve for the upstream portion as shown in Fig. 5.26. Equation (5.40) gives the shape of the crest for $H = 1$, and all values of Y and X must be multiplied by whatever value of H is selected for the design head. It has been shown that there is an orderly variation of the coefficient of discharge for heads above and below the design head for weirs of this type. Kirkpatrick[24] presented a curve relating C/C_0 to H/H_0 for seven dams of the Tennessee Valley Authority (TVA) having shapes similar to the standard crest, C being the coefficient of discharge for any head H, and C_0 and H_0 the corresponding values for the design head. His curve is reproduced in Fig. 5.27. A similar relationship between C/C_0 and H/H_0 developed by Randolph[23] is also shown in Fig. 5.27. Values of the coefficients of discharge C_0 for the design head, obtained for the seven TVA dams, varied from 3.71 to 3.93, whereas the one reported by Randolph was 3.97. All the tests reported here were made on scale models and include the effect of contractions caused by piers at the ends of the crests. Kirkpatrick's tests were made through one bay of a spillway with the adjacent bays closed. If, for example, a standard crest is designed for $H_0 = 15$ ft and it is assumed that the value of C_0 is 3.90, then it may be estimated from Fig. 5.27 that C for $H = 20$ ft is $1.05 \times 3.90 = 4.10$. Values obtained in this manner can be considered only as estimates, pending a laboratory calibration.

Table 5.9a Values of C in Formula $Q = CLH^{3/2}$ for Weirs of Trapezoidal Cross Section with Vertical Upstream Face and Sloping Downstream Face, U.S. Customary System

Slope of upstream face, horizontal: vertical	Width of crest, ft	Head H, ft								
		1.0	1.5	2.0	2.5	3.0	3.5	4.0	4.5	5.0
2:1	0.33	3.85	3.82	3.79	3.77	3.75	3.73	3.70	3.67	3.64
2:1	0.66	3.41	3.57	3.65	3.70	3.72	3.72	3.73	3.73	3.73
3:1	0.66			3.57	3.57	3.57	3.57	3.57	3.57	3.57
4:1	0.66			3.48	3.48	3.48	3.48	3.48	3.48	3.48
5:1	0.66			3.39	3.39	3.39	3.39	3.39	3.39	3.39

Table 5.9b Values of C in Formula $Q = CLH^{3/2}$ for Weirs of Trapezoidal Cross Section with Vertical Upstream Face and Sloping Downstream Face, Metric System

Slope of upstream face, horizontal: vertical	Width of crest, cm	Head H, m						
		0.30	0.45	0.60	0.75	0.90	1.20	1.50
2:1	10	2.13	2.11	2.09	2.08	2.07	2.04	2.01
2:1	20	1.88	1.97	2.02	2.04	2.06	2.06	2.06
3:1	20			1.97	1.97	1.97	1.97	1.97
4:1	20			1.92	1.92	1.92	1.92	1.92
5:1	20			1.87	1.87	1.87	1.87	1.87

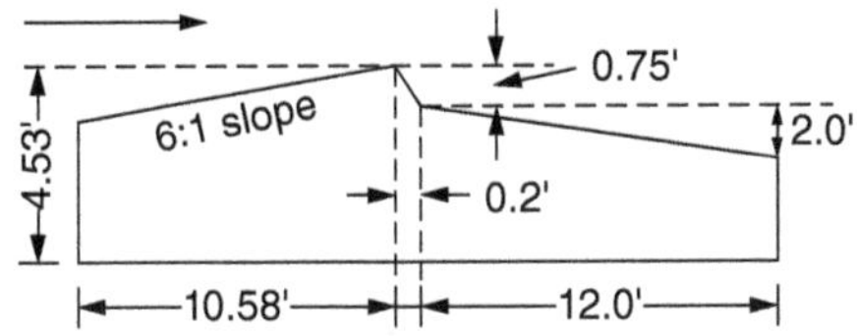

FIGURE 5.11 Irregular weir crest.

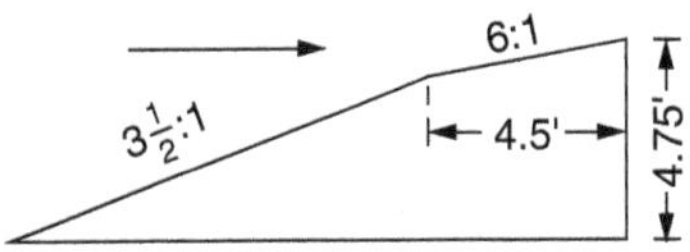

FIGURE 5.12 Irregular weir crest.

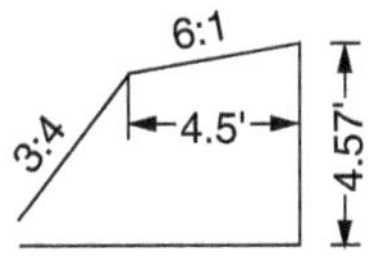

FIGURE 5.13 Irregular weir crest.

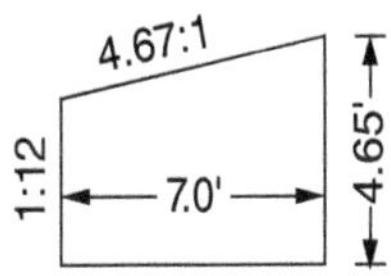

FIGURE 5.14 Irregular weir crest.

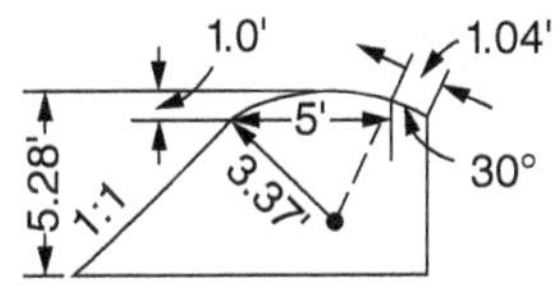

FIGURE 5.15 Irregular weir crest.

TABLE 5.10a Values of C in Formula $Q = CLH^{3/2}$ for Weirs of Irregular Cross Section, U.S. Customary System

	Head H, ft									
Figure	**1.0**	**1.5**	**2.0**	**2.5**	**3.0**	**3.5**	**4.0**	**4.5**	**5.0**	**5.5**
5.11	3.13	3.22	3.22	3.22	3.22	3.22	3.22	3.22	3.22	3.22
5.12		3.41	3.35	3.30	3.33	3.37	3.38	3.38	3.38	3.38
5.13	3.47	3.46	3.41	3.35	3.32	3.33	3.37	3.41	3.46	
5.14				3.44	3.39	3.38	3.38	3.39	3.41	
5.15	3.28	3.29	3.32	3.39	3.46	3.51	3.59	3.62	3.65	

TABLE 5.10b Values of C in Formula $Q = CLH^{3/2}$ for Weirs of Irregular Cross Section, Metric System

	Head H, m								
Figure	**0.30**	**0.45**	**0.60**	**0.75**	**0.90**	**1.05**	**1.20**	**1.50**	**1.70**
5.11	1.73	1.78	1.78	1.78	1.78	1.78	1.78	1.78	1.78
5.12		1.88	1.85	1.82	1.84	1.86	1.87	1.87	1.87
5.13	1.92	1.91	1.88	1.85	1.83	1.84	1.86	1.91	
5.14				1.90	1.87	1.87	1.87	1.88	
5.15	1.81	1.82	1.83	1.87	1.91	1.94	1.98	2.02	

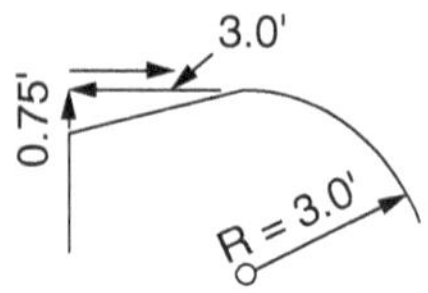

Figure 5.16 Irregular dam crest.

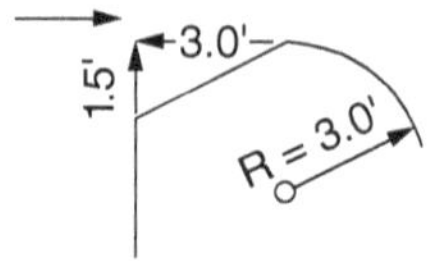

Figure 5.17 Irregular dam crest.

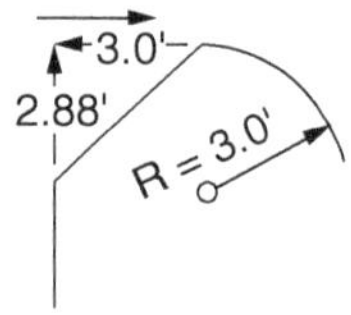

Figure 5.18 Irregular dam crest.

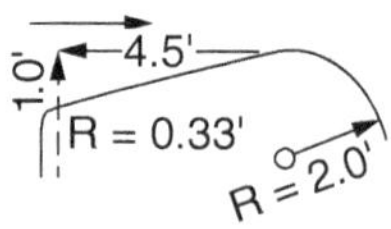

Figure 5.19 Irregular dam crest.

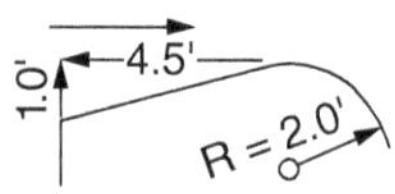

Figure 5.20 Irregular dam crest.

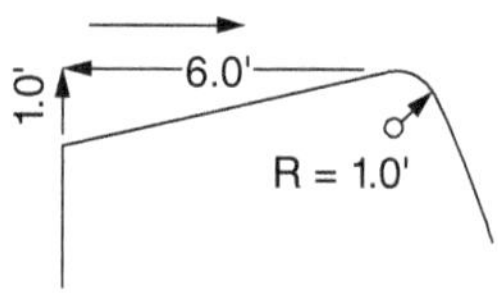

Figure 5.21 Irregular dam crest.

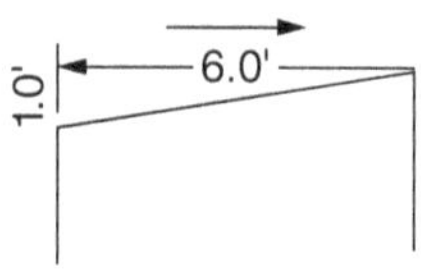

Figure 5.22 Irregular dam crest.

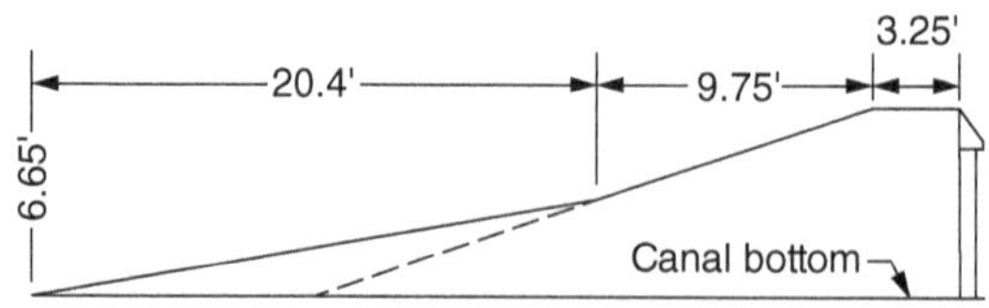

Figure 5.23 Irregular dam crest.

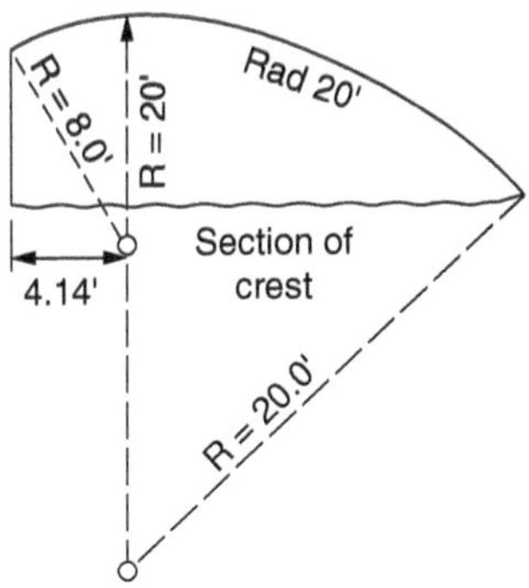

FIGURE 5.24 Austin, Texas, dam.

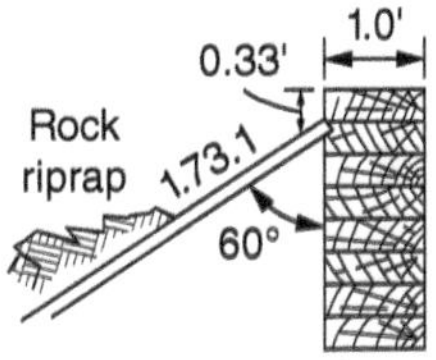

FIGURE 5.25 Blackstone River dam.

Selection and Use of Weirs

Weir selection for a particular situation depends on the range of discharges to be measured, the accuracy desired, and whether or not the weir can be calibrated after installation. Most weir equations for sharp-edged weirs are not accurate for heads less than 0.2 ft. Therefore, for very small discharges, a V-notch weir should be selected. On the other hand, few discharge measurements have been made on sharp-edged weirs for heads greater than 2 ft. Therefore, if a sharp-crested weir is to be used, it might be suggested that it be selected so that H will be between 0.2 and 2.0 ft. Weirs with rounded crests are used with much higher heads. If the weir is so located that it can be calibrated by means of volumetric or weighing tanks in the position that it will be used, exceedingly great accuracy can be expected, especially for sharp-crested weirs, and the errors will be principally those resulting from the determination of the head. When weirs are used in locations where they cannot be calibrated in place, the errors are likely to be much greater, as indicated by the difference in results obtained by various investigators for similar weirs. If the head is measured too far upstream, there may be some drop in the water surface due to friction. In order to avoid errors due to drawdown at the nappe, the head should be measured at least $4H$ upstream from horizontal crested weirs and $2.5H$ upstream from V-notch weirs. A detailed discussion of various types of errors which may affect weir measurements have been presented by Thomas.[25]

Heads are usually measured in a stilling well by means of a hook gauge, pressure transducer, float gaguge, ultrasonic level sensor or radar level sensors. The pipe connecting well and channel should not project beyond the channel surface. The gauge reading corresponding to zero head should be determined with great care and checked at

Table 5.11a Values of C in Formula $Q = CLH^{3/2}$ from Experiments at Cornell University,* U.S. Customary System

Figure	Length of model, ft	Head H, ft									
		0.5	1.0	1.5	2.0	2.5	3.0	3.5	4.0	4.5	5.0
5.16	7.94		3.30	3.32	3.36	3.40	3.43	3.48	3.53	3.62	3.72
5.16	15.97	3.32	3.44	3.46	3.42	3.41	3.46	3.50			
5.17	7.98		3.38	3.46	3.51	3.55	3.58	3.62	3.68	3.74	3.83
5.17	15.97	3.22	3.48	3.61	3.67	3.70	3.72				
5.18	15.97	3.15	3.45	3.64	3.75	3.82	3.87	3.88			
5.19	15.97	3.23	3.34	3.43	3.52	3.59	3.64				
5.20	15.97	3.18	3.30	3.37	3.42	3.46	3.49	3.52	3.54		
5.21	15.97	3.28	3.50	3.54	3.52	3.36	3.31	3.30			
5.22	15.97	3.53	3.54	3.55	3.50	3.35	3.27	3.25	3.25		
5.23	15.93	3.13	3.14	3.10	3.14	3.20	3.26	3.31	3.37		
5.24		3.09	3.11	3.33							
5.25			3.80								

*Experiments were made on models resembling existing dams (except for the last two experiments, which were actual dams).

TABLE 5.11b Values of C in Formula $Q = CLH^{3/2}$ from Experiments at Cornell University,* Metric System

Figure	Length of model, m	Head H, m									
		0.15	0.30	0.45	0.60	0.75	0.90	1.05	1.20	1.35	1.50
5.16	2.42		1.82	1.83	1.86	1.88	1.90	1.92	1.95	2.00	2.06
5.16	4.87	1.83	1.90	1.91	1.89	1.88	1.91	1.93			
5.17	2.43		1.87	1.91	1.94	1.96	1.98	2.00	2.03	2.07	2.12
5.17	4.87	1.78	1.92	1.99	2.03	2.04	2.06				
5.18	4.87	1.74	1.91	2.01	2.07	2.11	2.14	2.14			
5.19	4.87	1.78	1.85	1.90	1.94	1.98	2.01				
5.20	4.87	1.76	1.82	1.86	1.89	1.91	1.93	1.94	1.96		
5.21	4.87	1.81	1.93	1.96	1.94	1.86	1.83	1.82			
5.22	4.87	1.95	1.96	1.96	1.93	1.85	1.81	1.80	1.80		
5.23	4.86	1.73	1.73	1.71	1.73	1.77	1.80	1.83	1.86		
5.24		1.71	1.72	1.84							
5.25			2.10								

*Experiments were made on models resembling existing dams (except for the last two experiments, which were actual dams).

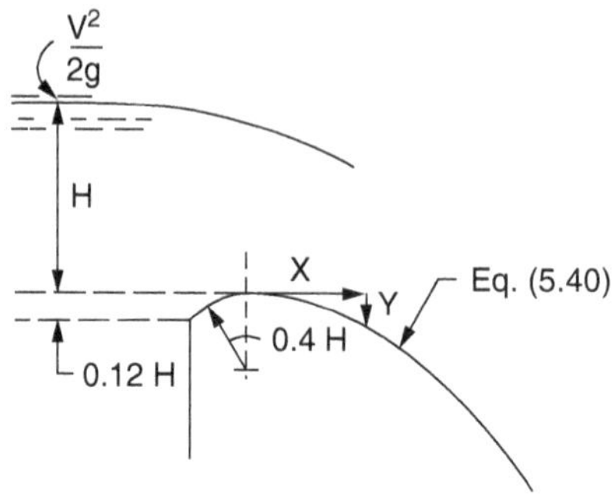

FIGURE 5.26 Spillway section "standard" crest.

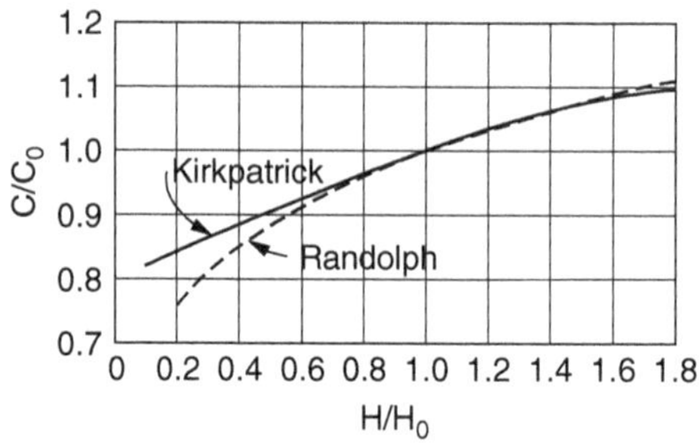

FIGURE 5.27 Coefficients for spillway sections.

frequent intervals. If accuracy is essential, the head used to compute the discharge should be the mean of at least 10, and preferably 20, separate measurements made at equal intervals of about 30 s.

Weirs for Measuring Stream Flow

Weirs of many types are used to measure the discharge of small streams. Sharp-crested weirs have the advantage that laboratory calibration conditions can be quite accurately reproduced in the field. Sharp-edged metal blades, however, require considerable maintenance, and in some cases they may be damaged by floating debris. The U.S. Geological Survey has standardized and calibrated a number of rounded-crested weirs specifically for stream-gauging purposes. Their Trenton and Columbus types[26] have cross sections similar to that shown in Figs. 5.1*d* and 5.26, thus avoiding the disadvantages of sharp-crested weirs. The Columbus type of deep-notch weir (Fig. 5.28) has the same cross section as the Columbus type round-crested one, and, in addition, the crest slopes downward toward the center to form a notch. This weir combines a large

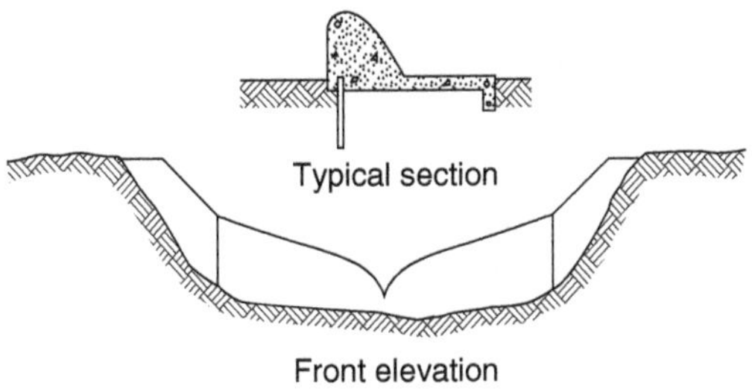

Figure 5.28 Columbus deep-notch–type weir.

maximum capacity with good sensitivity at low discharges. It must be carefully constructed to specifications in order to ensure that the rating curve derived in the laboratory will also apply in the field. In some cases standardized metal forms are used to ensure exact duplication.

When weirs are used to measure natural streams, a recording gauge is used. Methods of developing such a gauging station are discussed elsewhere.[27]

References

1. See, for example, V. L. Streeter, *Fluid Mechanics*, McGraw-Hill, New York, pp. 413–417, 1962.
2. R. E. Horton, "Weir Experiments, Coefficients, and Formulas," U.S. Geol. Survey Water Supply and Irrigation Paper 200, 1907.
3. J. B. Francis, *Lowell Hydraulic Experiments*, 4th ed., 1883; also *Trans. ASCE*, vol. 13, p. 303, 1884.
4. A. Fteley and F. P. Stearns, "Flow of Water over Weirs," *Trans. ASCE*, vol. 12, pp. 1–118, 1883.
5. H. Bazin, *Ann. Ponts et Chaussées*, Oct. 1888; trans. by Marichal and Trautwine, *Proc. Eng. Club*, Philadelphia, Jan. 1890; also *Ann. Ponts et Chaussées*, first quarter, 1894.
6. F. Frese, "Versuche über den Abfluss des Wassers bei vollkommenen Überfällen," *Z. Ver. deut. Ing*., S. 1285. 1890.
7. H. W. King, *Handbook of Hydraulics*, McGraw-Hill, New York, 1918.
8. T. Rehbock, Discussion of Schoder and Turner, "Precise Weir Measurements," *Trans. ASCE*, vol. 93, pp. 1143–1162, 1929; also *Handbuch der Ingenieurwissenschaften*, pt. 3, vol. 2, Leipzig, sec. 1, p. 58, 1912.
9. E. W. Schoder and K. B. Turner, "Precise Weir Measurements," *Trans. ASCE*, vol. 93, pp. 1143–1162, 1929.
10. C. E. Kindsvater and R. W. Carter, "Discharge Characteristics of Rectangular Thin-Plate Weirs," *Trans. ASCE*, vol. 124, pp. 772–822, 1959.
11. A. D. Flinn and C. W. D. Dyer, "The Cipolletti Trapezoidal Weir," *Trans. ASCE*, vol. 32, 1894.
12. A. T. Lenz, "Viscosity and Surface Tension Effects on V-Notch Weir Coefficients," *Trans. ASCE*, vol. 69, pp. 759–802, 1943.

13. J. Barr, "Experiments upon the Flow of Water over Triangular Notches," *Engineering*, 89, 435., Apr. 8, 15, 1910.
14. R. A. Hertzler, "Determination of a Formula for the 120-deg. V-Notch Weir," *Civil Eng. (N.Y.)*, p. 756, Nov. 1938.
15. J. K. Vennard and R. F. Weston, "Submergence Effect on Sharp-Crested Weirs," *Eng. News-Rec.*, vol. 130, p. 818, Jun. 3, 1943.
16. J. R. Villemonte, "Submerged-Weir Discharge Studies," *Eng. News-Rec.*, vol. 139, p. 866, pp. 54–57, Dec. 25, 1947.
17. F. T. Mavis, "How to Calculate Flow over Submerged Thin-Plate Weirs," *Eng. News-Rec.*, vol. 143, p. 65, Jul. 7, 1949.
18. J. C. Stevens, "Experiments on Small Weirs and Modules," *Eng. News*, vol. 64, p. 171, Aug. 18, 1910.
19. J. G. Woodburn, "Tests on Broad Crested Weirs," *Trans. ASCE*, vol. 96, pp. 387–408, 1932.
20. T. U. Taylor, "The Austin Dam," U.S. Geol. Survey Water Supply and Irrigation Paper 40, 1900.
21. W. P. Creager, *Engineering for Masonry Dams*, Wiley, New York, 1929.
22. W. P. Creager, J. D. Justin, and J. Hinds, *Engineering for Dams*, vol. 11, Wiley, New York, 1945.
23. R. R. Randolph, "Hydraulic Tests on the Spillway of Madden Dam," *Trans. ASCE*, vol. 103, 1938; also discussion by E. Scimeni, *ibid.*, p. 1113.
24. H. W. Kirkpatrick, "Discharge Coefficients for Spillways of TVA Dams," *Trans. ASCE*, vol. 22, pp. 190–210, 1957.
25. C. W. Thomas, "Errors in Measurement of Irrigation Water," *Trans. ASCE*, vol. 124, pp. 319–340, 1959.
26. "Equipment for River Measurements, Preliminary Report on Studies of Artificial Control," U.S. Geol. Survey Mimeographed Circ., 1935.
27. C. O. Wisler and E. F. Brater, *Hydrology*, 2d ed., Wiley, New York, 1959.

SECTION 6
Pipes

As commonly understood in hydraulics, a pipe is any closed conduit that carries water under pressure. Usually pipes are of circular cross section. Closed conduits that normally do not flow full are generally classed as open channels. Sewers and drainage tile come under this classification.

Since the factors affecting losses of head in conduits are independent of pressure, the same laws apply to the flow of water in both pipes and open channels.

Formulas for loss of head due to friction may involve the hydraulic radius r, which is the area divided by the wetted perimeter. For circular conduits flowing full, $r = d/4$, d being the diameter. It has usually been found more convenient to write d directly into formulas for circular conduits than to use r and later substitute its value in terms of d. For this reason the common practice is to express open-channel formulas as functions of r and pipe formulas as functions of d. For convenience in tabulating formulas and arranging tables, all matter pertaining to conduits flowing full is included in this section, and all matter pertaining to conduits flowing partially full is included in Sec. 7 on open channels. The word *pipe* as here used, therefore, includes all types of conduits flowing full.

The following symbols will be used in this section:

L = length of pipe, ft (m)
d = diameter of pipe, ft (m)
a = area of cross section of pipe, ft^2 (m^2)
p = wetted perimeter of pipe, ft (m)
$r = a/p$ = hydraulic radius ($d/4$ for round pipe flowing full), ft (m)
V = average velocity, ft/s (m/s)
$Q = aV$ = discharge, ft^3/s (m^3/s)
h = energy loss due to friction in length of pipe, ft (m)
$s = h/l$ = energy loss per foot (meter) of pipe
f, n, C = friction-loss coefficients
h_1, h_2, h_3, etc. = minor losses in pipes, ft (m)

K_1, K_2, K_3, etc. = minor-loss coefficients
μ = viscosity, slugs / ft·s (N·s/m²)
ρ = density, slugs/ft³ (kg/m³)
$\upsilon = \mu/\rho$ = kinematic viscosity, ft²/s (m²/s) (Table 1.2)
p = pressure intensity, lb/ft² (N/m²)
w = specific weight, lb/ft³ (N/m³) (Table 1.2)
g = gravitational acceleration, ft/s² (m/s²)
ϵ = magnitude of wall roughness, ft (m)

Fundamental Principles

Problems that involve steady flow in pipes require the application of the principles of continuity and of energy that were developed in Sec. 3. The determination of forces on pipe bends or nozzles using the momentum principle was also discussed in Sec. 3. The energy-loss term in the Bernoulli equation becomes of primary importance in pipe problems as distinct from orifices and weirs, in which energy losses were shown to be relatively small. The application of the basic principles, discussed in Sec. 3, to pipes will be demonstrated for the pipe systems shown in Figs. 6.1 and 6.2. Figure 6.1 shows a pipe carrying water from one reservoir to another. The pipe diameter is uniform in pipe *a* from point 2 to 5 and in pipe *b* from point 5 to 6, there being an increase in diameter at point 5. The general application of the Bernoulli equation is from point 1 to point 7 as follows:

$$\alpha_1 \frac{V_1^2}{2g} + \frac{p_1}{w} + z_1 = \alpha_7 \frac{V_7^2}{2g} + \frac{p_7}{\omega} + z_7 + \Sigma h_l \tag{6.1}$$

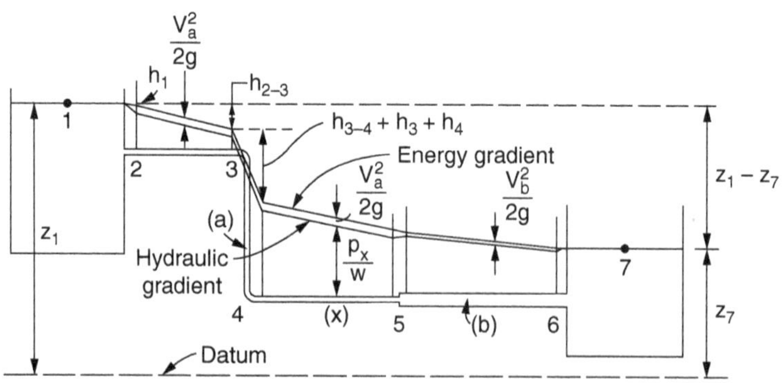

FIGURE 6.1 Pipe system.

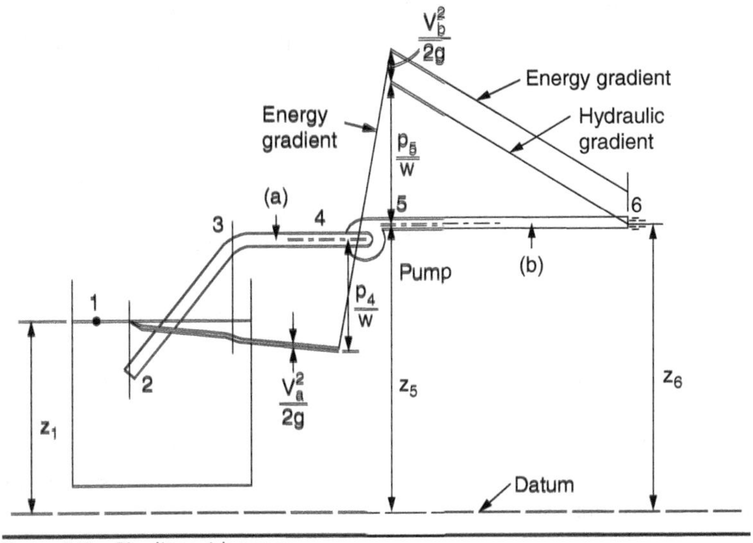

FIGURE 6.2 Pipeline with pump.

Because points 1 and 7 are in the large tanks, the kinetic energy terms may be neglected, and because they are located in the atmosphere, the pressure terms are zero. Thus Eq. (6.1) can be written

$$z_1 - z_7 = \Sigma h_l \tag{6.2}$$

where

$$\Sigma h_l = h_2 + h_{2-3} + h_3 + h_{3-4} + h_4 + h_{4-5} + h_5 + h_{5-6} + h_6 \tag{6.3}$$

and

h_2 = entrance loss
$h_{2-3} + h_{3-4} + h_{4-5}$ = total friction loss in straight sections of pipe a
$h_3 = h_4$ = loss at bends
h_5 = loss due to sudden enlargement
h_{5-6} = friction loss in pipe b
h_6 = loss at outlet

Losses at entrances, bends, enlargements, and contractions and other similar local losses are called *minor losses*, as distinguished from the friction loss in the straight sections. All the energy-loss terms can be expressed in terms of the local average velocities, as will be shown later. From continuity (Sec. 3),

$$Q = a_a V_a = a_b V_b \tag{6.4}$$

It may be seen from Eqs. (6.2) to (6.4) that a solution can be made for $z_1 - z_7$ if the discharge is known, or since all the energy losses can be expressed in terms of the velocities, the discharge can be determined if $z_1 - z_7$ and the pipe sizes are known. Example 6.3 is a numerical example based on Fig. 6.1.

Two lines shown in Fig. 6.1 have not been previously mentioned. The *energy gradient* is the line showing the total energy at any point in the pipe. The *hydraulic gradient*, or pressure *gradient*, is the line showing the pressure head, or piezometric head, at any point in the pipe. Sketching these lines is often very helpful in solving pipe problems. The energy gradient must always drop as it is traced in the direction of flow, whereas the pressure gradient may rise at locations such as point 5 in Fig. 6.1, where there is a conversion of kinetic energy to potential energy. For pipes of uniform diameter the energy loss per meter of pipe is equal to the drop in pressure gradient per meter.

Once the discharge and $z_1 - z_7$ are known in the pipe system shown in Fig. 6.1, the pressure at any point in the pipe can be determined either by placing numerical values on the graph or by writing the Bernoulli equation to the desired point. For example, to obtain the pressure at point x, the Bernoulli equation would be written from 1 to x as follows:

$$\alpha_1 \frac{V_1^2}{2g} + \frac{p_1}{w} + z_1 = \alpha_a \frac{V_a^2}{2g} + \frac{p_x}{w} + z_x + \Sigma h_l \tag{6.5}$$

Again, neglecting V_1, noting that $p_1 = 0$, and letting $\alpha_a = 1$,

$$\frac{p_x}{w} = z_1 - \left(\frac{V_a^2}{2g} + z_x + \Sigma h_l\right) \tag{6.6}$$

In Eq. (6.6) z_1, V_a, and z_x are known. The energy-loss term for this application consists of the following losses:

$$\Sigma h_l = h_2 + h_{2-3} + h_3 + h_{3-4} + h_4 + h_{4-x} \tag{6.7}$$

all of which have been defined previously, except h_{4-x}, which is the friction loss in pipe a for a length extending from point 4 to x. This problem is solved numerically in Example 6.3.

The pipe arrangement shown in Fig. 6.2 includes a pump in the system. The Bernoulli equation written from the water surface of the supply reservoir 1 to the center of the emerging stream 6, omitting V_1, α, and p_1 as in the previous illustration, is

$$z_1 + h_p = \frac{V_6^2}{2g} + \frac{p_6}{w} + z_6 + \Sigma h_l \tag{6.8}$$

In Eq. (6.8) p_6 is zero and h_p is the energy supplied by the pump in meter-newtons per newton. The energy losses are

$$\Sigma h_l = h_2 + h_{2-4} + h_3 + h_{5-6}$$

which can be identified by reference to Fig. 6.2. In this case, the gauge pressure is negative in pipe a, as shown by the location of the pressure gradient. The meaning of h_p may be further illustrated by writing the Bernoulli equation from point 4 to 5,

$$\frac{V_a^2}{2g} + \frac{p_4}{w} + z_4 + h_p = \frac{V_b^2}{2g} + \frac{p_5}{w} + z_5 \tag{6.9}$$

Then

$$h_p = (z_5 - z_4) + \frac{V_b^2}{2g} - \frac{V_a^2}{2g} + \frac{p_5}{w} - \frac{p_4}{w} \tag{6.10}$$

where p_4/w will be a negative quantity.

Usually, in problems involving pumps, the discharge and the geometric layout are known, and the size of pipe and the pump capacity must be determined. The solution is best made by solving for the required pump capacity for various selected pipe sizes.

Because all the terms in this form of the Bernoulli equation (see Sec. 3) have the unit of foot-pounds per pound or meter-newtons per newton, the rate at which energy E must be supplied by the pump is

$$E = Qwh_p \ ft-lb/s(\mathrm{m \cdot N/s}) \tag{6.11}$$

Loss of Energy due to Friction

Many investigators have sought to determine the laws governing the flow of fluids in pipes. One of the earliest expressions for energy loss in a pipe was developed by Chezy in 1775 (see later this section). Many other empirical formulas have been developed from test data. Most of these formulas were based on the assumption that the energy loss depends only on the velocity, the dimensions of the conduit, and the wall roughness. The works of Hagen (1839), Poiseuille (1840), and Reynolds (1883) showed that the density and viscosity of the fluid also affect the energy loss. Later, principally as the result of the work of Nikuradse (1933), it became generally recognized that the effect of roughness does not depend on the actual magnitude of the roughness, but on the ratio of the roughness size to the diameter of the pipe. Modern developments are oriented chiefly toward finding the effect of the form and spacing of roughness on energy losses.

Laminar Flow in Pipes

The laws of laminar flow in pipes were determined experimentally, independently by Hagen and by Poiseuille. The Hagen-Poiseuille law can be developed from fundamental principles as follows. Consider the forces acting on a cylinder of fluid of length l and radius y, as shown in Fig. 6.3. If steady motion exists, the force caused by the difference in pressure on the ends of the cylinder must be exactly balanced by the force resulting from the shear stress at the boundaries, as expressed by the equation

$$(p_1 - p_2)\pi y^2 = \tau 2\pi yl \tag{6.12}$$

Simplifying and introducing the value of τ from Eq. (1.3), Eq. (6.12) becomes

$$(p_1 - p_2)y = -2\mu \frac{du\,l}{dy}$$

the minus sign being included because increments of u and y are opposite in sign, as shown in Fig. 6.3. The preceding equation may be solved for du and integrated,

$$u = \int du = \frac{-(p_1 - p_2)y^2}{4\mu l} + C$$

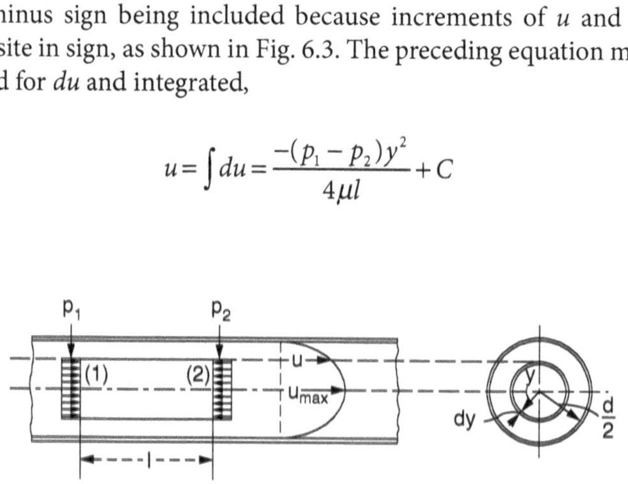

FIGURE 6.3 Laminar flow in pipes.

The value of C may be obtained from the boundary condition that $y = d/2$ when $u = 0$. Then

$$C = \frac{(p_1 - p_2)d^2}{16\,\mu\, l}$$

and

$$u = \frac{p_1 - p_2}{4\,\mu l}\left(-y^2 + \frac{d^2}{4}\right) \tag{6.13}$$

This equation shows the velocity distribution for laminar flow in circular pipes to be a paraboloid of revolution. The value of u_{max} may be determined by letting $y = 0$,

$$u_{max} = \frac{(p_1 - p_2)d^2}{16\,\mu\, l} \tag{6.14}$$

The discharge is obtained by integration as follows:

$$Q = \int dQ = \int u\,dA = \int u 2\pi y\,dy$$

Substituting the value of u from Eq. (6.13),

$$Q = \frac{\pi(p_1 - p_2)}{2\,\mu l}\int_0^{d/2}\left(-y^3 dy + \frac{d^2 y\,dy}{4}\right)$$

and

$$Q = \frac{\pi(p_1 - p_2)d^4}{128\,\mu l} \tag{6.15}$$

The average velocity V may be obtained from Eq. (6.15) by dividing by the area of the pipe,

$$V = \frac{(p_1 - p_2)d^2}{32\,\mu l} \tag{6.16}$$

It may be seen from Eqs. (6.14) and (6.16) that for laminar flow the maximum velocity is twice the average velocity. The expression for the change in piezometric head in a length l may be obtained by solving for $p_1 - p_2$ and dividing by the specific weight,

$$s\frac{p_1 - p_2}{w} = h = \frac{32\,\mu l V}{d^2 \rho g} \tag{6.17}$$

Introducing the Reynolds number [Eq. (3.1)], Eq. (6.17) becomes

$$h = \frac{64}{R}\frac{l}{d}\frac{V^2}{2g} \tag{6.18}$$

If in Eq. (6.18) the roughness coefficient f is used in place of $64/R$, the expression becomes

$$h = f\frac{l}{d}\frac{V^2}{2g} \tag{6.19}$$

which is the well-known Darcy-Weisbach[1,2] formula for flow in pipes, h being the energy loss per unit length of pipe. It is clear from the preceding that for laminar flow f is completely independent of roughness but varies only with the Reynolds number, that is, with the relative strength of the viscous and inertial forces. The straight line shown in Fig. 6.4 is a graphic representation of the relation

$$f = \frac{64}{R} \tag{6.20}$$

The experimental work of Hagen and Poiseuille and tests by many later investigators have established the accuracy of this relationship beyond question. So well does the relation hold that the application of Eq. (6.17) to circular tubes is one of the most basic methods of determining viscosity.

The Hagen-Poiseuille law [Eq. (6.15)], or the corresponding relation between f and R [Eq. (6.20)], applies when R is less than 2000. In the range of Reynolds numbers from 2000 to 4000, flow changes from laminar to turbulent. Values of f are uncertain in this range. If a pipeline were to be designed for flow in this range, the only safe procedure would be to assume that flow is turbulent and to select f by extending the curves shown in Fig. 6.4.

Turbulent Flow in Pipes

When flow occurs at Reynolds numbers greater than 4000, values of f in the Darcy-Weisbach formula [Eq. (6.19)] vary with roughness as well as with viscosity and density. Turbulent flow may be divided into three categories: flow in smooth pipes, flow in relatively rough pipes at high velocities, and flow in the transition zone between the first two categories.

For the case of flow in very smooth pipes, values of f vary with R, as shown by the lower curve of Fig. 6.4. It may be seen that this curve never becomes horizontal, which shows that the fluid properties

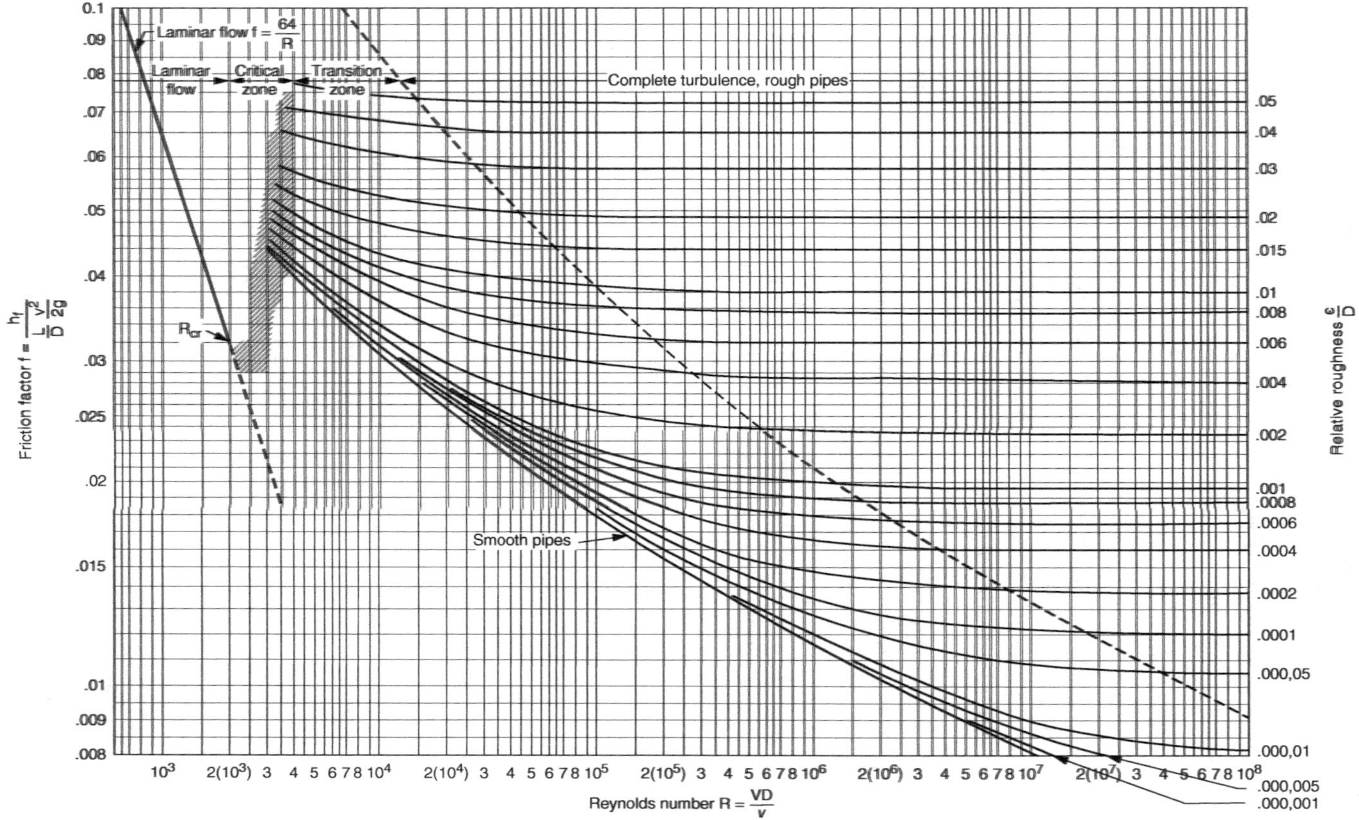

FIGURE 6.4 Moody diagram.

influence the flow throughout the entire range of Reynolds numbers. Glass pipes, very smooth drawn tubing, and reasonably smooth pipes having very large diameters would fall into this category.

Flow in rough pipes at large values of R is illustrated in Fig. 6.4 by the zone above and to the right of the broken line. This is the zone of fully developed turbulence. In this zone, the f curves become horizontal, thus showing that flow is entirely independent of the fluid properties. Nikuradse[3] showed that for pipes artificially roughened with uniform sand grains, values of f in this zone depend only on the relative roughness ϵ/d, where ϵ is the dimension of the roughness and d the pipe diameter in the same units. He determined this by testing pipes which had been artificially roughened with sand of uniform size.

The third category of turbulent flow in pipes occurs when values of f fall in the zone between the curve for smooth pipes and the broken line in Fig. 6.4. Flow in commercial pipe usually occurs in this category. In this zone, the f curves for various values of relative roughness depart from the smooth-pipe curves at successive locations and become horizontal as they enter the zone of fully developed turbulence. It was difficult at first to apply the principle of relative roughness to commercial pipe because such a small number of experimental points extended into the zone of fully developed turbulence. This difficulty was overcome by Colebrook,[4] who developed a relationship between f, ϵ/d, and R, which applies reasonably well to all the transition curves,

$$\frac{1}{\sqrt{f}} = -2\log\left(\frac{\epsilon}{3.7d} + \frac{2.51}{R\sqrt{f}}\right) \tag{6.21}$$

With the aid of this equation, Colebrook[4] was able to assign roughness values to various commercial pipe materials by comparing values of f for these pipes with Nikuradse's values for artificially roughened pipe. The broken line separating the transition zone from the zone of fully developed turbulence in Fig. 6.4 was suggested by Rouse.[5] The various curves shown in Fig. 6.4 were presented in this form by Moody.[6] Equations representing the curves in the turbulent flow portion of Fig. 6.4, developed by Wood,[7] are presented in Sec. 13 [Eqs. (13.11) and (13.12)]. Additional values of ϵ are presented in Table 6.1.

It must be recognized that this method of analyzing pipe friction data has certain inherent weaknesses because it relates all types of energy dissipation caused by wall friction to that resulting from closely spaced sand grains. The turbulence mechanism near rough walls is undoubtedly affected by the longitudinal and circumferential spacing and the form of the roughness, as well as by the actual

Table 6.1 Roughness Size ϵ for Large Pipes*

Material	Condition	ϵ, ft	ϵ, mm
Concrete	Unusually rough, poor alignment at joints	0.003–0.002	0.9–0.6
	Rough, visible form marks, spalling	0.003–0.0012	0.6–0.4
	Wood-floated or brushed surface in good condition, good joints	0.0012–0.0006	0.4–0.2
	Centrifugally cast, new, smooth, steel forms	0.0005	0.2
	Average workmanship, smooth joints, new, very smooth, steel forms	0.0006–0.0002	0.2–0.06
	First-class workmanship, smooth joints	0.0002–0.00005	0.06–0.02
Butt-welded steel	Severe tuberculation and incrustation	0.02–0.008	6.1–2.4
	General tuberculation	0.008–0.0031	2.4–0.9
	Heavy brush-coated enamels and tars	0.0031–0.0012	0.9–0.4
	Light rust	0.0012–0.0005	0.4–0.2
	Hot-asphalt-dipped	0.0005–0.0002	0.2–0.06
	New smooth pipe, centrifugally applied enamels	0.0002–0.00003	0.06–0.01
Girth-riveted steel	Severe tuberculation and incrustations	0.035–0.012	10.7–3.6
	General tuberculation	0.012–0.0044	3.6–1.3
	Rusted	0.0044–0.0020	1.3–0.6
	New, smooth, centrifugally applied enamels	0.002–0.0005	0.6–0.2
	Heavy brush-coated asphalts and tars	0.006–0.003	1.8–0.9
	Hot-asphalt-dipped, brush-coated graphite	0.003–0.001	0.9–0.3
Fully riveted steel	Severe tuberculation	0.03–0.02	9.1–6.1
	General tuberculation	0.02–0.007	6.1–2.1
	Fairly smooth, 3 rows longitudinal rivets	0.007–0.0034	2.1–1.0
	Fairly smooth, 2 rows longitudinal rivets	0.005–0.002	1.5–0.6
	Fairly smooth, 1 row longitudinal rivets	0.0034–0.001	1.0–0.3
Wood stave	Excessive growth on walls, rough projecting staves at joints	0.08–0.001	24–0.3
	Used condition	0.001–0.0009	0.3
	New, first-class construction	0.0004–0.0001	0.1–0.03
Steel†	Enamel-coated steel, 51 in. (1.30 m) diameter	0.000016	0.005
Corrugated metal‡	Depth of corrugations, ½ in. (1.3 cm) Spacing of corrugations, $2^2/_3$ in. (6.8 cm)	0.015–0.018	55–46
Rock§	Unlined	2.0	600

*Values were obtained from Bradley and Thompson,[8] except as otherwise noted.

†From Burke.[9]

‡From Webster and Metcalf.[10] Value of ϵ estimated by the authors.

§From Elder.[11]

size of the roughness.[12,13] This aspect of the problem has been elucidated by Morris.[14,15] Evidence of the nonsimilarity of some types of commercial-pipe roughness and sand-grain roughness is the fact that tests sometimes show increasing values of *f* in the transition zone whereas the Colebrook-White curves show decreasing values of *f* in this zone. Other tests, interpreted solely on the basis of relative roughness and the Reynolds number, are not entirely independent of the pipe diameter. These conditions are illustrated by results obtained from corrugated-metal pipes (see Selection of Energy-Loss Equation and Friction Factor).[10,16–18] The range of values of ϵ given in Fig. 6.4 and in Table 6.1 is undoubtedly due to these factors as well as to the condition of the pipe surface.

Roughness increases with age in pipes that are subject to corrosion. Ippen[19] reported one case with observations made on galvanized steel pipe in which it was found that the value of ϵ doubled within 3 years as the result of moderate conditions of use. Freeman[20] tested a number of new wrought-iron pipes and some "old, rusty" wrought-iron pipes of the same size. Evaluating the roughness of his pipes in comparison with Nikuradse's data. Freeman obtained the following results:

Type of pipe	ϵ, ft	ϵ, mm
2-in. (5-cm) new wrought iron	0.00014	0.04
2-in. (5-cm) old, rusty wrought iron	0.003	0.9
3-in. (7.5-cm) new wrought iron	0.00015	0.05
3-in. (7.5-cm) slightly rusty wrought iron	0.00014	0.13
3-in. (7.5-cm) very rusty, old wrought iron	0.0031	0.95
4-in. (10-cm) new wrought iron	0.00016	0.05
4-in. (10-cm) old, rusty wrought iron	0.0095	2.9

It may be noted that the roughness size is from 20 to 60 times larger for very old, rusty pipe than for new pipe. If the relative roughness is obtained, however, and inserted in Fig. 6.4, it will be seen that the values of *f* for the rusty pipe are only two or three times as great as those for new pipe.

Values of *f* for fire hose, computed from test data of Underwriters' Laboratories, Inc., are given in Table 6.2.

The Chezy Formula

This formula was the earliest attempt to express energy loss in conduits algebraically,

$$V = c\sqrt{rs} \tag{6.22}$$

TABLE 6.2 Values of f in Darcy-Weisbach Formula for 6-cm Fire Hose

	Velocity, ft/s				
	4	6	10	15	20
	Velocity, m/s				
Description	1	2	3	4	5
Unlined canvas	0.038	0.038	0.037	0.035	0.034
Rough rubber-lined cotton	0.032	0.031	0.031	0.030	0.029
Smooth rubber-lined cotton	0.024	0.023	0.022	0.019	0.018

If in Eq. (6.22) r is replaced by $d/4$ and s by h/l, the value of h is given by the expression

$$h = \frac{4}{c^2}\frac{l}{d}V^2 \tag{6.23}$$

from which it can be seen that the Darcy-Weisbach formula [Eq. (6.19)] is a rearrangement of the Chezy formula, the roughness coefficients being related as follows:

$$f = \frac{8g}{c^2} \tag{6.24}$$

From the previous discussion concerning Fig. 6.4 it may be concluded that the Chezy formula would give excellent results for flow in rough conduits at large Reynolds numbers, where the exponent of V is approximately 2. When later investigators found that this formula did not adhere to test results for smooth pipe with low velocities, other empirical formulas were devised to satisfy each particular group of tests. Only in recent years has it become generally recognized that all such tests can be unified by means of the Reynolds number.

The Manning Formula

Manning[21] concluded that the c in Chezy's formula should vary with $r^{1/6}$ as follows:

$$c = \frac{r^{1/6}}{n} \tag{6.25}$$

where n is the roughness coefficient. Substitution of Eq. (6.25) in the Chezy formula [Eq. (6.22)] yields the following equation, which is in the form usually referred to as the Manning formula:

$$V = \frac{r^{2/3} s^{1/2}}{n} \text{(metric units)} \tag{6.26}$$

Since the Manning coefficient, n, is not dimensionless, a conversion factor is required to use the formula in English units and the Manning formula [Eq. (6.26)] becomes:

$$V = 1.486 \frac{r^{2/3} s^{1/2}}{n} \text{(English units)}$$

For round pipes flowing full, $d/4$ may be substituted for r, and the Manning formula becomes

$$V = \frac{0.0.590}{n} d^{2/3} s^{1/2} \text{ (English units)}$$

$$V = \frac{0.40}{n} d^{2/3} s^{1/2} \text{ (metric units)} \tag{6.26a}$$

A comparison of the Manning equation [Eq. (6.26)] with the Darcy-Weisbach equation [Eq. (6.19)] reveals the following relationship between n and f:

$$f = \frac{185 n^2}{d^{1/3}} \text{(English units)}$$

$$f = \frac{122.5 n^2}{d^{1/3}} \text{(metric units)} \tag{6.27}$$

It is convenient to express the Manning formula, when used for pipes, in one of the following forms:

$$V = \frac{0.590}{n} d^{2/3} s^{1/2} \text{ (English units)}$$

$$V = \frac{0.40}{n} d^{2/3} s^{1/2} \text{ (metric units)} \tag{6.26a}$$

$$Q = \frac{0.463}{n} d^{8/3} s^{1/2} \text{ (English units)}$$

$$Q = \frac{0.314}{n} d^{8/3} s^{1/2} \text{ (metric units)} \tag{6.26b}$$

$$h = 2.87n^2 \frac{lV^2}{d^{4/3}} \text{ (English units)}$$

$$h = 6.25n^2 \frac{lV^2}{d^{4/3}} \text{ (metric units)} \tag{6.26c}$$

$$h = 4.66n^2 \frac{lQ^2}{d^{16/3}} \text{ (English units)}$$

$$h = 10.14n^2 \frac{lQ^2}{d^{16/3}} \text{ W (metric units)} \tag{6.26d}$$

$$d = \left(\frac{2.159Qn}{s^{1/2}}\right)^{3/8} \text{ (English units)}$$

$$d = \left(\frac{3.185Qn}{s^{1/2}}\right)^{3/8} \text{ (metric units)} \tag{6.26e}$$

$$s = \left(\frac{2.159Qn}{d^{8/3}}\right)^{1/2} \text{ (English units)}$$

$$s = \left(\frac{3.185Qn}{d^{8/3}}\right)^{1/2} \text{ (metric units)} \tag{6.26f}$$

The use of these formulae in solving three types of problems is described next.

1. Determine d, with s, n, and Q being given. Equation (6.26*e*) can be used to compute the diameter directly.
2. Determine s, with d, n, and Q being given. Equation (6.26*f*) can be used to compute the slope of the energy grade line directly.
3. Determine Q, with s, n, and d being given. Equation (6.26*f*) can be used to compute the diameter directly.

Table 6.3 lists recommended values of n. Practically all the experimental results that have been published lie between the extremes of these values.

The Hazen-Williams Formula

This formula[22] is widely used in waterworks design,

$$V = 1.318C_H r^{0.63} s^{0.54} \quad \text{(English)}$$

$$V = 0.85C_H r^{0.63} s^{0.54} \quad \text{(metric)} \tag{6.28}$$

It is an empirical equation based on laboratory and field observation.

TABLE 6.3 Values of n to Be Used with Manning Formula

Pipe	Variation		Use in designing	
	From	To	From	To
Clean uncoated cast-iron pipe	0.011	0.015	0.013	0.015
Clean coated cast-iron pipe	0.010	0.014	0.012	0.014
Dirty or tuberculated cast-iron pipe	0.015	0.035		
Riveted steel pipe	0.013	0.017	0.015	0.017
Lockbar and welded pipe	0.010	0.013	0.012	0.013
Galvanized iron pipe	0.012	0.017	0.015	0.017
Brass and glass pipe	0.009	0.013		
Wood-stave pipe	0.010	0.014		
Wood-stave pipe, small diameter			0.011	0.012
Wood-stave pipe, large diameter			0.012	0.013
Concrete pipe	0.010	0.017		
Concrete pipe with rough joints			0.016	0.017
Concrete pipe, "dry mix," rough forms			0.015	0.016
Concrete pipe, "wet mix," steel forms			0.012	0.014
Concrete pipe, very smooth			0.011	0.012
Vitrified sewer pipe	0.010	0.017	0.013	0.015
Common clay drainage tile	0.011	0.017	0.012	0.014
Corrugated metal (6.7 × 1.3 cm)	0.023	0.026		
Corrugated metal (7.5 × 2.5 and 15 × 2.5 cm)	0.026	0.029		
Corrugated metal (15 × 5 cm structural plate)	0.030	0.033		
Rock, unlined	0.038	0.041		
Enameled steel	0.009	0.010		

Typical values of the Hazen-Williams discharge coefficient C_H are given in the following table:

Type of pipe	C_H
Extremely smooth and straight	140
Very smooth	130
New riveted steel	110
Old riveted steel	100
Old cast iron	95
Old pipes in poor condition	60–80

Equation (6.28) has the advantage that the coefficient C_H is not a function of the Reynolds number, but it has the disadvantage that it is valid only for water and is not able to account for temperature or viscosity variations.

Selection of Energy-Loss Equation and Friction Factor

Only two energy-loss equations have been presented in detail, the Darcy-Weisbach equation [Eq. (6.19)] and the Manning equation [Eq. (6.26)]. Many other formulas have been used for this purpose, notably those of Williams-Hazen,[22] Tutton,[23] Unwin,[24] Scobey,[25-27] Barnes,[28] and Lea.[29]

The Darcy-Weisbach equation, along with the concept of the variation of f with R and $\in/d$, represents a basic approach to the solution of pipe friction losses. It is dimensionally correct, its general form can be derived analytically for laminar flow, and it can be used for any fluid. The Manning formula is the only one of the empirical type of energy-loss equations that is still used extensively. Numerous test results are available for the establishment of values of n. Because the Manning equation includes the hydraulic radius, it can be applied to conduits of any shape. Its principal weakness is that it can be used only for water at normal temperatures. Nevertheless, when dealing with water, results obtained with the Manning equation are undoubtedly just as accurate as those obtained from the Darcy-Weisbach equation. Indeed, because of the tendency of n to vary over a smaller range than f, there are situations in which it may be more dependable to use the Manning equation. For example, tests previously referred to[10,16] for corrugated metal pipes resulted in values of f varying from 0.051 to 0.078, a range of 53 percent based on the smaller value, whereas for the same tests the values of n varied from 0.0229 to 0.0248, a range of only 8 percent based on the smaller value. Obviously, it is simpler and perhaps more accurate to use an n of 0.024 for all corrugated metal pipes in the range of diameters and Reynolds numbers tested than to use the somewhat more cumbersome Darcy-Weisbach equation. Furthermore, extrapolation to other pipe sizes might also be more safely accomplished in this case by using the Manning equation. However, particularly for smoother pipe and large values of the Reynolds number, more dependable results can be obtained by means of the Darcy-Weisbach equation.

Although the developments described in the preceding section have vastly improved the understanding of the mechanics of energy dissipation in conduits, the problem is not yet fully understood, and the engineer should realize that the solution of all problems outside the laminar range must depend on the interpolation or extrapolation of experimental data determined under conditions which may have been dissimilar to that in which they will be used. For materials and pipe sizes commonly used, the curves of f and tables of n will give good results. When unusual surfaces or very large pipe sizes are involved, and especially for very long conduits in which the energy loss becomes a vital design consideration, the literature should be searched for tests on similar conduits. The U.S. Bureau of Reclamation has presented much information on friction factors.[8] Other test results of interest have been reported by Burke[9] and Elder.[11]

The computation of energy loss will be illustrated by the following examples. Section 13, Examples 13.6 and 13.7, illustrate similar computations using a digital computer.

Example 6.1 Determine the energy loss in 1000 ft of new 24 in. concrete pipe when water at 70°F is flowing at an average velocity of 6 ft/s (1) using the Darcy-Weisbach equation and (2) using the Manning equation.

1. Obtaining the kinematic viscosity from Table 1.2,

$$R = \frac{dV}{\nu} = \frac{24/12 \times 6.0}{1.06 \times 10^{-5}} = 1.13 \times 10^{6}$$

From Table 6.1, $\epsilon = 0.0005$ ft. Then $\epsilon/d = 0.0005/2 = 0.00025$. From Fig. 6.4, $f = 0.015$. Then, from Eq. (6.19),

$$h = f\frac{l}{d}\frac{V^2}{2g} = 0.015 \times \frac{1000}{2} \times \frac{36}{64.4} = 4.2 \text{ ft}$$

2. From Table 6.3, minimum $n = 0.011$. Then from Eq. (6.26*c*),

$$h = \frac{2.87 n^2 l V^2}{d^{4/3}} = \frac{2.87 \times 0.011^2 \times 1000 \times 36}{(2)^{4/3}} = 4.10 \text{ ft} \quad ▲$$

Example 6.2 Determine the energy loss in 15 m of 6-mm tubing for a liquid having a unit weight of 9580 N/m^3 and a viscosity of 1.50 N·s/m^2 when the velocity is 3 cm/s.

$$R = \frac{d\upsilon\rho}{\mu} = \frac{0.006 \times 0.03 \times 9580}{1.5 \times 9.8} = 0.117$$

and from Eqs. (6.19) and (6.20),

$$h = f\frac{l}{d}\frac{V^2}{2g} = \frac{64}{R}\frac{l}{d}\frac{V^2}{2g} = \frac{64}{0.117} \times \frac{15}{0.006} \times \frac{0.03^2}{19.6} = 62.8\,\text{m} \quad \blacktriangle$$

Minor Losses

Energy losses resulting from rapid changes in the direction or magnitude of the velocity are called *minor* losses, or *local* losses. The term minor loss is appropriate for pipelines that include long reaches of uniform straight pipe. However, for short pipes it is a misnomer, because the minor losses may be greater than the friction losses.

Minor losses are usually expressed in terms of the kinetic energy and a coefficient, as illustrated by the equation

$$h_m = K_m \frac{V^2}{2g} \tag{6.29}$$

When a change in pipe size occurs, two velocities are involved, and both may be included in the expression for energy loss, as will be shown for the case of enlargements and contractions. Under certain conditions, particularly for the compound-pipe problem (see Compound Pipes), it may be convenient to express the energy loss in terms of equivalent length l_e of straight pipe. If the energy-loss term for minor losses given by Eq. (6.29) is equated to the loss for a straight pipe of length l_e, from Eq. (6.19),

$$K_m \frac{V^2}{2g} = f\frac{l_e}{d}\frac{V^2}{2g}$$

and

$$l_e = \frac{K_m d}{f} \tag{6.30}$$

Then the actual length of pipe can be increased by the amount l_e and the total loss will be given by Eq. (6.19).

Loss at Entrance

When water enters a pipe from a reservoir, the loss is expressed as

$$h_1 = K_1 \frac{V^2}{2g} \tag{6.31}$$

Values of K_1 for various conditions are given in the following table:

Type of entrance	K_1
Square-cornered entrance flush with wall	0.5
Rounded entrance	0.04–0.2
Inward-projecting square-cornered entrance	0.8–0.9

Additional information on entrance losses is given in the discussion of culverts in Sec. 4.

Loss of Head due to Enlargement

Borda has found that the loss in pipes due to *sudden* enlargement may be represented by the theoretical formula

$$h_2 = \frac{(V_1 - V_2)^2}{2g} \tag{6.32}$$

where h_2 is the head lost and V_1 and V_2 are the velocities in the smaller and larger pipes, respectively.

This loss was also investigated experimentally by Baer,[30] Brightmore,[31] Archer,[32] and others. These experiments indicated that Borda's formula gives values of h_2 too small for the lower velocities and smaller differences in diameter, and too large for the opposite conditions. Archer deduced the formula

$$h_2 = \frac{(V_1 - V_2)^{1.919}}{2g} = 0.051(V_1 - V_2)^{1.919} \quad \text{(metric)} \tag{6.33}$$

$$h_2 = 1.098\frac{(V_1 - V_2)^{1.919}}{2g} = 0.01705(V_1 - V_2)^{1.919} \quad \text{(English)}$$

This formula appears to be as satisfactory as any yet suggested. Experiments at the University of Michigan indicated that Archer's formula holds quite accurately in the limit when $V_2 = 0$.

Table 6.4 lists the values of K_2 for use in the formula

$$h_2 = K_2 \frac{V_1^2}{2g} \tag{6.34}$$

Losses due to *gradual* enlargement were investigated by Parker[33] from a study of experiments by Andres, Gibson, and others.

Table 6.4a Values of K_2 for Determining Loss of Head due to Sudden Enlargement in Pipes, from Formula $h_2 = K_2(V_1^2/2g)$, U.S. Customary System

	Velocity V_1,* ft/s											
d_2/d_1*	2	3	4	5	6	7	8	10	12	15	20	30	40
1.2	0.11	0.10	0.10	0.10	0.10	0.10	0.10	0.09	0.09	0.09	0.09	0.09	0.08
1.4	0.26	0.26	0.25	0.24	0.24	0.24	0.24	0.23	0.23	0.22	0.22	0.21	0.20
1.6	0.40	0.39	0.38	0.37	0.37	0.36	0.36	0.35	0.35	0.34	0.33	0.32	0.32
1.8	0.51	0.49	0.48	0.47	0.47	0.46	0.46	0.45	0.44	0.43	0.42	0.41	0.40
2.0	0.60	0.58	0.56	0.55	0.55	0.54	0.53	0.52	0.52	0.51	0.50	0.48	0.47
2.5	0.74	0.72	0.70	0.69	0.68	0.67	0.66	0.65	0.64	0.63	0.62	0.60	0.58
3.0	0.83	0.80	0.78	0.77	0.76	0.75	0.74	0.73	0.72	0.70	0.69	0.67	0.65
4.0	0.92	0.89	0.87	0.85	0.84	0.83	0.82	0.80	0.79	0.78	0.76	0.74	0.72
5.0	0.96	0.93	0.91	0.89	0.88	0.87	0.86	0.84	0.83	0.82	0.80	0.77	0.75
10.0	1.00	0.99	0.96	0.95	0.93	0.92	0.91	0.89	0.88	0.86	0.84	0.82	0.80
∞	1.00	1.00	0.98	0.96	0.95	0.94	0.93	0.91	0.90	0.88	0.86	0.83	0.81

*d_2/d_1—ratio of diameter of larger pipe to diameter of smaller pipe; V_1—velocity in smaller pipe.

Table 6.4b Values of K_2 for Determining Loss of Head due to Sudden Enlargement in Pipes, from Formula $h_2 = K_2(V_1^2/2g)$, Metric System

	Velocity V_1,* m/s										
d_2/d_1*	**0.5**	**1.0**	**2.0**	**3.0**	**4.0**	**5.0**	**6.0**	**7.0**	**8.0**	**9.0**	**10.0**
1.2	0.11	0.10	0.10	0.09	0.09	0.09	0.09	0.09	0.09	0.09	0.09
1.4	0.26	0.26	0.24	0.23	0.23	0.22	0.22	0.22	0.21	0.21	0.21
1.6	0.40	0.39	0.36	0.35	0.35	0.34	0.33	0.33	0.32	0.32	0.32
1.8	0.51	0.49	0.46	0.45	0.44	0.43	0.42	0.42	0.41	0.41	0.41
2.0	0.60	0.58	0.54	0.52	0.52	0.51	0.50	0.50	0.49	0.48	0.48
2.5	0.74	0.72	0.67	0.65	0.64	0.63	0.62	0.62	0.61	0.60	0.59
3.0	0.84	0.80	0.75	0.73	0.71	0.70	0.69	0.68	0.67	0.67	0.66
4.0	0.93	0.89	0.83	0.80	0.79	0.77	0.76	0.75	0.74	0.74	0.73
5.0	0.97	0.93	0.87	0.84	0.83	0.81	0.80	0.79	0.78	0.77	0.76
10.0	1.00	0.98	0.92	0.89	0.87	0.85	0.84	0.83	0.82	0.82	0.81
∝	1.00	1.00	0.94	0.91	0.89	0.87	0.86	0.85	0.84	0.83	0.82

*d_2/d_1—ratio of diameter of larger pipe to diameter of smaller pipe; V_1—velocity in smaller pipe.

The formula suggested by Andres for a conical enlargement may be written

$$h_2 = f\,\frac{V_1^2 - V_2^2}{2g} \tag{6.35}$$

where V_1 and V_2 are the velocities in the smaller and larger pipes, respectively, and f is an empirical coefficient depending for its value on the angle θ, θ being double the angle between the axis of the pipe and its side.

Andres gives values of f for smaller values of θ, and Gibson for values up to 90°. Their results are not entirely concordant, but this author has used them to plot a mean curve giving the results of Andres more weight for the smaller angles. The following are results obtained in this manner:

θ	2°	3°	4°	5°	6°	7°	8°	9°	10°	11°	12°
f	0.033	0.036	0.039	0.042	0.046	0.050	0.055	0.066	0.078	0.090	0.100
θ	15°	20°	25°	30°	35°	40°	45°	50°	60°	75°	90°
f	0.16	0.31	0.40	0.49	0.55	0.60	0.64	0.67	0.72	0.72	0.67

Using these values of f, Table 6.5 lists values of K_2 in the formula.

$$h_2 = K_2\,\frac{V_1^2}{2g} \tag{6.34}$$

Here V_1 is the velocity in the smaller pipe.

Loss of Head due to Contraction

Merriman[34] suggests the following formula for determining the loss of head due to sudden contraction:

$$h_3 = \left(\frac{1}{c} - 1\right)^2 \frac{V_2^2}{2g} \tag{6.36}$$

where V_2 is the velocity in the smaller pipe and

$$c = 0.582 + \frac{0.0418}{1.1 - r} \tag{6.37}$$

r being the ratio of the smaller to the larger diameter.

TABLE 6.5 Values of K_2 for Determining Loss of Head due to Gradual Enlargement in Pipes, from Formula $h_2 = K_2(V_1^2/2g)$

	Angle of Cone, Deg*													
d_2/d_1*	2	4	6	8	10	15	20	25	30	35	40	45	50	60
1.1	0.01	0.01	0.01	0.02	0.03	0.05	0.10	0.13	0.16	0.18	0.19	0.20	0.21	0.23
1.2	0.02	0.02	0.02	0.03	0.04	0.09	0.16	0.21	0.25	0.29	0.31	0.33	0.35	0.37
1.4	0.02	0.03	0.03	0.04	0.06	0.12	0.23	0.30	0.36	0.41	0.44	0.47	0.50	0.53
1.6	0.03	0.03	0.04	0.05	0.07	0.14	0.26	0.35	0.42	0.47	0.51	0.54	0.57	0.61
1.8	0.03	0.04	0.04	0.05	0.07	0.15	0.28	0.37	0.44	0.50	0.54	0.58	0.61	0.65
2.0	0.03	0.04	0.04	0.05	0.07	0.16	0.29	0.38	0.46	0.52	0.56	0.60	0.63	0.68
2.5	0.03	0.04	0.04	0.05	0.08	0.16	0.30	0.39	0.48	0.54	0.58	0.62	0.65	0.70
3.0	0.03	0.04	0.04	0.05	0.08	0.16	0.31	0.40	0.48	0.55	0.59	0.63	0.66	0.71
∞	0.03	0.04	0.05	0.06	0.08	0.16	0.31	0.40	0.49	0.56	0.60	0.64	0.67	0.72

*d_2/d_1—ratio of diameter of larger pipe to diameter of smaller pipe. Angle of cone is twice the angle between cone axis and its side.

Brightmore[31] experimented on pipes 6 in. (15 cm) in diameter contracted to 4 and 3 in. (10 and 7.5 cm), the mean of his results being represented approximately by the formula

$$h_3 = \frac{0.7(V_1 - V_2)^2}{2g} \tag{6.38}$$

Parker[33] suggests that Eq. (6.36) be used for higher velocities when the head lost is 1 ft (0.3 m) or more, whereas Eq. (6.38) is more reliable for smaller losses of head.

Following this suggestion, this author computed h_3 by both formulas for various velocities and diameter ratios. The results were then plotted and curves drawn through the points by gradually changing from results obtained by Eq. (6.38) for lower velocities to those from Eq. (6.36) for higher velocities. The corresponding values of K_3 for determining the loss of head due to sudden contraction in the formula

$$h_3 = K_3 \frac{V_2^2}{2g} \tag{6.39}$$

are given in Table 6.6.

Loss of Head due to Gate Valves

Average values of K_g for losses at gate valves, as determined by Corp and Ruble,[35] are given in Table 6.7.

Loss of Head due to Bends

The minor loss at a bend results from a distortion of the velocity distribution, thereby causing additional shear stresses within the fluid. The normal velocity distribution may not be restored for a distance as great as 25 diameters below the bend. The bend loss is considered to be the loss in excess of the loss for an equal length of straight pipe. It is expressed in terms of kinetic energy

$$h_b = K_b \frac{V^2}{2g} \tag{6.40}$$

Tests by Beij[36] indicated that the loss is a function of the relative radius R/d (bend radius/pipe diameter) and the roughness of the pipe. Throughout the range of his tests, for Reynolds numbers varying from 20,000 to 300,000, he found that the bend loss was independent of the Reynolds number. Figure 6.5 is the curve plotted by Beij through his test results. These tests were run on 4-in. (10-cm) pipe having a roughness such that the value of f in the Darcy-Weisbach formula

Table 6.6a Values of K_3 for Determining Loss of Head due to Sudden Contraction in Pipes, from Formula $h_2 = K_2(V_1^2/2g)$, U.S. Customary System

	Velocity V_2,* ft/s											
d_2/d_1^*	**2**	**3**	**4**	**5**	**6**	**7**	**8**	**10**	**12**	**15**	**20**	**30**	**40**
1.1	0.00	0.00	0.01	0.01	0.02	0.03	0.04	0.06	0.09	0.15	0.29	0.75	1.49
1.2	0.00	0.01	0.02	0.03	0.04	0.06	0.07	0.12	0.18	0.28	0.54	1.38	2.74
1.4	0.01	0.02	0.04	0.07	0.10	0.13	0.17	0.27	0.40	0.65	1.14	2.68	4.98
1.6	0.02	0.04	0.06	0.10	0.14	0.20	0.26	0.40	0.67	0.89	1.56	3.44	5.97
1.8	0.02	0.05	0.08	0.13	0.19	0.25	0.33	0.51	0.73	1.12	1.92	4.05	6.72
2.0	0.02	0.05	0.09	0.14	0.21	0.28	0.36	0.55	0.79	1.19	2.06	4.28	7.09
2.2	0.02	0.06	0.10	0.15	0.22	0.30	0.38	0.59	0.84	1.28	2.20	4.56	7.41
2.5	0.03	0.06	0.10	0.16	0.23	0.31	0.40	0.62	0.88	1.34	2.30	4.76	7.71
3.0	0.03	0.06	0.11	0.17	0.24	0.32	0.42	0.65	0.92	1.40	2.41	4.98	8.11
4.0	0.03	0.06	0.12	0.18	0.25	0.34	0.44	0.69	0.97	1.48	2.53	5.24	8.48
5.0	0.03	0.07	0.12	0.18	0.26	0.35	0.46	0.70	1.00	1.52	2.60	5.36	8.67
10.0	0.03	0.07	0.12	0.19	0.27	0.36	0.47	0.72	1.02	1.56	2.68	5.56	9.06
∝	0.03	0.07	0.12	0.19	0.27	0.36	0.47	0.72	1.03	1.58	2.71	5.68	9.36

*d_2/d_1—ratio of diameter of larger pipe to diameter of smaller pipe; V_2—velocity in smaller pipe.

Table 6.6b Values of K_3 for Determining Loss of Head due to Sudden Contraction in Pipes, from Formula $h_2 = K_2(V_1^2/2g)$, Metric System

	Velocity V_2,* m/s										
d_2/d_1*	0.5	1.0	2.0	3.0	4.0	5.0	6.0	7.0	8.0	9.0	10.0
1.1	0.03	0.04	0.04	0.04	0.04	0.04	0.05	0.05	0.05	0.05	0.05
1.2	0.07	0.07	0.07	0.08	0.08	0.08	0.09	0.09	0.10	0.10	0.10
1.4	0.17	0.17	0.17	0.18	0.18	0.18	0.18	0.19	0.19	0.19	0.19
1.6	0.26	0.26	0.26	0.26	0.26	0.26	0.25	0.25	0.25	0.25	0.24
1.8	0.34	0.34	0.34	0.33	0.32	0.31	0.31	0.30	0.29	0.29	0.28
2.0	0.38	0.38	0.37	0.36	0.35	0.34	0.33	0.33	0.32	0.31	0.30
2.2	0.40	0.40	0.39	0.38	0.37	0.36	0.35	0.35	0.34	0.33	0.32
2.5	0.42	0.42	0.41	0.40	0.39	0.38	0.37	0.36	0.35	0.34	0.33
3.0	0.44	0.44	0.43	0.42	0.41	0.40	0.39	0.38	0.37	0.36	0.35
4.0	0.47	0.46	0.45	0.44	0.43	0.42	0.41	0.40	0.38	0.37	0.36
5.0	0.48	0.48	0.46	0.45	0.45	0.44	0.42	0.41	0.39	0.38	0.37
10.0	0.49	0.48	0.47	0.46	0.46	0.44	0.43	0.42	0.41	0.40	0.39
∝	0.49	0.49	0.47	0.47	0.46	0.45	0.44	0.43	0.42	0.41	0.40

*d_2/d_1—ratio of diameter of larger pipe to diameter of smaller pipe; V_2—velocity in smaller pipe.

Table 6.7 Values of K_g for Determining Loss of Head due to Gate Valves, from Formula $h_2 = K_2(V_1^2/2g)$

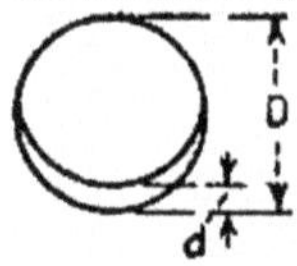

Nominal diameter of valve, cm	in.	Ratio of height *d* of valve opening to diameter *D* of full valve opening					
		1/8	¼	3/8	½	¾	1
1.3	½	450	60	22	11	2.2	1.0
1.9	¾	310	40	12	5.5	1.1	0.28
2.5	1	230	32	9.0	4.2	0.90	0.23
3.8	1½	170	23	7.2	3.3	0.75	0.18
5	2	140	20	6.5	3.0	0.68	0.16
10	4	91	16	5.6	2.6	0.55	0.14
15	6	74	14	5.3	2.4	0.49	0.12
20	8	66	13	5.2	2.3	0.47	0.10
30	12	56	12	5.1	2.2	0.47	0.07

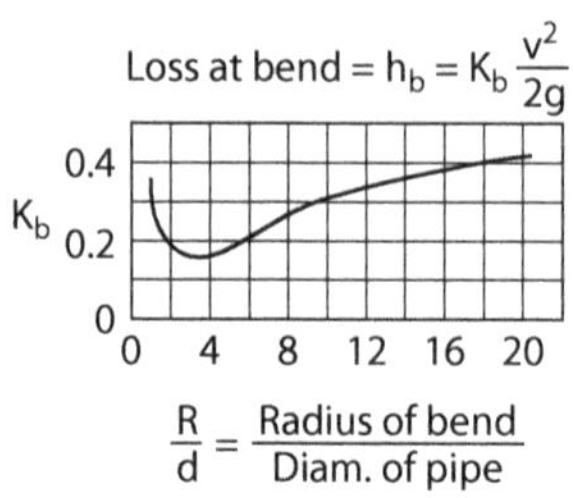

Figure 6.5 Energy loss in pipe bends.

varied from 0.020 to 0.025. For very smooth pipe or for very large pipes the values of K_b are smaller than those given by Fig. 6.5, and the following values of K_b are suggested[37]:

R/d	K_b
1	0.23
2	0.13
4 or larger	0.08

For very rough pipe, values of K_b are likely to be larger than those given by Fig. 6.5.

For bends of less than 90°, Fuller[38] gave the following approximate rules:

- For loss of head due to 45° bends, use three-fourths of that due to 90° bends of the same radius.
- For loss of head due to 22.5° bends, use one-half of that due to 90° of the same radius.

A curve of reduction factors for bends of less than 90° given by the U.S. Bureau of Reclamation shows that the 90° bend loss should be multiplied by 0.83 for 60° bends, by 0.70 for 45° bends, and by 0.42 for 22.5° bends.

The solution of a pipe problem involving minor losses is illustrated by the following example.

Example 6.3 In Fig. 6.1 pipes *a* and *b* are steel pipes having diameters of 6 and 12 in., respectively. The discharge is 3 ft³/s, and the fluid is water at 70°F. The radii of the bends are twice the respective pipe diameters. Pipe lengths are

2–3	200 ft
3–4	250 ft
4–5	300 ft
5–6	300 ft

1. Determine the difference in elevations of the water surfaces in the two reservoirs $(z_1 - z_7)$.

$$V_a = 3.0/0.196 = 15.3 \text{ ft/s}$$
$$V_b = 3.0/0.785 = 3.82 \text{ ft/s}$$

Then

$$\frac{V_a^2}{2g} = 3.63 \text{ ft} \qquad \text{and} \qquad \frac{V_b^2}{2g} = 0.227 \text{ ft}$$

The various elements in the total energy loss are shown in Eq. (6.3) and identified below Eq. (6.3). They will now be evaluated.

From Eq. (6.34),

$$h_2 = 0.5\frac{V_a^2}{2g} = 1.8 \text{ ft}$$

From the section on turbulent flow,

$$h_{2-3} = f\frac{l}{d}\frac{V^2}{2g} = f\frac{200}{0.5} \times 3.63$$

The value of viscosity is obtained from Table 1.2. Then

$$R_a = \frac{dV}{\nu} = \frac{0.5 \times 15.3}{1.06 \times 10^{-5}} = 720{,}000$$

From Table 6.1,

$$\epsilon = 0.00015$$

and therefore

$$\frac{\epsilon}{d} = \frac{0.00015}{0.5} = 0.0003$$

From Fig. 6.4, $f = 0.016$. Then, from the preceding,

$$h_{2-3} = 0.016 \times 200/0.5 \times 3.63 = 23.2 \text{ ft}$$

and, by similarity,

$$h_{3-4} = 29.0 \text{ ft} \quad \text{and} \quad h_{4-5} = 34.8 \text{ ft}$$

From Fig. 6.5,

$$h_3 = h_4 = 0.2\frac{V_a^2}{2g} = 0.7 \text{ ft}$$

From Table 6.4,

$$h_5 = 0.5 \times 3.63 = 1.9 \text{ ft}$$

The Reynolds number for pipe b is

$$R_b = \frac{dv}{\nu} = \frac{1.0 \times 3.82}{1.06 \times 10^{-5}} = 360{,}000$$

and from Fig. 6.4, $f = 0.0156$ and

$$h_{5-6} = 0.0156 \times 300/1.0 \times 0.227 = 1.1 \text{ ft}$$

The outlet loss from Table 6.5 is

$$h_6 = \frac{V_b^2}{2g} = 0.2 \text{ ft}$$

Then, as shown by Eq. (6.2), the difference in the elevation of the water surfaces can be found by summation of all the losses as follows:

$$z_1 - z_7 = 1.8 + 23.2 + 0.7 + 29.0 + 0.7 + 34.8 + 1.9 + 1.1 + 0.2$$
$$= 93.4 \text{ ft·m}$$

It should be noted that several of the so-called minor losses are greater than the friction loss in pipe *b*.

2. Find the pressure at x if the distance from point 4 to x is 150 ft, $z_1 = 200$ ft, and $z_x = 50$ ft.
The solution of this problem is discussed under Fundamental Principles. The summation of the energy losses shown in Eq. (6.7) can be evaluated from the values in part 1 of this example, except for h_{4-x}, which is

$$h_{4-x} = 0.016 \times 150 / 0.5 \times 3.63 = 17.4 \text{ ft}$$

Then

$$\Sigma h_l = 1.8 + 23.2 + 0.7 + 29.0 + 0.7 + 17.4 = 72.8 \text{ ft}$$

Then, from Eq. (6.6),

$$\frac{p_x}{w} = z_1 - \left(\frac{V_a^2}{2g} + z_x + \Sigma h_l \right)$$
$$= 200 - (3.6 + 50 + 72.8) = 73.6 \text{ ft}$$

and

$$p_x = 73.6 \times 62.4/144 = 31.9 \text{ lb/in}^2 \quad \blacktriangle$$

Compound Pipes

Two types of compound-pipe problems may be encountered. One type occurs in pipe networks when two or more paths are available for water flowing between two points, such as points 1 and 5 in Fig. 6.6. A second type, illustrated by Fig. 6.7, occurs when water can flow to or from a junction of three or more pipes from independent sources or outlets.

The first type, often called the *parallel-pipe problem*, can be solved by making use of the fact that the total energy loss through the system is the same for all paths followed by the water. This will be illustrated by writing the Bernoulli equation from the water surface of the reservoir

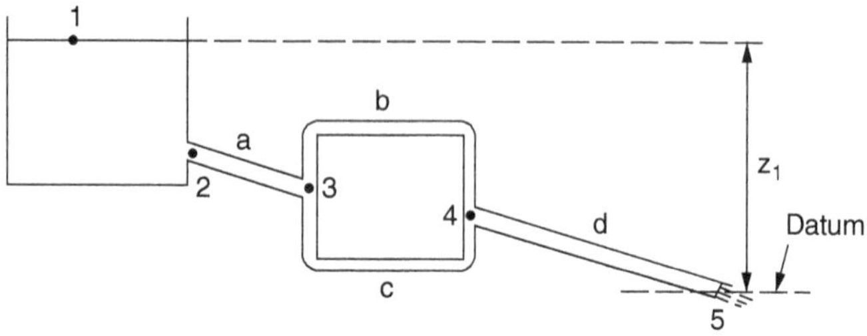

FIGURE 6.6 Parallel pipes.

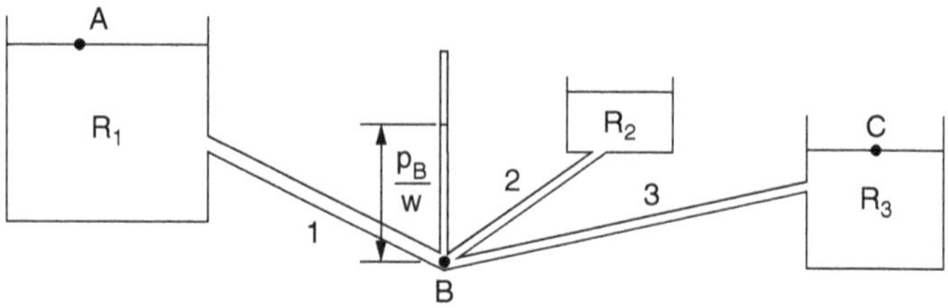

FIGURE 6.7 Branching pipe system.

to the outlet of pipe *d* in Fig. 6.6, taking the datum through point 5 at the outlet of pipe *d*:

$$\frac{V_1^2}{2g} + \frac{p_1}{w} + z_1 = \frac{V_5^2}{2g} + \frac{p_5}{w} + z_5 + \Sigma h_l \tag{6.41}$$

and omitting terms which are zero,

$$z_1 = \frac{V_5^2}{2g} + \Sigma h_t \tag{6.42}$$

The expression for the energy loss may now be evaluated, first considering the path through pipe *b* and then that through pipe *c*,

$$\Sigma h_a = h_2 + h_{2-3} + h_{3(a)} + h_{3-4(a)} + h_{4(a)} + h_{4-5} \tag{6.43}$$

and

$$\Sigma h_b = h_2 + h_{2-3} + h_{3(b)} + h_{3-4(b)} + h_{4(b)} + h_{4-5} \tag{6.44}$$

where h_2, h_3, and h_4 are minor losses at the points indicated, and h_{2-3}, h_{3-4}, and h_{4-5} are friction losses. Noting from Eq. (6.42) that, because

z_1 and $V_5^2/2g$ are the same no matter which path the water takes, $\Sigma h_a = \Sigma h_b$, it follows from Eqs. (6.43) and (6.44) that

$$h_{3(a)} + h_{3-4(a)} + h_{4(a)} = h_{3(b)} + h_{3-4(b)} + h_{4(b)} \tag{6.45}$$

All the terms in Eq. (6.45) can be expressed in terms of velocity, using Eq. (6.19) for the friction loss and an appropriate minor-loss equation. Then the only two unknowns will be V_b and V_c. A second equation involving continuity can then be written,

$$Q_a = Q_b + Q_c = Q_d \tag{6.46}$$

or

$$A_a V_a = A_b V_b + A_c V_c = A_d V_d \tag{6.46a}$$

If then the total discharge is known, Eqs. (6.45) and (6.46) can be solved for the discharge in pipes *b* and *c*, and the total energy drop across the system can be obtained from Eq. (6.42).

Pipe systems of the type shown in Fig. 6.7 usually require solutions by successive trials because it may not be obvious in which direction the flow occurs in one of the pipes. For example, in Fig. 6.7 it is clear that water flows out of reservoir R_1 and into R_3, but flow may be either into or out of R_2. A convenient method of starting is to assume that there is no discharge in pipe 2. This establishes the pressure head at point *B* as being that of the water surface in R_2. Then, by writing the Bernoulli equation from *A* to *B* and from *B* to *C*, the discharges in pipes 1 and 2 can be determined. If then $Q_1 > Q_3$, it follows that the assumed value of p_B/w must be increased, and therefore that flow must be into R_2 and

$$Q_1 = Q_2 + Q_3 \tag{6.47}$$

If, on the other hand, $Q_1 < Q_3$, it follows that the assumed value of p_B/w must be reduced, thus causing the water to flow out of R_2 and

$$Q_1 + Q_2 = Q_3 \tag{6.48}$$

Once the proper selection of these continuity equations has been made, the problem can be solved by assuming various pressures at *B* until the appropriate equation is satisfied.

Distribution Networks

Most waterline distribution networks are a complex combination of looping and branching pipelines. A direct analytical solution of pressures and flow distribution in such a network is not possible, and a trial and error procedure must be used. The most widely used procedure is that developed by Cross.[39] Other methods have also been developed to handle distribution networks, such as electronic network analyzers[40,41] and various techniques for solution on the digital computer.[42–45] In this section the Hardy Cross method[39] and the nodal method[45] are briefly discussed.

Hardy Cross Solution

This method provides a procedure whereby successive corrections are systematically determined for assumed flows or pressure losses in individual pipes. The complicated network is viewed as a combination of simple circuits or loops, with each balanced in turn until compatible flow conditions exist throughout the system.

The following concepts, established in solving parallel-pipe problems (see Compound Pipes), are applied to individual loops:

1. Continuity must be satisfied at all junctions.
2. Energy loss must be the same for all paths followed by the water, or the algebraic sum of the energy losses around any loop must be zero.

The energy losses are computed by means of one of the standard energy-loss equations discussed previously, and minor losses and kinetic energy are usually neglected.

All the energy-loss equations can be expressed in the following form:

$$h = kQ^n \tag{6.49}$$

where k is a function of pipe geometry and fluid viscosity. If the Darcy-Weisbach equation [Eq. (6.19)] is used, $n = 2$ and $k = dfl/(2g\ da^2)$. If the Manning formula [Eq. (6.26)] is used, $n = 2$ and $k = In^2/2.028r^{4/3}a^2$ in English units or $k = In^2/r^{4/3}a^2$ in metric units. If the Hazen-Williams formula [Eq. (6.28)] is used, $n = 1.85$ and $k = l/(1.318C_H r^{0.63}a)^{1.85}$ in English units or $k = l/(0.85C_H r^{0.63}a)^{1.85}$ in metric units, where C_H is the Hazen-Williams roughness factor.

If the initial set of assumed values of discharge is designated as Q_1 and the first correction factor as ΔQ_1, then the second set of discharges is

$$Q_2 = Q_1 + \Delta Q_1 \tag{6.50}$$

The sum of the energy losses around a loop based on the first assumptions is

$$h_1 = \Sigma k Q_1^n \tag{6.51}$$

and after the first correction is applied, the sum of the energy losses is

$$h_2 = \Sigma k(Q_1 + \Delta Q_1)^n \tag{6.52}$$

It is convenient to expand the right-hand term into a series and retain only the first two terms of the series as follows:

$$h_2 = \Sigma k(Q_1^n + nQ_1^{n-1}\,\Delta Q_1) \tag{6.53}$$

Because the algebraic sum of the losses around a loop must be zero, the right-hand terms of Eq. (6.53) can be set equal to zero and then solved for ΔQ,

$$\Sigma k\left(Q_1^n + nQ_1^{n-1}\,\Delta Q_1\right) = 0 \tag{6.54}$$

$$\Delta Q_1 = \frac{\Sigma k Q_1^n}{\Sigma\left|knQ_1^{n-1}\right|} \tag{6.55}$$

By use of the absolute value of the denominator in Eq. (6.55), the sign of ΔQ automatically becomes opposite to the sign of ΣkQ^n. In solving a problem, the user must adopt a sign convention for the direction of flow in the pipes around a loop. Each loop in the system is handled in the same way so that continuity is satisfied at each junction.

A step-by-step procedure is outlined as follows and illustrated in Example 6.4.

1. Establish the geometric configuration of the system and the coefficients for Eq. (6.49) for each pipeline in the system.
2. Assume reasonable flow rates in all pipes in such a manner as to satisfy continuity at each junction.
3. For each elementary loop in the system:
   a. Evaluate the loss in each pipe with Eq. (6.49).
   b. Find the algebraic sum of the head loss around the loop, after establishing a sign convention.
   c. Compute the sum of the quantities nkQ_1^{n-1} from each pipe in the loop without regard to the direction.
   d. Calculate the flow adjustment for the loop from Eq. (6.55).
4. Apply step 3 to each circuit in the system, make the flow adjustments to each pipe, and repeat the process until the desired accuracy is obtained.

Example 6.4 Determine the distribution of flow in the network of Fig. 6.8 with the inflow and outflow shown. Also, determine the elevation of the hydraulic grade line at points 1 and 5 if the reservoir elevation is 10 m. There is no flow into or out of the reservoir.

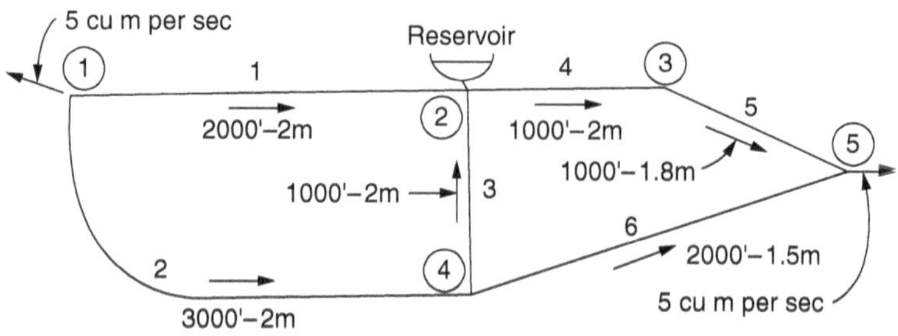

FIGURE 6.8 Distribution network.

This network contains six pipes and two loops. The Hazen-Williams formula is used, and the coefficient is assumed to be 100 in each pipe. The lengths, diameters, and values of k for each pipe are listed in the following table:

Pipe	Length, m	Diameter, m	k	Assumed Q	ΔQ	Corrected Q
1	2000	2.0	0.146	−2.5	−0.3	−2.8
2	3000	2.0	0.218	+2.5	−0.3	+2.2
3	1000	2.0	0.073	+1.0	−0.3	+0.7
				−1.0	+0.3	−0.7
4	1000	2.0	0.073	−3.5	+0.3	−3.2
5	1000	1.8	0.122	−3.5	+0.3	−3.2
6	2000	1.5	0.590	+1.5	+0.3	+1.8

The exponent in Eq. (6.49) is $n = 1.85$. Flows in the counterclockwise direction were taken as positive. The assumed flows with the corresponding signs for the flow directions shown in Fig. 6.8 are also listed in the table. Note that the discharge in pipe 3 has a plus sign for the left-hand loop and a minus sign for the right-hand loop. Flow directions are computed for the left and right loops in accordance with steps 3*a* to 3*d* in the procedure outlined. The computed correction factors and the corrected discharges are also listed. The correction factors for both loops are applied to pipe 3, and both decrease the discharge.

The process is illustrated in the following table. It should be repeated until the pipe flows do not change significantly.

Pipe	ΣkQ^n	ΣnkQ^{n-1}
	Left loop	
1	$0.146 \times 2.5^{1.85} = -0.80$	$1.85 \times 0.146 \times 2.5^{0.85} = 0.59$
2	$0.218 \times 2.5^{1.85} = +1.19$	$1.85 \times 0.218 \times 2.5^{0.85} = 0.88$
3	$0.073 \times 1.0^{1.85} = +0.073$	$1.85 \times 0.073 \times 1.0^{0.85} = \frac{0.14}{1.61}$
	$\Delta Q = \frac{-(-0.80+1.19+0.073)}{1.61} = -0.3$	
	Right loop	
3	$0.073 \times 1.0^{1.85} = -0.073$	$1.85 \times 0.073 \times 1.0^{0.85} = 0.14$
4	$0.073 \times 3.5^{1.85} = -0.74$	$1.85 \times 0.073 \times 3.5^{0.85} = 0.39$
5	$0.122 \times 3.5^{1.85} = -1.24$	$1.85 \times 0.122 \times 3.5^{0.85} = 0.65$
6	$0.59 \times 1.5^{1.85} = +1.25$	$1.85 \times 0.59 \times 1.5^{0.85} = \frac{1.54}{2.72}$
	$\Delta Q = \frac{-(-0.073-0.74-1.24+1.25)}{2.72} = +0.3$	

In this example, the flows are balanced after the first adjustment since the originally assumed flows were quite close.

The pressure level at junction 3 is $10.0 - 0.073 \times 3.2^{1.85} = 9.37$ m, and at the outflow junction it is $9.37 - 0.122 \times 3.2^{1.85} = 8.32$ m. At node 1 it is $10.0 + 0.146 \times 2.8^{1.85} = 10.98$ m. ▲

Nodal Method

In the previous example if a pump with a known head discharge curve were located at junction 1 (Fig. 6.8) whereas the reservoir still remains at junction 2, the method used here acquires another level of complexity. If the system becomes even more complex with additional connecting reservoirs, interior pumps, or valves, the previous method becomes too complicated to be useful. A junction-oriented method is suggested as an alternative approach.[45] The computations are quite lengthy for hand calculation, but the method is easily programmed for solution by digital computer.

The basic concept of the nodal method involves the evaluation of corrections to the originally assumed elevation of the hydraulic grade line at each junction in the system. Corrections are computed and applied repeatedly until the steady-state relations are satisfied throughout the system. The procedure begins with assumed values of the head at each junction in the system. Flow directions in pipes are also indicated, but flow estimates are not needed. External boundaries must be described by a fixed flow rate, valve relationship, pump characteristic curve, fixed reservoir elevation, or other relationship. A linearized form of the pipe flow equation is then applied to each pipe at a junction, and a head correction *DH* is computed so that continuity is satisfied at the junction. Each node is treated independently, the corrections are applied, and the process is repeated. Convergence to the final result is much slower than with the Hardy Cross method. The primary motivation for using this method is to provide a capability to handle many different types of controls and boundary conditions in a system. Although the procedure is much better suited to computer solution, details are presented here along with an example.

Equation (6.49) is rewritten in the form

$$Q = Rh^{RN} \tag{6.56}$$

where $RN = 1.0/n$ and $R = 1.0/k^{RN}$. In terms of the elevations of the hydraulic grade line, this equation can be written for each pipe at a node as follows:

$$Q = R(H_j - H_i)^{RN} \tag{6.57}$$

where i indicates the node for which the equation is written, j the node at the opposite end of the pipe, and $H_j > H_i$ for flow into the node. If the initially assumed values of H were correct, the sum of the Q's would be zero at each node. Since this is not the case, a correction reaction factor *DH* is introduced, which should satisfy continuity at each node. This is done by adding *DH* to H_i. Then Eq. (6.57) becomes

$$Q = R[H_j - (H_i + DH)]^{RN}$$

Expanding the right-hand term and retaining only the first two terms yields

$$Q = R[(H_j - H_i)^{RN} - RN{\cdot}DH(H_j - H_i)^{RN-1}] \tag{6.58}$$

Equation (6.58) can be written

$$Q = A - C{\cdot}DH \tag{6.59}$$

where A and C have the following values:

$$A = R(H_j - H_i)^{RN-1} \tag{6.60}$$

$$C = R \cdot RN(H_j - H_i)^{RN-1} \tag{6.61}$$

If flow is out of the node, $H_i > H_j$ and the quantities A and C are redefined as follows:

$$A = -R(H_i - H_j)^{RN} \tag{6.62}$$

$$C = R \cdot RN(H_i - H_j)^{RN-1}\, H_j)^{RN-1} \tag{6.63}$$

At an interior junction of two or more pipes, continuity requires that the sum of all inflows to the junction must be zero. This can be expressed as

$$\Sigma A - DH\Sigma C = 0 \tag{6.64}$$

The adjustment to the hydraulic grade line is given by

$$DH = \frac{\Sigma A}{\Sigma C} \tag{6.65}$$

If there is a known inflow QV_i to the junction, Eq. (6.64) becomes

$$\Sigma A - DH\Sigma C + QV_i = 0 \tag{6.66}$$

and

$$DH = \frac{\Sigma A + QV_i}{\Sigma C} \tag{6.67}$$

A junction at a reservoir has a fixed elevation and therefore does not require a calculation. A pump providing flow to the system is described by its head discharge curve. In the vicinity of the operating point a straight line

$$Q = C_1 + C_2\,\Delta H \tag{6.68}$$

or a parabola

$$Q = C_1 + C_2\Delta H + C_1\,\Delta H^2 \tag{6.69}$$

may be used to describe the characteristic curve, where ΔH is the head rise across the pump. If the linear Eq. (6.68) is used to describe the flow from the pump to the junction, and the suction reservoir is assumed to be at zero elevation, continuity at the junction can be expressed as

$$\Sigma A - DH\Sigma C + C_1 + C_2(H_i + DH) = 0 \tag{6.70}$$

The head adjustment at the node is given by

$$DH = \frac{\Sigma A + C_1 + C_2 H}{\Sigma C - C_2} \tag{6.71}$$

The following example illustrates one correction to each node in a simple system.

Example 6.5 Determine the distribution of flow in the network of Fig. 6.8 with the addition of a pump to provide the inflow at junction 1. The characteristic curve of the pump is shown in Fig. 6.9, and the suction reservoir is at elevation zero. The same outflow of 5 m^3/s is required at node 5, and the reservoir elevation at node 2 is 10.0 m. However, because of the specified relation between head and discharge at node 1, the input at node 1 may differ from 5 m^3/s, and inflow or outflow may be required at the reservoir. It is assumed that this could occur without changing the elevation of the water surface.

The lengths and diameters of the pipes, the values of k for use in Eq. (6.49), and values of R for use in Eq. (6.56) are given in the following table. Values of k are the same as in Example 6.4, and $RN = 1/n = 0.54$. The node numbers with the original estimated heads are also tabulated along with the corrected heads after one iteration.

Pipe	Length, m	Diameter, m	*k*	*R*	Node	Estimated *H*, m	Corrected *H*, m
1	2000	2.0	0.146	2.83	1	11.0	11.47
2	3000	2.0	0.218	2.28	2	10.0	10.0
3	1000	2.0	0.073	4.11	3	9.0	9.43
4	1000	2.0	0.073	4.11	4	10.5	9.89
5	1000	1.8	0.122	3.11	5	8.5	8.17
6	2000	1.5	0.590	1.33			

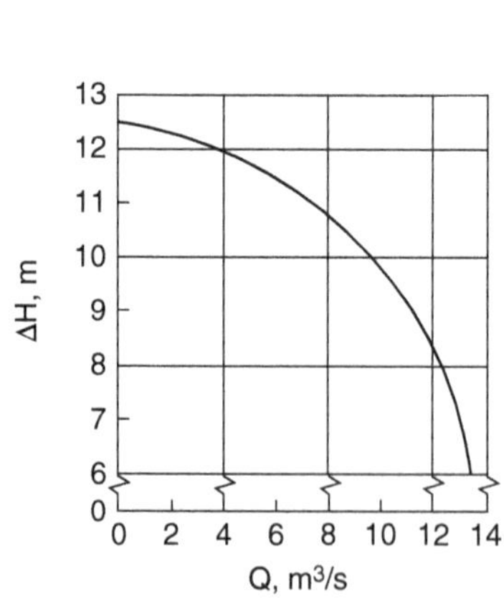

FIGURE 6.9 Pump characteristic curve.

The computations for the first iteration are as follows:

Computations for first iteration		
Pipe no.	***A***	***C***
	Node 3	
4	$4.11 \times 1.0^{0.54} = 4.11$	$\frac{4.11 \times 0.54 \times 1.0^{0.54}}{1.0} = 2.22$
5	$-3.11 \times 0.5^{0.54} = \underline{-2.14}$	$\frac{3.11 \times 0.54 \times 0.5^{0.54}}{0.5} = \underline{2.31}$
	$\Sigma A = 197$	$\Sigma C = 4.53$
	$DH = \frac{\Sigma A}{\Sigma C} = \frac{1.97}{4.53} = 0.43$	
	Node 4	
2	$2.28 \times 0.5^{0.54} = 4.57$	$\frac{2.28 \times 0.54 \times 0.5^{0.54}}{0.5} = 1.69$
3	$-4.11 \times 0.5^{0.54} = -2.83$	$\frac{4.11 \times 0.54 \times 0.5^{0.54}}{0.5} = 3.05$
6	$-1.33 \times 2.0^{0.54} = \underline{-1.93}$	$\frac{1.33 \times 0.54 \times 2.0^{0.54}}{2.0} = \underline{0.52}$
	$A = -3.19$	$C = 5.26$
	$DH = \frac{\Sigma A}{C} = \frac{-3.19}{5.26} = -0.61$	
	Node 5	
5	2.14	2.31
	<u>1.93</u>	<u>0.52</u>
	$\Sigma A = 4.07$	$\Sigma C = 2.83$
	$DH = \frac{\Sigma A + QV_5}{\Sigma C} = \frac{4.07 - 5.0}{2.83}$	$= -0.33$
	Node 1	
1	$2.83 \times 1.0^{0.54} = -2.83$	$\frac{2.83 \times 0.54 \times 1.0^{0.54}}{1.0} = 1.53$
	$= \underline{-1.57}$	$= \underline{1.69}$
	$\Sigma A = -4.40$	$\Sigma C = -3.22$
Pump equation [Eq. (6.68)]	$Q = 7.6$	$H = 11.0$
	$Q = 5.8$	$H = 11.5$
	$C_1 = 47.2$	$C_2 = -3.6$
	$DH = \frac{-4.40 + 47.2 - 3.6 \times 11.0}{3.22 + 3.6} = \frac{3.2}{6.82} = 0.47$	

At node 3, Eq. (6.65) is solved with values of A and C determined from Eqs. (6.60) and (6.61), respectively, for pipe 4 and from Eqs. (6.62) and (6.63), respectively, for pipe 5. The same equations are used at node 4. At node 5 Eq. (6.67) is used in place of Eq. (6.63), and at node 1 Eq. (6.71) is applied. The level of the hydraulic grade line at node 2 is fixed at the elevation of the water surface (10.0 m), and therefore no correction is needed.

The corrected heads are shown in the table. The same procedure should be repeated until the pressure head at each node does not change significantly. The flow in each pipe can then be computed by use of Eq. (6.56). By use of the corrected heads given in the table, the flows in pipes 1 to 6 are 3.48, 2.92, −1.25, 3.03, 3.52, and 1.78 m^3/s. A check of continuity shows that a balance has not been achieved. For example, the inflow at the pump is $3.48 + 2.92 = 6.40$ m^3/s. The outflow at node 5 is $3.52 + 1.78 = 5.30$ m^3/s. The inflow from the reservoir is $3.03 + 1.25 - 3.48 = 0.8$ m^3/s, giving a total inflow of $6.4 + 0.8 = 7.2$ m^3/s, which exceeds the outflow by 1.9 m^3/s. Several more iterations are required before continuity is satisfied at all nodes as well as in regard to the total inflow and outflow. ▲

References

1. M. H. Darcy, *Recherches expérimentales relatives au mouvement de l'eau dans les tuyaux*, Mallet-Bachelier, Paris, 1857.
2. J. Weisbach, *Die Experimental-Hydraulik*, Engelhardt, Freiburg, Germany, 1855.
3. J. Nikuradse, "Strömungsgesetze in rauhen Rohren," *VDI-Forschungsh.*, no. 361, 1933.
4. C. F. Colebrook, "Turbulent Flow in Pipes, with Particular Reference to the Transition Region between Smooth and Rough Pipe Laws," *J. Inst. Civil Eng. (London)*, vol. 11, no. 4, pp. 133–156, Feb. 1939.
5. H. Rouse, "Evolution of Boundary Roughness," *Proc. 2nd Hydraulics Conf.*, Univ. Iowa Bull. 27, 1943.
6. L. F. Moody, "Friction Factors for Pipe Flow," *Trans. ASME*, vol. 66, no. 4, pp. 671–684, Nov. 1944.
7. D. V. Wood, "An Explicit Friction Factor Relationship," *Civil Eng. Div., ASCE*, 102 (HY5), pp. 60–61, Dec. 1966.
8. J. N. Bradley and L. R. Thompson, "Friction Factors for Large Conduits Flowing Full," U.S. Bureau of Reclamation, Eng. Monograph 7, 1965.
9. M. F. Burke, "High Velocity Tests in a Penstock," *Trans. ASCE*, vol. 120, pp. 863–896, 1955.
10. M. J. Webster and L. R. Metcalf, "Friction Factors in Corrugated Metal Pipe," *J. Hydraul. Div., ASCE*, 85(HY 9): pp. 35–74, Sep. 1959.
11. R. Elder, "Friction Factors in the Appalachia Tunnel," *Trans. ASCE*, vol. 123, pp. 1249–1274, 1958.
12. C. W. Harris, "An Engineering Concept of Flow in Pipes," *Trans. ASCE*, vol. 115, p. 909, 1950.

13. V. L. Streeter and H. Chu, "Fluid Flow and Heat Transfer in Artificially Roughened Pipes," Armour Research Foundation Project 4918, Chicago, 1949.
14. H. M. Morris, "Flow in Rough Conduits," *Trans. ASCE*, vol. 120, pp. 373–410, 1955.
15. H. M. Morris, "Design Methods for Flow in Rough Conduits," *Trans. ASCE*, pt. 1, vol. 126, pp. 454–490, 1961.
16. L. G. Straub and H. M. Morris, "Hydraulic Tests on Corrugated Metal Culvert Pipes," St. Anthony Falls Hydraulic Lab., Tech. Paper 5, ser. B, 1950.
17. C. R. Neill, "Hydraulic Roughness of Corrugated Pipe," *J. Hydraul. Div., ASCE*, May 1962.
18. J. M. Normann and H. G. Bossy, "Hydraulic Flow Resistance Factors for Corrugated Metal Conduits," *Fed. Highway Admin.*, Sep. 1970.
19. A. T. Ippen, discussion of Moody.[6]
20. J. R. Freeman, *Flow of Water in Pipes and Pipe Fittings*, ASME, New York, 1944.
21. R. Manning, "Flow of Water in Open Channels and Pipes," *Trans. Inst. Civil Eng. (Ireland)*, vol. 20, pp. 161–207, 1890.
22. G. S. Williams and A. Hazen, *Hydraulic Tables*, 3rd ed., John Wiley, New York, 1933.
23. C. H. Tutton, "The Flow of Water in Pipes," *J. Assoc. Eng. Soc.*, vol. 23, p. 151, 1899.
24. W. C. Unwin, *A Treatise on Hydraulics*, A. & C. Black, London, p. 217, 1907.
25. F. C. Scobey, "The Flow of Water in Concrete Pipe," U.S. Dept. Agriculture Bull. 852, 1920.
26. F. C. Scobey, "The Flow of Water in Wood Stave Pipe," U.S. Dept. Agriculture Bull. 376, 1916, rev. 1926.
27. F. C. Scobey, "The Flow of Water in Riveted Steel and Analogous Pipe," U.S. Dept. Agriculture Bull. 150, 1930.
28. A. A. Barnes, *Hydraulic Flow Reviewed*, Spon and Chamberlain, New York, 1916.
29. F. C. Lea, *Hydraulics*, Edward Arnold, London, 1909.
30. H. Baer, "Experiments on Hydraulic Losses", *Dinglers Polytech. J.*, vol. 88, pp. 177–179, Mar. 23, 1907.
31. A. W. Brightmore, *"Lloss of Pressure in Water Flowing Through Straight and Curved Pipes", Minutes of the Proceedings of the Proc. Inst. Civil Eng.*, vol. 169, p. 323, 1906–1907.
32. W. H. Archer, "Loss of Head due to Enlargements in Pipes," *Trans. ASCE*, vol. 76, pp. 999–1026, 1913.
33. P. à Morley Parker, *The Control of Water*, Routledge & Kegan Paul, London, pp. 796–800, 1925.
34. M. Merriman, *Treatise on Hydraulics*, 10th ed., Wiley, New York, p. 183, 1916.
35. C. I. Corp and R. O. Ruble, "Experiments on Loss of Head in Valves and Pipes of One-Half to Twelve Inch Diameter," *Bull. Univ. Wis., Eng. Ser.*, vol. 9, no. 1, p. 59, 1922.
36. K. H. Beij, "Pressure Losses for Fluid Flow in 90° Pipe Bends," *J. Res., Nat. Bur. Standards*, vol. 21, pp. 1–18, Jul. 1938.
37. A. G. Anderson and L. G. Straub, "Hydraulics of Conduit Bends," St. Anthony Falls Hydraulic Lab., Bull. 1, Dec. 1948.
38. W. E. Fuller, "Loss of Head in Bends," *J. New England Water Works Assoc.*, vol. 27, no. 4, pp. 509–521, Dec. 1913.
39. H. Cross, "Analysis of Flow in Networks of Conduits or Conductors," Univ. Illinois Bull. 286, Nov. 1946.

40. M. S. McIlroy, "Direct Reading Electric Analyzer for Pipeline Networks," *J. Am. Water Works Assoc.*, vol. 42, p. 347, Apr. 1950.
41. M. B. McPherson and J. V. Rodzidul, "Water Distribution Design and the McIlroy Network Analyzer," *J. Hydraul. Div., ASCE*, vol. 84, pp. 1–15, Apr. 1958.
42. J. H. Dillingham, "Computer Analysis of Water Distribution Systems, pt. II," *Water Sewage Works*, pp. 43–45, Feb. 1967.
43. M. McCormick and C. J. Bellamy, "A Computer Program for the Analysis of Networks of Pipes and Pumps," *J. Inst. Eng. (Sydney, Australia)*, Mar. 1968.
44. U. Shamir and C. D. Howard, "Water Distribution Systems Analysis," *J. Hydraul. Div., ASCE*, vol. 94, pp. 219–234, Jan. 1968.
45. V. L. Streeter, "Water Hammer Analysis of Distribution Systems," *J. Hydraul. Div., ASCE*, vol. 93, pp. 185–201, Sep. 1967.

SECTION 7

Steady Uniform Flow in Open Channels

Reference to Sec. 3 will show that the conditions specified in the title of this section require that the discharge in the channel be constant with respect to time and that the cross-sectional area remain the same from place to place in the channel. For subcritical flow (see Sec. 8), this condition can exist throughout the full length of the channel only if the outlet end of the channel is controlled so that there will be no drawdown or backwater. For supercritical flow, uniform flow can occur throughout the channel only if the water enters the channel at the uniform-flow depth from a pressure chamber and if no obstruction exists at the outlet end of the channel. Strictly speaking, this type of flow can occur only in parallel-walled channels, thus precluding all natural streams. Practically speaking, however, there are often reaches of natural streams in which flow is nearly uniform, and in many cases flow can be considered as steady in rivers for short-time intervals.

The principles governing the relationship between depth, slope, and discharge for uniform flow depend entirely on the rate of energy dissipation due to friction. Consequently, this section deals entirely with this aspect of flow in open channels. However, because the rate of energy dissipation for gradually varied flow depends on the same variables as in the case of uniform flow, the material presented here will also be used in Sec. 8. The problems involved in steady nonuniform flow are discussed in Secs. 8 and 9, and unsteady flow in open channels is treated in Secs. 10 and 11.

Elements of a Cross Section

The more important elements of cross sections, together with the symbols that will be used to designate them, as illustrated in Fig. 7.1, are as follows:

- *Area a* always means the cross-sectional area of the stream.
- *Wetted perimeter p* is the length of the wetted surface of the channel, the line *abc*.
- *Hydraulic radius $r = a/p$* is the area divided by the wetted perimeter.
- *Depth D*, if not specified otherwise, refers to the maximum depth of water in the cross section.
- *Top width T* is the width of the cross section at the water surface.
- *Mean depth $D_m = a/T$* is the area of cross section divided by the top width. This term will be applied generally to all sections, but it is not literally descriptive of channels with overhanging sides, such as circular conduits flowing more than half full.
- *Depth to center of gravity $\bar{y}$* is the depth to the center of gravity of the cross section of the stream.

These terms and symbols are all used in the discussions and formulas contained in this and the following sections. The hydraulic radius enters into formulas involving velocity or discharge. The mean depth occurs in the criterion for indicating when flow in a channel is at critical depth. The depth to the center of gravity of a cross section is employed in determining hydrostatic pressures in problems involving sudden changes in depth of flow.

Sectional Forms

Most of the sectional forms used for open channels are shown in Fig. 7.2. It is convenient to have tables that facilitate the determination of numerical values of the elements of a cross section. The equations

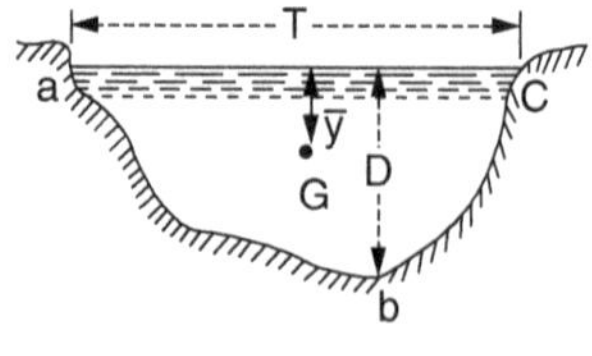

FIGURE 7.1 Cross section of open channel.

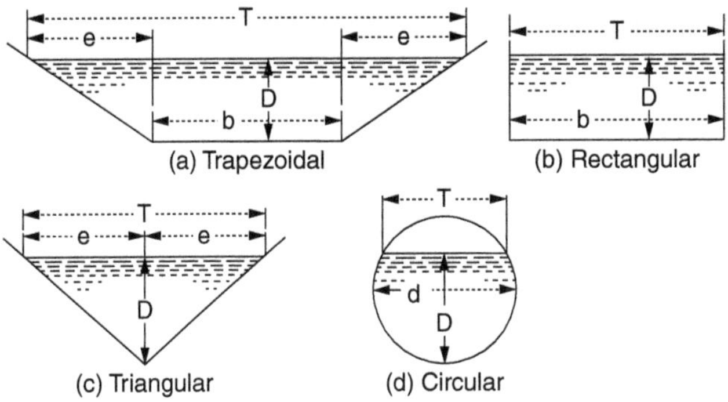

FIGURE 7.2 Simple forms of channel sections.

used in deriving these tables are shown for trapezoidal sections only. Similar equations were derived and used for developing the tables for circular sections.

The *trapezoidal section* (Fig. 7.2*a*) is always used for earth canals. Ordinary earth sections have relatively flat side slopes, usually not steeper than 1:1 in cut and 1½:1 in fill. In rock, hardpan, or other hardened material and for lined canals, trapezoidal sections with very steep side slopes are often employed. It is not uncommon to have different side slopes on the two sides of a canal. Often the uphill side of an earth canal will have a steeper slope than the downhill side.

As indicated in Fig. 7.2*a*, D is the maximum depth of water and b is the bottom width of the canal. The area of the trapezoidal section is

$$a = eD + bD \tag{7.1}$$

Letting

$$z = \frac{e}{D} \tag{7.2}$$

and

$$x = \frac{D}{b} \tag{7.3}$$

then

$$a = \left(z + \frac{1}{x} \right) D^2 \tag{7.4}$$

The wetted perimeter is

$$p = b + 2(e^2 + D^2)^{1/2} \tag{7.5}$$

or, substituting for b and e as before,

$$p=\left[\frac{1}{x}+2(z^2+1)^{1/2}\right]D \tag{7.6}$$

Then

$$r=\frac{a}{p}=\frac{1/x + z}{1/x+2(z^2+1)^{1/2}}D \tag{7.7}$$

The top width is

$$T=b+2e=\left(\frac{1}{x}+2z\right)D \tag{7.8}$$

Then the expression for mean depth becomes

$$D_m=\frac{a}{T}=\frac{1/x+z}{1/x+2z}D \tag{7.9}$$

The distance down from the water surface to the center of gravity is obtained by taking moments as follows:

$$a\overline{y}=bD\frac{D}{2}+eD\frac{D}{3}$$

Substituting for a, b, and e, using Eqs. (7.4), (7.3), and (7.2), respectively,

$$\overline{y}=\frac{1/2x+z/3}{z+1/x} \tag{7.10}$$

If the slopes of the two sides of the channel are different, an average value of z used in Eqs. (7.4), (7.9), and (7.10) will give correct values of a, D_m, and $\overline{y}$, respectively, but Eq. (7.7), used with an average z, will not give exact values of r. For example, if $D=5$, $b=10$, and $z=1$ and 2, using z (average) $=1.5$, then $r=3.12$, whereas the correct result is 3.10. For smaller differences in z, the error will be relatively less.

The *rectangular* and *triangular* sections are special cases of the trapezoidal section. The former has $z=0$, and the latter has $b=0$. The rectangular section (Fig. 7.2*b*) is used for wooden flumes and for various types of lined conduits. Triangular cross sections (Fig. 7.2*c*) are seldom encountered, but channels of this form have interesting hydraulic properties.

For the rectangular section, $a=bD$, $r=bD/(b+2D)$, $\overline{y}=D/2$, and $D_m=D$. For the triangular section, $a=zD^2$, $r=zD/2(z^2+1)^{1/2}$, $\overline{y}=D/3$, and $D_m=D/2$.

Circular conduits, designed to flow partially full under normal operating conditions, are commonly used for sewers and drains and for other purposes where underground channels are required. Semicircular sections are sometimes employed for flumes and lined canals.

As indicated in Fig. 7.2*d*, D is the maximum depth of a partially filled circular conduit and d is the diameter. The area is

$$a=\left(\frac{360-\theta}{360}\frac{\pi}{4}+\frac{\text{Sin}\,\theta}{8}\right)\frac{D^2}{x^2} \tag{7.11}$$

in which θ is the angle between the radii subtending the water surface and $x = D/d$.

The wetted perimeter of a circular channel flowing part full is

$$p=\left(\frac{360-\theta}{360}\right)\frac{\pi D}{x} \tag{7.12}$$

in which θ and x are as defined before. Table 7.1 below will assist in the calculation of the area and wetted perimeter as well as the factor $ar^{2/3}/d^{8/3}$ which is useful in solving the Manning formula, which will be discussed later.

Most Efficient Channel Section

Considered purely from the standpoint of the hydraulics of a channel, it can be seen from the Manning equation [Eq. (7.33)] that for a given area the most efficient channel will be the one with the minimum wetted perimeter. Therefore, if in the expression for the wetted perimeter given by Eq. (7.6) the depth is replaced in terms of a, z, and x from Eq. (7.4), p can be differentiated first with respect to x and then with respect to z. The first operation yields the following expression:

$$\frac{1}{x}=\frac{b}{D}=2\left[(z^2+1)^{1/2}-z\right] \tag{7.13}$$

which yields the best ratio b/D for any z. The second differentiation gives $z=1/\sqrt{3}$, which, when substituted into Eq. (7.13), gives $b/D=2/\sqrt{3}$, thus showing that the most efficient trapezoidal section is a half hexagon. The most efficient rectangular section is that for which $b/D = 2$, as shown by letting $z = 0$ in Eq. (7.13). It may be seen from Eq. (7.13) that for flat side slopes, as would be used in earth canals, the most efficient sections have small bottom widths. All cross sections which satisfy Eq. (7.13) have forms such that semicircles can be inscribed in them, and the most efficient open-channel cross-sectional shape of all is a semicircle. Steep-sided parabolas also have a high efficiency.

Table 7.1 Factors to Assist Solving Manning's Equation in Circular Conduits

D/d	a/A	p/P	$ar^{2/3}/d^{8/3}$	D/d	a/A	p/P	$ar^{2/3}/d^{8/3}$
0.00	0.000	0.000	0.000	0.52	0.53	0.51	0.17
0.02	0.005	0.090	0.000	0.54	0.55	0.53	0.18
0.04	0.013	0.128	0.001	0.56	0.58	0.54	0.19
0.06	0.024	0.158	0.002	0.58	0.60	0.55	0.20
0.08	0.037	0.183	0.004	0.60	0.63	0.56	0.21
0.10	0.052	0.205	0.007	0.62	0.65	0.58	0.22
0.12	0.068	0.225	0.010	0.64	0.68	0.59	0.23
0.14	0.085	0.244	0.013	0.66	0.70	0.60	0.24
0.16	0.103	0.262	0.017	0.68	0.72	0.62	0.25
0.18	0.122	0.279	0.022	0.70	0.75	0.63	0.26
0.20	0.142	0.295	0.027	0.72	0.77	0.65	0.27
0.22	0.163	0.311	0.033	0.74	0.79	0.66	0.28
0.24	0.185	0.326	0.039	0.76	0.82	0.67	0.29
0.26	0.207	0.341	0.046	0.78	0.84	0.69	0.30
0.28	0.229	0.355	0.053	0.80	0.86	0.70	0.30
0.30	0.252	0.369	0.061	0.82	0.88	0.72	0.31
0.32	0.276	0.383	0.069	0.84	0.90	0.74	0.32
0.34	0.300	0.396	0.078	0.86	0.91	0.76	0.32
0.36	0.324	0.410	0.086	0.88	0.93	0.77	0.33
0.38	0.349	0.423	0.096	0.90	0.95	0.80	0.33
0.40	0.374	0.436	0.105	0.92	0.96	0.82	0.33
0.42	0.399	0.449	0.115	0.94	0.98	0.84	0.34
0.44	0.424	0.462	0.125	0.96	0.99	0.87	0.33
0.46	0.449	0.475	0.135	0.98	1.00	0.91	0.33
0.48	0.475	0.487	0.145	1.00	1.00	1.00	0.31
0.50	0.500	0.500	0.156				

Not only does Eq. (7.13) give the channel with the smallest wetted perimeter for a given area but it therefore also gives the channel that provides a given discharge with the minimum area. Therefore, the section determined from Eq. (7.13) will be the most economical section to build insofar as both excavation and canal lining are concerned. However, earth canals may involve other considerations, such as construction difficulties or maintenance problems, which make it desirable to depart from the ideal section.

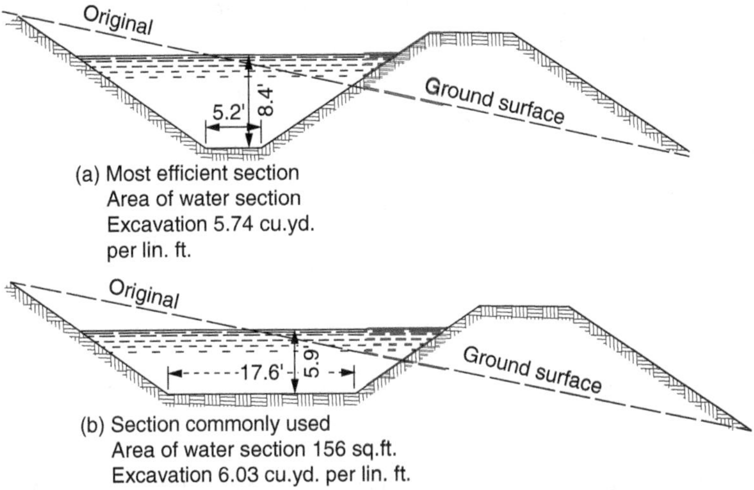

FIGURE 7.3 Trapezoidal canal sections. $Q = 200$ ft³/s; $z = 1.5{:}1$, $n = 0.0225$; $s = 0.000134$; freeboard = 2 ft; slope of ground = 20%; top width of embankment = 2.4 m.

Two earth-canal sections with balanced cut and fill are illustrated in Fig. 7.3. Each has the same capacity and all conditions except the ratio of depth to bottom width and the area of the section are the same for each. Figure 7.3*a* is the theoretically most efficient section, whereas Fig. 7.3*b* has dimensions approximating those commonly employed for canals of this capacity. For the indicated ground slope of 20 percent, there will be about 5 percent more excavation in section *b* than in *a*, but the former section will be easier to construct and maintain than the latter. For canals in flatter country the difference in excavation for the two sections will be relatively less than that indicated in the figure. As the ground slope increases, the excavation saved by using the more efficient section increases.

Energy Losses in Open Channels

As in pipes, the energy losses are of two types, those due to friction and those due to sudden changes in the direction and magnitude of the velocity, which are called minor losses. However, because this section of the handbook deals with uniform flow, only friction losses will be discussed here, and minor losses will be covered in Sec. 8.

The mechanics of uniform flow can be illustrated by first considering laminar flow in wide channels. Thereafter turbulent flow will be discussed.

As was shown by Eq. (3.11), the Bernoulli equation for an open channel takes on the following form:

$$z_1 + D_1 + \alpha_1 \frac{V_1^2}{2g} = z_2 + D_2 + \alpha_2 \frac{V_2^2}{2g} + h$$

where h is the energy loss between points 1 and 2, as illustrated in Fig. 7.4. For *uniform* flow the depth and velocity terms are the same at all points, so the Bernoulli equation reduces to

$$z_1 - z_2 = h \tag{7.14}$$

thus showing that the energy loss h (Fig. 7.4) is equal to the drop in the bottom, or that the energy gradient, water surface, and bottom are parallel.

If s is the energy loss per unit of length, then, for any open channel,

$$s = \frac{h}{l} = \sin \theta_e \tag{7.15}$$

where θ_e is the angle between the energy gradient and a horizontal plane and

$$h = sl \tag{7.16}$$

The relation between the bottom slope s_0, the angle of the bottom with the horizontal θ, the length l, and $z_1 - z_2$ is expressed by the equations

$$s_0 = \frac{z_1 - z_2}{l \cos\theta} = \tan\theta \tag{7.17}$$

and

$$z_1 - z_2 = s_0 l \cos \theta \tag{7.18}$$

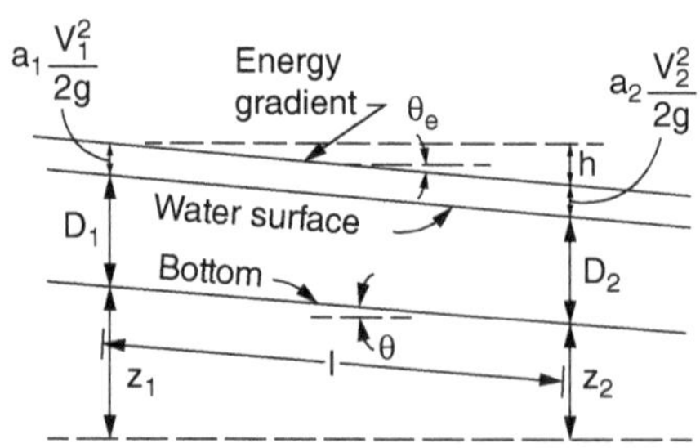

Figure 7.4 Uniform flow.

Then, from Eqs. (7.14) and (7.16) to (7.18), *for uniform flow only*,

$$s = s_0 \cos\theta = \sin\theta \tag{7.18a}$$

When θ is very small, as is often the case for open channels, $\cos\theta \to 1$ or $\sin\theta \approx \tan\theta$, and

$$s \approx s_0$$

Again it should be emphasized that this relation applies only to uniform flow. In nonuniform flow the bottom slope has no relation to s. In conclusion it may be stated that, under all conditions, s is the energy loss per unit of length, and for open channels with very small slopes, it may also be defined as the slope of the energy gradient. For uniform flow in a prismatic channel, s is also the drop in the channel per unit of length, or $\sin\theta$, and for very small slopes, it becomes nearly equal to the slope of the channel, s_0 or $\tan\theta$.

Laminar Flow with Free Surface

The law of laminar flow with a free surface for the case of wide rectangular channels may be developed in the same manner as for pipes. Consider the free body of fluid shown in Fig. 7.5, having a width of unity, a length l, and a height $D - y$. The summation of forces in the direction of flow gives

$$\tau l = wl(D - y)\sin\theta \tag{7.19}$$

Replacing τ with its value from Eq. (1.3), rearranging the terms, and simplifying, Eq. (7.19) becomes

$$du = \frac{w\sin\theta}{\theta}(D\,dy - y\,dy)$$

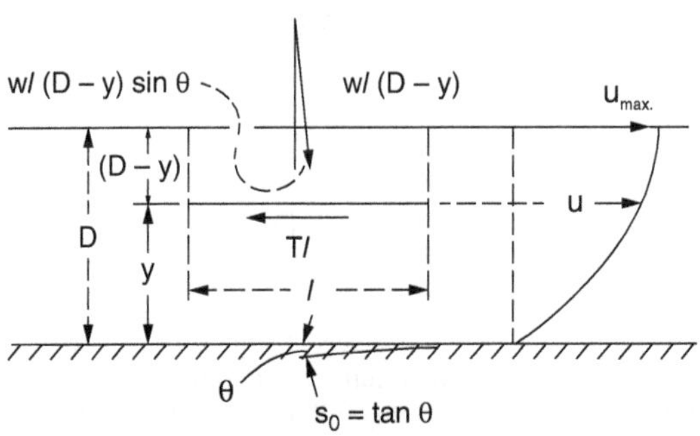

FIGURE 7.5 Laminar flow in open channels.

and integration gives

$$u = \frac{w \sin \theta}{\theta}\left(Dy - \frac{y^2}{2}\right) + C \tag{7.20}$$

The value of C in Eq. (7.20) must be zero to satisfy the condition that $u = 0$ when $y = 0$. The maximum velocity may be obtained by letting $y = D$. Then

$$u_{\text{max}} = \frac{w \sin \theta \, D^2}{2\mu} \tag{7.21}$$

The discharge is obtained from a summation of small elements of discharge, utilizing the value of u from Eq. (7.20),

$$q = \int dq = \int u \, dy = \frac{W \sin \theta}{\mu} \int_o^D \left(Dy \, dy - \frac{y^2 dy}{2} \right)$$

from which

$$q = \frac{w \sin \theta \, D^3}{3\,\mu} \tag{7.22}$$

The average velocity is

$$V = \frac{q}{D} = \frac{w \sin \theta \, D^2}{3\mu} \tag{7.23}$$

and

$$\sin \theta = \frac{3\mu V}{wD^2} \tag{7.24}$$

Finally, the Reynolds number $R = DV\rho/\mu$[a] may be introduced to give

$$\sin \theta = \frac{3V^2}{RgD} \tag{7.25}$$

From Eq. (7.15),

$$s = \frac{h}{l} = \frac{3V^2}{RgD} \tag{7.26}$$

---

[a]The Reynolds number for open channels is sometimes written with $4r$ as the length parameter, where r is the hydraulic radius. This makes the magnitudes of R correspond to those for pipes flowing full in which the length parameter is the diameter d. Because for wide channels $r = D$, all values of R in Figs. 7.6 and 7.7 would have been four times as large if $4r$ had been used.

and rearranging terms,

$$h = \frac{6}{R}\frac{l}{D}\frac{V^2}{2g} \tag{7.27}$$

Letting

$$f = \frac{6}{R} \tag{7.28}$$

then

$$h = f\frac{l}{D}\frac{V^2}{2g} \tag{7.29}$$

In this form the equation resembles closely the Darcy-Weisbach equation for pipes [Eq. (6.19)], and Eq. (7.28) is the counterpart of the similar relationship for pipes given by Eq. (6.20). Because the Manning equation is commonly used for turbulent flow in open channels, it is of interest to relate f to n. This can be done by equating the value of s obtained from Eq. (7.29) with its value from the Manning equation [see Eq. (7.33)],

$$f = \frac{1}{D}\frac{V^2}{2g} = \frac{n^2V^2}{r^{4/3}} \quad \text{(metric units)}$$

$$f = \frac{n^2V^2}{2.208r^{4/3}} \quad \text{(English units)}$$

Because this derivation is for very wide channels, r may be replaced with D to obtain the following relationships:

$$\begin{aligned} f &= 2g\frac{n^2}{D^{1/3}} \quad \text{(metric units)} \\ f &= \frac{2g}{2.21}\frac{n^2}{D^{1/3}} \quad \text{(English units)} \end{aligned} \tag{7.30}$$

and

$$\begin{aligned} n &= \frac{f^{1/2}D^{1/6}}{\sqrt{2g}} \quad \text{(metric units)} \\ n &= \frac{1.49f^{1/2}D^{1/6}}{\sqrt{2g}} \quad \text{(English units)} \end{aligned} \tag{7.31}$$

Plotting $\log f$ against $\log R$ in accordance with Eq. (7.28) yields a straight line having a slope of −1, as shown in Figs. 7.6 and 7.7. Numerous tests of laminar flow in wide channels with smooth surfaces have

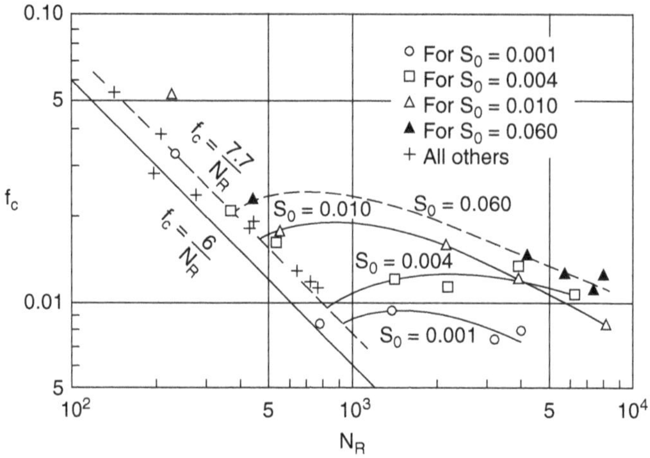

Figure 7.6 *f* versus *R* for rough side of masonite.

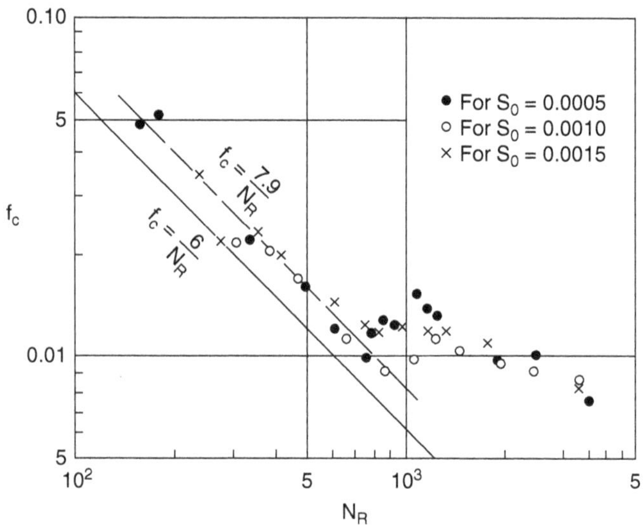

Figure 7.7 *f* versus *R* for cement.

verified Eq. (7.28).[b] However, for rough surfaces the results follow trends parallel to that of Eq. (7.28), but with a larger value of *f*. Such relationships can be represented by the equation

$$f = \frac{C}{R} \tag{7.32}$$

[b]References to research on this topic, together with a more detailed discussion, are presented in Woo and Brater.[1]

In Fig. 7.6, the value of C from tests by Woo and Brater[1] for the rough side of masonite is 7.7, and in Fig. 7.7 tests conducted at the U.S. Waterways Experiment Station[2] on cement surfaces yield $C = 7.9$. Tests at the University of Michigan[1] with a surface of glued sand having an average size of 0.04 in gave a value of $C = 9.8$ for slopes equal to or less than 0.003. For greater slopes, C increased consistently with s_0, reaching a value of 25.0 for a slope of 0.060. Tests at Vicksburg[12] with a movable sand surface showed that a sand diameter of 0.0081 in acted like a smooth surface ($C = 6.0$), but that C began to increase for a sand diameter of 0.019 in. All the Vicksburg tests were made with relatively small slopes, so that no variation of C with slope could be detected.

Data plotted in Figs. 7.6 and 7.7 show clearly how the relationship between f and R changes when turbulence begins. It may be noted from Fig. 7.6 that the slope of the channel affects the point at which turbulence begins, there being a tendency for turbulent flow to begin at lower values of R for steeper slopes. A review of published data[1] indicates that laminar flow ceases at values of R ranging from 400 to 900. These values appear to be consistent with those of pipes flowing full, because the depth term used in the Reynolds number is approximately equivalent to the hydraulic radius, and for a pipe flowing full, $r = d/4$. Consequently if values of R for pipes were based on r instead of d, they would be one-fourth as large, and therefore the transition into turbulent flow would begin in about the same range of values of R as shown here.

The relationships derived and discussed here are for rectangular channels which are sufficiently wide so that the effects of the sidewall will be negligible. Based on an analytical expression for laminar flow in rectangular conduits first presented by Boussinesq,[3] it can be shown[1] that the theoretical value of C in Eq. (7.32) varies with the ratio of width to depth b/D, according to the following tabulation:

b/D	C
∞	6.0
50	6.2
10	6.8
5	8.0

Laminar flow in channels other than rectangular has been investigated by Straub et al.[4]

Turbulent Flows in Open Channels

It is quite probable that future research and analysis will lead to the development of practical methods of determining energy losses in open channels in the transitional and fully turbulent ranges using an equation in the form of Eq. (7.29) and the concept of relating f to relative roughness and Reynolds number, as is done for round pipes flowing full (see Turbulent Flow in Pipes, Sec. 6). However, the problem is exceedingly difficult for open channels because of the variety of shapes used and because relative roughness varies with depth. Furthermore, in earth canals or natural rivers the problem is further complicated when scour begins. An outline of research and writing on this topic is presented in a report of the American Society of Civil Engineers.[5] An example of the variation of f with R is provided by test data from nine large concrete-lined canals by the U.S. Bureau of Reclamation.[6] The values of R were computed from the expression $R = 4rV\rho/\mu$, and the values of f were computed from Eq. (7.29) after replacing D with $4r$. The report states that the 52 plotted test points "were chosen carefully to minimize the effects of curvature, algal growth, pier losses, and all other effects except canal size," A straight-line least-squares fit, through a logarithmic plotting of the points, showed f varying from about 0.0148 for $R = 3 \times 10^6$ to 0.0117 for $R = 3 \times 10^7$. Although there is a scattering of the values of f about the fitted line over a range varying from 10 percent below to 30 percent above the line, the trend along the line is clearly evident. In this case the chief variable in the Reynolds number appears to be the hydraulic radius, and the ranges of values of the Manning number n derived from these tests for various values of the hydraulic radius are tabulated in the following subsection.

The Manning Equation

Fortunately, most practical problems fall in the fully turbulent range of flow at high values of the Reynolds number, where the Manning equation gives satisfactory results. The historical relation of the Manning equation to the Chezy equation was discussed in Sec. 6 under Turbulent Flow in Pipes and under Selection of Energy-Loss Equation and Friction Factor.

The Manning formula is written

$$V = \frac{1.486 r^{2/3} s^{1/2}}{n} \text{ (English units) } \quad V = \frac{r^{2/3} s^{1/2}}{n} \text{ (metric units)} \qquad (7.33)$$

where V is in feet per second (meters per second) and r in feet (meters). The corresponding expressions for discharge are obtained by multiplying by the cross-sectional area a to obtain

$$Q = \frac{1.486 a r^{2/3} s^{1/2}}{n} \text{ (English units) } \quad Q = \frac{a r^{2/3} s^{1/2}}{n} \text{ (metric units)} \qquad (7.34)$$

Q being in cubic feet per second (cubic meters per second).

The use of these equations is illustrated in Example 7.1.

Example 7.1 Flow is uniform in a trapezoidal channel at a depth of 1.5 m. The bottom width is 15 m, side slopes are 1:1, the bottom slope is 0.00001, and n is 0.020. Determine the discharge.

$$a = 15 \times 1.5 + 2 \times \frac{1.5 \times 1.5}{2} = 24.8\ \text{m}^2$$

$$p = 15 + 2 \times \sqrt{2} \times 1.5 = 19.2\ \text{m}$$

$$r = \frac{24.8}{19.2} = 1.29\ \text{m}$$

From Eq. (7.34),

$$Q = \frac{24.8 \times 1.29^{2/3} \times 0.00001^{1/2}}{0.020}$$
$$= 4.64\ \text{m}^3/\text{s}$$

The most common type of open-channel problem is the design situation in which the channel size is required for a given Q and s. This problem must be solved by successive approximations. ▲

Example 7.2 What depth is required if a channel is to convey 200 ft³/s in a trapezoidal channel having a bottom width of 50 ft, side slopes of 1:1, $n = 0.020$, and a bottom slope of 0.00001?

Using Eq. (7.34)

$$Q = \frac{1.486 a r^{2/3} s^{1/2}}{n} = \frac{1.486 \times 0.00001^{1/2}}{0.020} a r^{2/3} = 200$$

$$a r^{2/3} = 851.2$$

Using Eqs. (7.4) and (7.7) to get the relationships for a and r, successive approximations of D can be made until the above equation is satisfied, as shown in the table below.

D	*a*	*r*	$ar^{2/3}$
1	51.0	0.97	49.8
3	159.0	2.72	309.7
5	275.0	4.29	725.7
6	336.0	5.02	984.7
5.5	305.3	4.66	851.1

The required depth is 5.5 ft. ▲

Example 7.3 Determine the dimensions of a semicircular channel, which will carry 3.0 m^3/s with a slope of 0.0001 if $n = 0.013$. The channel is to have 0.3 m of free board. Then,

$$D = \frac{d}{2} - 0.3$$

and

$$\frac{D}{d} = 0.5 - \frac{0.3}{d}$$

Using Eqs. (7.11) and (7.12) to compute the area (a) and the wetted perimeter (p) of the section and Eq. (7.34) to compute the discharge the solution can be found through an iterative process as shown in the table below.

d	D	$ar^{2/3}$	Q
1	0.20	0.03	0.02
3	1.20	1.97	1.51
4	1.70	4.73	3.64
3.5	1.45	3.16	2.43
3.8	1.60	4.05	3.12
3.75	1.58	3.90	3.00

The diameter of the semicircular channel is 3.75 m. ▲

Roughness Coefficients

Values of n to be used in the Manning equation are given in Table 7.2. These values have been selected from published literature. The U.S. Bureau of Reclamation report[6] on large concrete-lined canals, described in Turbulent Flow in Open Channels, provides a graph showing 52 test values of n plotted against the hydraulic radius. The author has drawn envelope curves which include all but one of the 52 test points. The following tabulation shows the ranges of n for three selected values of the hydraulic radius r, as taken from these envelope curves.

r, ft	*r*, m	*Values of n for large concrete-lined canals*
4	1.2	0.013–0.015
8	2.4	0.014–0.016
12	3.6	0.015–0.017

This same report showed that values of n were increased as much as 40 percent by algal growth during the summer season.

TABLE 7.2 Values of n to Be Used with Manning Formula

Surface	Best	Good	Fair	Bad
Uncoated cast-iron pipe	0.012	0.013	0.014	0.015
Coated cast-iron pipe	0.011	0.012*	0.013*	
Commercial wrought-iron pipe, black	0.012	0.013	0.014	0.015
Commercial wrought-iron pipe, galvanized	0.013	0.014	0.015	0.017
Smooth brass and glass pipe	0.009	0.010	0.011	0.013
Smooth lockbar and welded "OD" pipe	0.010	0.011*	0.013*	
Riveted and spiral steel pipe	0.013	0.015*	0.017*	
Vitrified sewer pipe	0.011	0.013*	0.015	0.017
Common clay drainage tile	0.011	0.012*	0.014*	0.017
Glazed brickwork	0.011	0.012	0.013*	0.015
Brick in cement mortar, brick sewers	0.012	0.013	0.015*	0.017
Neat cement surfaces	0.010	0.011	0.012	0.013
Cement mortar surfaces	0.011	0.012	0.013*	0.015
Concrete pipe	0.012	0.013	0.015*	0.016
Wood stave pipe	0.010	0.011	0.012	0.013
Plank flumes:				
Planed	0.010	0.012*	0.013	0.014
Unplaned	0.011	0.013*	0.014	0.015
With battens	0.012	0.015*	0.016	
Concrete-lined channels	0.012	0.014*	0.016*	0.018
Cement-rubble surface	0.017	0.020	0.025	0.030
Dry-rubble surface	0.025	0.030	0.033	0.035
Dressed-ashlar surface	0.013	0.014	0.015	0.017
Semicircular metal flumes, smooth	0.011	0.012	0.013	0.015
Semicircular metal flumes, corrugated	0.0225	0.025	0.0275	0.030
Canals and ditches:				
Earth, straight and uniform	0.017	0.020	0.023*	0.025
Rock cuts, smooth and uniform	0.025	0.030	0.033*	0.035
Rock cuts, jagged and irregular	0.035	0.040	0.045	
Winding sluggish canals	0.023	0.025*	0.028	0.030
Dredged earth channels	0.025	0.028*	0.030	0.033
Canals with rough stony beds, weeds on earth banks	0.025	0.030	0.035*	0.040
Earth bottom, rubble sides	0.028	0.030*	0.033*	0.035

(Continued)

Table 7.2 Values of *n* to Be Used with Manning Formula (*Continued*)

Surface	Best	Good	Fair	Bad
Natural stream channels:				
(1) Clean, straight bank, full stage, no rifts or deep pools	0.025	0.028	0.030	0.033
(2) As (1), but some weeds and stones	0.030	0.033	0.035	0.040
(3) Winding, some pools and shoals, clean	0.033	0.035	0.040	0.045
(4) As (3), lower stages, more ineffective slope and sections	0.040	0.045	0.050	0.055
(5) As (3), some weeds and stones	0.035	0.040	0.045	0.050
(6) As (4), stony sections	0.045	0.050	0.055	0.060
(7) Sluggish river reaches, rather weedy or with very deep pools	0.050	0.060	0.070	0.080
(8) Very weedy reaches	0.075	0.100	0.125	0.150

*Values commonly used in design.

Values of n given in these tabulations were nearly all obtained for very large values of the Reynolds number. When the Reynolds number is small enough to approach the transition range ($rV/v < 15{,}000$ or $4rV/v < 60{,}000$), Eq. (7.29) should be used, or Eq. (7.31) may be used to compute appropriate values of n for the Manning equation.

Noneroding Velocities

Fortier and Scobey[7] pointed out that there is no sharp line of demarcation between the velocities that can no longer maintain silt in movement and those that will scour a canal bed. It is believed that there is a broad belt of velocities between these two "critical" velocities within which silt already loosened or brought in through a head gate will remain in suspension while the bed will not be scoured. In general, old and well-seasoned canals will stand much higher velocities than new ones. This is true particularly if the canal bed or the silt conveyed by the stream contains colloidal matter. Colloids, as applied to soils, give the properties of plasticity, cohesion, toughness when wet, and hardness when dry that are essential to an erosion resisting soil.

Table 7.3 contains canal velocities that are recommended. The following conclusions were drawn by Fortier and Scobey.

1. The laws of hydraulics governing the movement of loose silt and detritus in open channels are only distantly related to the

TABLE 7.3 Permissible Canal Velocities After Aging,* ft/s (m/s)

Original material excavated	Clear water, no detritus	Water transporting colloidal silts	Water transporting noncolloidal silts, sands, gravels, or rock fragments
Fine sand, noncolloidal	1.50 (0.50)	2.50 (0.75)	1.50 (0.50)
Sandy loam, noncolloidal	1.75 (0.55)	2.50 (0.75)	2.00 (0.60)
Silt loam, noncolloidal	2.00 (0.60)	3.00 (0.90)	2.00 (0.60)
Alluvial silts, noncolloidal	2.00 (0.60)	3.50 (1.05)	2.00 (0.60)
Ordinary firm loam	2.50 (0.75)	3.50 (1.05)	2.25 (0.68)
Volcanic ash	2.50 (0.75)	3.50 (1.05)	2.00 (0.60)
Fine gravel	2.50 (0.75)	5.00 (1.50)	3.75 (1.15)
Stiff clay, very colloidal	3.75 (1.15)	5.00 (1.50)	3.00 (0.90)
Graded, loam to cobbles, noncolloidal	3.75 (1.15)	5.00 (1.50)	5.00 (1.50)
Alluvial silts, colloidal	3.75 (1.15)	5.00 (1.50)	3.00 (0.90)
Graded, silt to cobbles, colloidal	4.00 (1.20)	5.50 (1.65)	5.00 (1.50)
Coarse gravel, noncolloidal	4.00 (1.20)	6.00 (1.80)	6.50 (2.00)
Cobbles and shingles	5.00 (1.50)	5.50 (1.65)	6.50 (2.00)
Shales and hardpans	6.00 (1.80)	6.00 (1.80)	5.00 (1.50)

*Recommended in 1926 by Special ASCE Committee on Irrigation Research.

laws governing the scouring of a canal bed and are not directly applicable.

2. The material of seasoned canal beds is composed of particles of different sizes, and when the interstices of the larger are filled by the smaller, the mass becomes more dense, stable, and less subject to the erosive action of water.
3. The velocity required to ravel and scour a well-bedded canal in any material is much greater than the velocity required to maintain movement of particles of that same material before becoming bedded or that have been raveled off by higher velocities than the bed would stand.
4. Colloids in either the material of the canal bed or the water conveyed by it, or in both, tend to cement particles of clay, silt, sand, and gravel in such a way as to resist erosive effects.
5. The grading of material running from fine to coarse, coupled with the adhesion between particles brought about by colloids,

makes possible high mean velocities without any appreciable scouring effect.

6. Irrigation canals may be designed for the velocity that is permissible when seasoned by age, as the demand for water grows with the age of the canal and the maximum mean velocity grows with the supply necessary to satisfy this demand.
7. Power canals are likely to be placed under peak load as soon as feasible after completion. For this reason a more conservative velocity should be chosen; otherwise scour may take place before seasoning.
8. Canals, when new, may be operated with velocities less than the maximum permissible by the use of check structures.
9. A slight excess of velocity is preferable to insufficient velocity. Check-drop structures will correct a slope causing erosion, but there is no method within reasonable cost by which velocities may be increased.
10. In the past, engineers have been led astray by a too close adherence to the results of experimental data on the transporting action of water on such materials as clay, silt, sand, and gravel, considered separately rather than on cohesive or mechanical combinations, and in consequence many canals have been designed and built on too flat slopes with correspondingly low velocities.
11. The growth of aquatic plants is only partly connected with velocities. Canals designed for the highest permissible velocities, from the standpoint of erosion, will be as free from plant growth as design alone can affect.

The practical difficulty that will be encountered in attempting to apply the foregoing conclusions or in selecting suitable velocities from Table 7.3 lies in properly defining the soil. Even with numerous borings and test pits it will be difficult to determine the real character of the soil along the line of a proposed canal. This is particularly true of locations in glacial lands, where the character of the soil may change at frequent intervals. Also, the definitions of soils given in Table 7.3 will doubtlessly convey widely varying meanings to different engineers.

Criteria for scour based only on velocity neglect the fact that the shear stress on the bottom is also related to the depth. Reference to Eq. (7.19) and Fig. 7.5 will show that the expression for average shear stress in a wide rectangular channel is

$$\tau = wD \sin \theta \tag{7.35}$$

A similar derivation for a channel of any shape yields

$$\tau = wr \sin \theta \tag{7.36}$$

Lane[8] has taken the critical velocities given in Table 7.3, together with other test results, and developed criteria for noncohesive materials based on critical tractive force rather than velocity. For materials having a median diameter (a size such that 50 percent by weight is larger) greater than 5 mm, his design criterion can be expressed by the equation

$$\tau_c = 0.40\, d \text{ (units of feet)} \quad \tau_c = 0.77\, d \text{ (units of meters)} \tag{7.37}$$

where τ_c is the limiting shear stress in pounds per square foot (newtons per square meter), and d is the particle size in inches (millimeters), such that 25 percent by weight is larger. For material having a median size smaller than 5 mm, criteria are in the form of the three curves reproduced in Fig. 7.8. Curve a is for canals carrying 2 percent or more of silt and clay two or three times per year. Curve b is for canals in which the silt and clay content reaches 0.2 percent two or three times per year, and curve c is for clear water. Lane also presents methods of taking into account the fact that the shear stress varies from place to place in the wetted perimeter and that the sides are more vulnerable than the bottom.

Because of the rather large scatter of experimental results, and also because the greater vulnerability of the sides is compensated for to some extent by the lower shear stresses on the sides, an approximate solution for the critical tractive force for channels in which $b/D \geqq 2$ and $z \geqq 1.5$ can be obtained from the criteria given by Eq. (7.37)

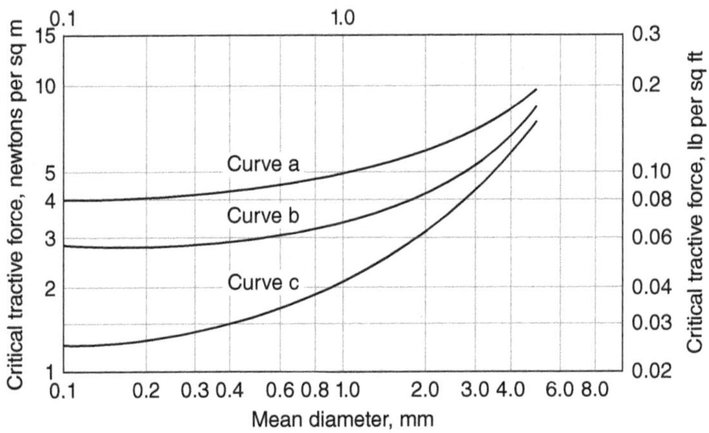

FIGURE 7.8 Criteria for scour, median size <5 mm.

and Fig. 7.8 without including the refinements mentioned. This can be illustrated by an example in which the 25 percent size is 25 mm and the depth is 1.5 m. From Eq. (7.37),

$$\tau = 0.77d = 19.3\ \text{N/m}^2$$

Then, from Eq. (7.35), the limiting slope is

$$\sin\theta = \frac{\tau}{wD} = \frac{19.3}{9800\times 1.5} = 0.0013$$

For this case the more refined solution given by Lane resulted in a value of 0.0011 for sin θ. For many practical cases, the approximate solution suggested here may be sufficiently accurate.

Another problem frequently encountered in designing canals is *prevention of sedimentation* when water enters a canal with a heavy load of suspended material. Much progress is being made on this phase of flow in open channels, but the state of development does not lend itself to a condensed presentation of simple criteria.[9,10]

Air Entrainment

When water flows at high velocities, the turbulence produces a very rough water surface which tends to entrap air. The vertical components of the velocity distribute the air through the water, creating the "white" water that is associated with air entrainment. This phenomenon has been observed and studied by many hydraulic engineers, and only a representative group of references is presented here.[11–14] Detailed investigations[14] have revealed that near the bottom of a channel the water contains many small air bubbles in suspension, whereas near the top the water mixture appears as if it were predominantly air transporting globules of water. Because of the rapid movement of the small globules of water, it is difficult to establish the location of the surface of the mixture. An arbitrary location of the surface such that 98 or 99 percent of the water is included is sometimes used. When the water enters a channel from a reservoir through a smooth entrance, entrainment does not occur in the central regions until the boundary layer has spread to the full depth. However, at the sidewalls entrainment begins sooner, because the boundary layer is exposed to the surface as soon as it begins to develop.

The water discharge in a channel with air entrainment can be expressed as

$$Q = \theta_w AV \tag{7.38}$$

where A is the total cross-sectional area of the mixture, V the average velocity of the mixture, and θ_w the specific gravity of the mixture.

The average velocity of the mixture can be estimated from open channel equations such as the Manning equation, using the same values of n as for ordinary flow. The value of θ_w must be known in order to solve for the area of cross section and the depth of the mixture. The value of θ_w depends on the velocity of the mixture, depth, roughness, entrance conditions, form of the channel, wind velocity and direction, and fluid properties. If only water within a limited range of temperatures is considered, the effect of two important fluid properties, viscosity and surface tension, may be considered as constant. Hall[13] has shown that for five channels which he investigated, the value of θ_w tends to vary with the Froude number,

$$F = \frac{V^2}{g r_c} \tag{7.39}$$

where r_e is the hydraulic radius that would occur if only water were flowing. All but one of the five flumes produced relationships of the following form:

$$\frac{1-\theta_w}{\theta_w} = K\frac{V^2}{g r_c} \tag{7.40}$$

For smooth concrete chutes a value of $K = 0.006$ is recommended. The K value increases with roughness. Using the same data, Douma[15] has developed a relationship which, after revision to include θ_w, takes the form

$$(1-\theta_w)^2 = \frac{0.002V^2}{gr} - 0.01 \tag{7.41}$$

where r is the hydraulic radius of the mixture. Either of these expressions may be used to obtain a rough estimate regarding the degree of bulking that might be expected in reasonably straight and smooth concrete channels.

References

1. D. C. Woo and E. F. Brater, "Laminar Flow in Rough Rectangular Channels," *J. Geophys. Res.*, vol. 66, p. 4207, Dec. 1961.
2. "Studies of River Bed Materials and Their Movement, with Special Reference to the Lower Mississippi River," U.S. Waterways Experiment Station, Paper 17, Jan. 1935.

3. M. J. Boussinesq, "Memoire sur l'influence des frottements dans lesmouvements réguliers des fluides," *J. Math. Pures Appl.*, ser. II, vol. 13, p. 377, 1868.
4. L. G. Straub, E. Silberman, and H. C. Nelson, "Open-Channel Flow at Small Reynolds Numbers," *Trans. ASCE*, vol. 123, p. 685, 1958.
5. Task Force on Friction Factors in Open Channels, Committee on Hydromechanics of the Hydraulics Division, ASCE, "Friction Factors in Open Channels," *J. Hydraul. Div., ASCE*, vol. 89, pt. I, p. 97, Mar. 1963.
6. P. J. Tilp and M. W. Scrivner, "Analysis and Description of Capacity Tests on Large Concrete-Lined Canals," U.S. Bureau of Reclamation, Tech. Memo 661, 1964.
7. S. Fortier and F. C. Scobey, "Permissible Canal Velocities," Special Committee on Irrigation Hydraulics, final report on this subject, *Trans. ASCE*, vol. 89, 1926.
8. E. W. Lane, "Design of Stable Channels," *Trans. ASCE*, vol. 120, p. 1234, 1955.
9. H. A. Einstein, "The Bed-Level Function for Sediment Transportation in Open Channel Flows," U.S. Dept. of Agriculture Tech. Bull. 1026, 1950.
10. C. B. Brown, *Engineering Hydraulics*, Chap. 12, Wiley, New York, 1950.
11. E. W. Lane, "Entrainment of Air in Swiftly Flowing Water," *Civil Eng. (N.Y.)*, p. 89, Feb. 1939.
12. G. H. Hickox, "Air Entrainment on Spillway Faces," *Civil Eng. (N.Y.)*, p. 562, Dec. 1945.
13. L. S. Hall, "Entrainment of Air in Flowing Water: A Symposium: Open Channel Flow at High Velocities," *Trans. ASCE*, vol. 108, p. 1394, 1943.
14. L. G. Straub and A. G. Anderson, "Self-Aerated Flow in Open Channels," *Trans. ASCE*, vol. 125, p. 456, 1960.
15. J. H. Douma, Discussion of Hall,[13] p. 1462.

SECTION 8

Open Channels with Nonuniform Flow

This section deals with the action of water in open channels under special conditions, which are usually associated with sudden or gradual changes in the cross-sectional area of the stream. The terms "rapidly varied flow" and "gradually varied flow" are used to describe these two types of nonuniform flow. The first part of this section is devoted to a discussion of the principles of rapidly varied flow. The effects of constrictions and enlargements on the water-surface profile are described. The principles of flow at critical depth are developed, and expressions for flow at critical depth are derived for all shapes of channel cross sections. Critical-depth meters and other examples of flow at critical depth are discussed. The principles of the hydraulic jump are derived, and methods of designing transition sections are described.

The second part of this section deals with gradually varied flow. The general differential equations of gradually varied flow are derived, and generalized water-surface profiles are presented. Methods of computing water-surface profiles under all conditions are described. Procedures for locating the position of a hydraulic jump are outlined, and methods of solving such special problems as flow in short channels, flow in chutes, and flow down very steep spillways are presented.

The velocity head, which appears in many of the equations presented in this section, is assumed to be $V^2/2g$, that is, α [see Eq. (3.11)] is assumed to be unity. The approximation thus introduced will not ordinarily be of importance, but it should be kept in mind because in some cases it may be necessary to modify expressions to include an estimated value of α.

Rapidly Varied Flow

The principles of rapidly varied flow may be derived by considering the specific energy of the water flowing in an open channel. The general Bernoulli constant for an open channel is

$$H = z + D + \frac{V^2}{2g} \tag{8.1}$$

in which H is the energy of the fluid with respect to an arbitrarily chosen datum plane, as illustrated in Fig. 8.1. The specific energy H_e is obtained by letting the datum plane pass through the bottom of the channel. Thus z in Eq. (8.1) becomes zero, and the expression for specific energy is as shown by the equation

$$H_e = D + \frac{V^2}{2g} \tag{8.2}$$

If the value of V from the equation

$$Q = aV \tag{8.3}$$

is substituted into Eq. (8.2), the following equation for specific energy is obtained:

$$H_e = D + \frac{Q^2}{2ga^2} \tag{8.4}$$

For the special case of discharge in a rectangular channel,

$$a = bD \tag{8.5}$$

and thus

$$Q = bDV \tag{8.6}$$

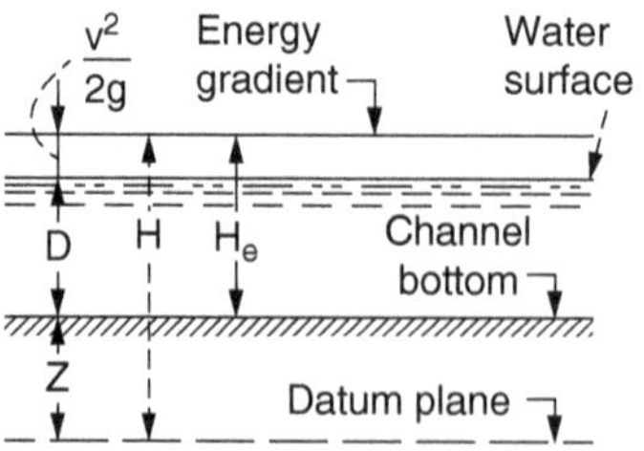

FIGURE 8.1 Energy of open-channel flow.

If the discharge per unit of width Q/b is denoted by q, insertion of Eqs. (8.5) and (8.6) into Eq. (8.4) yields the following expression for specific energy in a rectangular channel:

$$H_e = D + \frac{q^2}{2gD^2} \tag{8.7}$$

This equation, or the general expression, Eq. (8.4), may be used to show what happens when a channel cross section changes rapidly. For the sake of simplicity, the discussion will be based on the case of flow in a rectangular channel where Eq. (8.7) will apply. The equation may be studied from two points of view, first by keeping the discharge constant while H_e and D are permitted to vary, and second by holding H_e constant while D and q are the variables.

Constant-Discharge Relations

If the value of q is taken as a constant, Eq. (8.7) will yield three values of D for each value of H_e. One of the values will be negative, so that only two of the values have practical significance. Figure 8.2*a* was obtained by taking q as 2 ft^3 per foot of width (0.19 m^3/s per meter of width) and solving for various values of D and H_e. It may be seen that there are two values of D for each value of H_e, except at the point of minimum energy H_m, where there is only a single value. This particular depth is called *critical depth* D_c and has great significance in the solution of many nonuniform flow problems.

The usefulness of graphs of Eq. (8.7), such as the one shown in Fig. 8.2*a*, in solving problems where there is a sudden change of

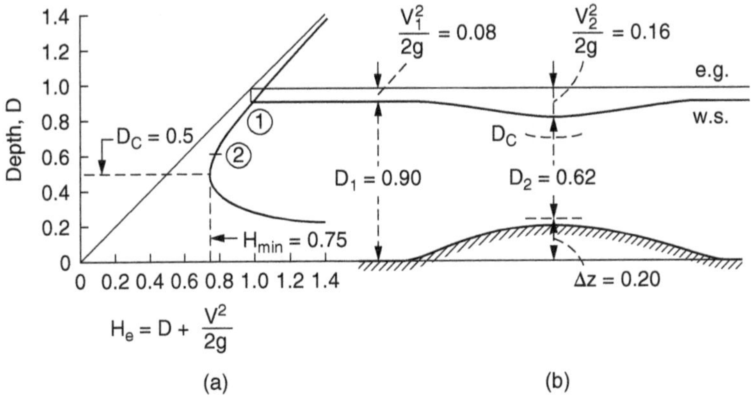

FIGURE 8.2 Relationship between specific energy and depth for constant discharge in rectangular channels.

bottom elevation is illustrated by the example shown in Fig. 8.2*b*. In this example, it was assumed that the depth of water was 0.9 ft and the bottom was raised 0.2 ft for a short distance. At the original depth, corresponding to point 1 in Fig. 8.2*a*, flow occurred with a specific energy of 0.98 ft. Raising the bottom 0.2 ft changes the specific energy to 0.78 ft, which corresponds to point 2 in Fig. 8.2*a*. The depth at this point is 0.62 ft, and the water-surface profile will be as shown in Fig. 8.2*b*. In this example, it was assumed that there was no energy loss due to the constriction.

The maximum distance that the bottom of this channel could be raised, when the original depth is 0.9 ft, without causing any backwater, is the one that will reduce the value of H_e to H_m, 0.75 ft in this case. The depth would then be D_c. This condition is shown by the broken lines in Fig. 8.2*b*. Any larger rise would attempt to reduce the specific energy to a value less than 0.75 ft, and the curve of Fig. 8.2*a* shows that a discharge of 2 ft^3 per foot of width cannot flow with H_e less than 0.75 ft. Consequently, the water would have to become deeper upstream in order to raise the energy gradient farther above the channel bottom or, in other words, to increase the specific energy with which the water approaches the constriction. The problem of lowering the channel bottom over a short distance may be solved in a similar manner.

Constant-Energy Relations

If the value of H_e in Eq. (8.7) is taken as constant, there will be two positive values of D for each value of q, as shown in Fig. 8.3*a*.

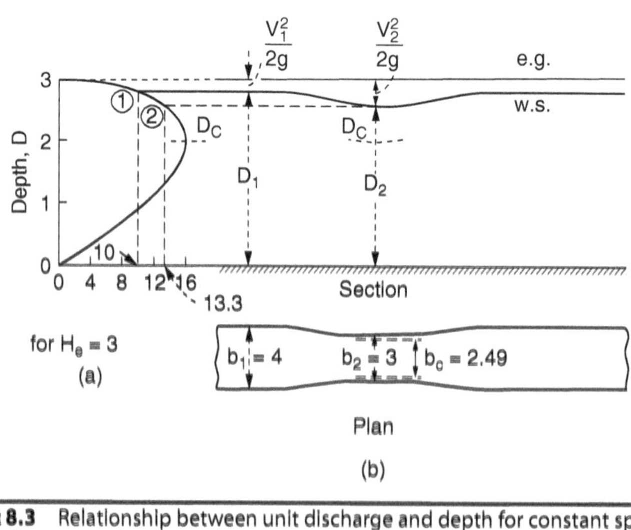

Figure 8.3 Relationship between unit discharge and depth for constant specific energy in rectangular channels.

The curve shown in the figure was obtained by taking the value of H_e as 3.0 ft (0.91 m) and computing the corresponding values of D and q from Eq. (8.7). Again, as in the case of Fig. 8.2*a*, there is a unique value of D, which in Fig. 8.3*a* occurs at the point of maximum discharge. This depth is the critical depth D_c.

The application of Fig. 8.3*a* to a channel constriction problem is illustrated by the example shown in Fig. 8.3*b*. The discharge in a rectangular channel 4 ft (1.22 m) wide (b_1) was assumed to be 10 ft^3 per foot of width (0.93 m^3/s per meter of width), corresponding to point 1 in Fig. 8.3*a*. The width (b_2) was assumed to be changed to 3 ft (0.91 m) for a short distance, thus increasing the discharge per unit of width to 13.3 ft^3/s (1.24 m^3/s) at the constriction. This value of q, point 2 in Fig. 8.3*a*, corresponds to a depth (D_1) of 2.59 ft (0.79 m). The water-surface profile will then be as shown in Fig. 8.3*a*.

The minimum width to which the channel may be constricted without causing backwater will be the width that will cause the maximum unit discharge to occur, 2.49 ft (0.76 m) in this case, and the depth at the constriction will then be the critical depth. This condition is illustrated by the broken lines in Fig. 8.3*b*. The graph of Eq. (8.7), Fig. 8.3*a*, shows that the discharge per unit of width cannot exceed 16.08 ft^3/s (1.49 m^3/s) with a specific energy of 3 ft (0.91 m). If, then, the width of the constriction is made less than 2.49 ft (0.76 m), thus increasing the value of q above 16.08 ft^3/s (1.49 m^3/s), the water must become deeper upstream from the constriction in order to gain the additional energy required for this larger discharge.

The case of an expansion in the channel may be solved in a similar manner with the aid of Fig. 8.3*a*. Had the original depth been less than D_c, the water surface would have risen in passing through the constriction.

Analytical Solutions of Constriction Problems

The problem of determining the depth of water in a constriction may also be solved without the aid of curves by writing the Bernoulli equation from a point above the constriction to the constriction. It is well to keep in mind, however, the general nature of the curves (Figs. 8.2*a* and 8.3*a*) when solving such problems because there are always two possible answers for the depth in the constriction. The correct one can be determined only by knowing in advance whether the depth will be greater or less than D_c. In the general case, the bottom may be raised or lowered, the width may be increased or decreased, and the energy loss must be included.

The Bernoulli equation written with reference to the symbols used in either Fig. 8.2*b* or Fig. 8.3*b* is

$$\frac{V_1^2}{2g} + D_1 = \frac{V_2^2}{2g} + D_2 + \Delta z + h_l \tag{8.8}$$

The value of the energy loss h_l may be determined in the manner that will be described under Minor Losses. The unknowns are V_2 and D_2, but V_2 may be related to Q, b_2, and D_2 by the use of Eq. (8.6), thus making D_2 the only unknown quantity. Arranged in this fashion, Eq. (8.8) may be solved by successive approximations.

Critical Depth—General Case

The expressions for flow at critical depth may be derived by setting $dH_e/dD = 0$, as suggested by Fig. 8.2*a*, or by setting $dq/dD = 0$, as suggested by Fig. 8.3*a*. The first-mentioned procedure will be used here to obtain general expressions for flow at critical depth, which are applicable to channels of any shape. The value of dH_e/dD will be obtained from

$$H_e = D + \frac{Q^2}{2ga^2} \tag{8.4}$$

The area of any cross section may be expressed as a function of depth,

$$a = f(D) \tag{8.9}$$

Replacing a in Eq. (8.4) with $f(D)$ and differentiating,

$$dH_e = dD - \frac{Q^2 f'(D)}{gf^3(D)} \tag{8.10}$$

From Eq. (8.9) we have

$$f'(D) = da \tag{8.11}$$

and from Fig. 8.4 it may be seen that

$$da = TdD \tag{8.12}$$

By replacing $f(D)$ with a and $f'(D)$ with TdD, the following value of dH_e/dD is obtained from Eq. (8.10):

$$\frac{dH_e}{dD} = 1 - \frac{TQ^2}{ga^3} \tag{8.13}$$

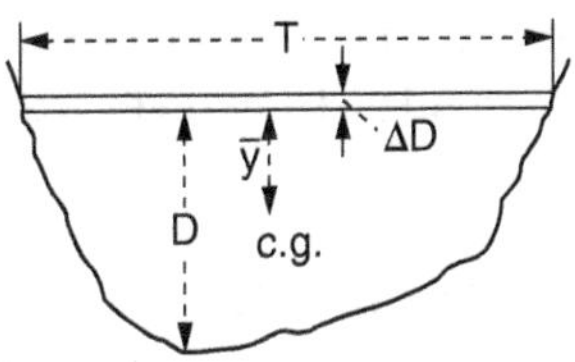

FIGURE 8.4 Channel cross section.

The following general expression for flow at critical depth is obtained by setting the right side of Eq. (8.13) equal to zero:

$$\frac{Q^2}{g}\frac{a^3}{T} \tag{8.14}$$

Equation (8.14) must be satisfied, no matter what the shape of the channel, when flow is at critical depth.

Other useful general expressions for flow at critical depth may be obtained by introducing the mean depth D_m, as defined by

$$D_m = \frac{a}{T} \tag{8.15}$$

The following critical-depth equations may be derived from Eq. (8.12) by substituting values from Eqs. (8.2), (8.3), and (8.15), keeping in mind that when $D = D_c$, $H_e = H_m$, where H_m is the minimum specific energy,

$$Q = a\sqrt{gD_m} \tag{8.16}$$

$$D_m = \frac{Q^2}{ga^2} \tag{8.17}$$

$$H_m = D_c + \frac{D_m}{2} \tag{8.18}$$

$$V_c = \sqrt{gD_m} \tag{8.19}$$

$$D_m = \frac{V_c^2}{g} \tag{8.20}$$

$$\frac{V_c^2}{gD_m} = 1 \tag{8.21}$$

The critical-depth equations for particular shapes of channels may be derived from these general equations, Eqs. (8.14) to (8.21), or by direct application of the basic principles. The equations for flow at critical depth in rectangular, triangular, trapezoidal, and circular channels are derived next.

Critical Depth in Rectangular Channels

In channels of rectangular cross section, the depth D_c is equal to the mean depth D_m, the bottom width b is equal to the top width T, and when the discharge is taken as the discharge per unit of width q, both b and T are equal to unity. By making appropriate substitutions, Eqs. (8.19) and (8.20) become

$$V_c = \sqrt{gD_c} \tag{8.22}$$

and

$$D_c = \frac{V_c^2}{g} \tag{8.23}$$

Equation (8.16) yields the following expressions for discharge in rectangular channels:

$$Q = \sqrt{g}\, bD_c^{3/2} \tag{8.24}$$

$$q = \sqrt{g} D_c^{3/2} \tag{8.25}$$

and Eq. (8.17) becomes

$$D_c = \sqrt[3]{\frac{q^2}{g}} \tag{8.26}$$

From Eq. (8.18),

$$H_m = \frac{3}{2} D_c \tag{8.27}$$

and

$$D_c = \frac{2}{3} H_m \tag{8.28}$$

If the value of D_c from Eq. (8.28) is substituted into Eq. (8.25), the following expression for discharge per unit of width is obtained:

$$q = \sqrt{g}\,(2/3)^{3/2}\, H_m^{3/2} \tag{8.29}$$

With the value of g taken as 32.16 ft/s², Eq. (8.29) becomes

$$q = 3.087 H_m^{3/2}$$

With the value of g taken as 9.8 m/s², Eq. (8.34) becomes

$$q = 1.706 H_m^{3/2} \tag{8.30}$$

Critical Depth in Triangular Channels

For the triangular section shown in Fig 8.5, the maximum depth is D_c, and the mean depth D_m is equal to ½D_c. From Eqs. (8.19) and (8.20),

$$V_c = \sqrt{\frac{gD_c}{2}} \tag{8.31}$$

and

$$D_c = \frac{2V_c^2}{g} \tag{8.32}$$

As shown in Fig. 8.5, z is the slope of the sides of the channel, expressed as the ratio of horizontal to vertical, and for symmetrical sections it is ½T/D_c. The area is $a = zD_c^2$. Substituting into Eqs. (8.16) and (8.17),

$$Q = \sqrt{\frac{g}{x}}\, zD_c^{5/2} \tag{8.33}$$

For $g = 32.16$ ft/s², Eq. (8.33) becomes

$$Q = 4.01 zD_c^{5/2}$$

or for $g = 9.8$ m/s², Eq. (8.33) becomes

$$Q = 2.21 zD_c^{5/2} \tag{8.34}$$

and

$$D_c = \sqrt[5]{\frac{2Q^2}{gz^2}} \tag{8.35}$$

From Eq. (8.18),

$$H_m = \frac{5}{4} D_c \tag{8.36}$$

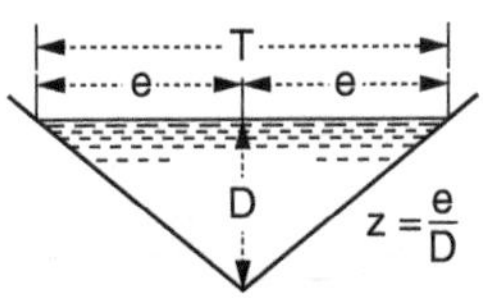

Figure 8.5 Triangular channel.

and
$$D_c = \frac{4}{5} H_m \tag{8.37}$$

Substituting this value of D_c into Eq. (8.33),

$$Q = \sqrt{\frac{g}{2}} \left(\frac{4}{5}\right)^{5/2} zH_m^{5/2} \tag{8.38}$$

For $g = 32.16$ ft/s²

$$Q = 2.295zH_m^{5/2}$$

or, for $g = 9.8$,

$$Q = 1.268zH_m^{5/2} \tag{8.39}$$

Critical Depth in Trapezoidal Sections

The trapezoidal section shown in Fig. 8.6 has a depth D_c and a bottom width b. The slope of the sides, horizontal divided by vertical, is z.

Expressing the mean depth D_m of Eqs. (8.19) and (8.20) in terms of channel dimensions, the following relations between critical depth D_c and average velocity V_c are obtained:

$$V_c = \sqrt{\frac{b + zD_c}{b + 2zD_c} gD_c} \tag{8.40}$$

and
$$D_c = \frac{V_c^2}{g} - \frac{b}{2z} + \sqrt{\frac{V_c^4}{g^2} + \frac{b^2}{4z^2}} \tag{8.41}$$

With mean depth D_m and the area a expressed in terms of channel dimensions, Eq. (8.16) yields the following expression for discharge at critical depth in a trapezoidal channel:

$$Q = \sqrt{g \frac{(b + zD_c)^3}{b + 2zD_c}} D_c^{3/2} \tag{8.42}$$

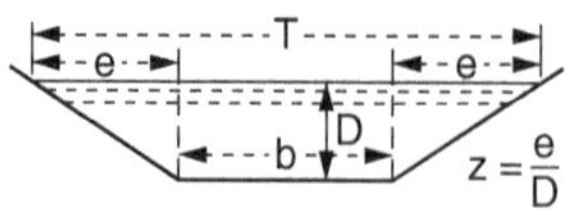

FIGURE 8.6 Trapezoidal channel.

The following expressions may be derived from Eq. (8.18):

$$H_m = \frac{3b + 5zD_c}{2b + 4zD_c} \tag{8.43}$$

and

$$D_c = \frac{4zH_m - 3b + \sqrt{16z^2H_m^2 + 16zH_m b + 9b^2}}{10z} \tag{8.44}$$

Critical Depth in Circular Channels

Referring to Fig. 8.7, the following expressions for the area, top width, and depth of a circular channel may be obtained:

$$a = \frac{d^2}{4}(\theta_r - \tfrac{1}{2}\sin 2\theta) \tag{8.45}$$

$$T = d\sin\theta \tag{8.46}$$

$$D_c = \frac{d}{2}(1 - \cos\theta) \tag{8.47}$$

where θ is in degrees and θ_r in radians. Values of a, T, and D_c from Eqs. (8.45) to (8.47) may be substituted into Eq. (8.14) to obtain the following expression:

$$Q = \frac{2^{5/2} g^{1/2} (\theta_r - \tfrac{1}{2}\sin 2\theta)^{3/2}}{8(\sin\theta)^{1/2}(1 - \cos\theta)} D_c^{5/2} \tag{8.48}$$

D_c can be determined for a given Q by successive approximations using Eqs 8.47 and 8.48.

Critical-Depth Meters

One of the most ideal methods of measuring the discharge in open channels is by means of constriction built for the purpose of causing critical depth to occur. This device, known as a critical-depth meter

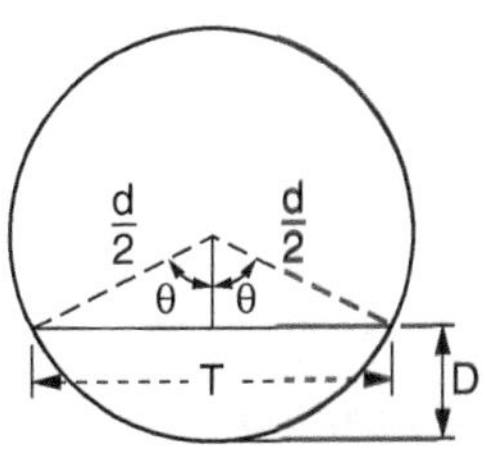

FIGURE 8.7 Circular channel.

or control meter, has the advantage over a weir in that it requires no stilling basin, is practically invulnerable to damage by floating debris, and does not require calibration.

Critical-depth meters may be made by constricting the width of the channel, by raising the bottom, or by doing both. The principles involved were fully set forth earlier in this section, where it was shown that raising the bottom a certain amount or constricting the width beyond a particular point causes backwater upstream from the constriction and critical depth in the constriction. The most important consideration in the design is that the throat be level, uniform in cross section, and of sufficient length to ensure that critical depth will occur in this section. A throat length only two or three times D_c is usually enough unless the meter ends in a fall or a very steep slope, in which case the throat length should be longer (see "Free Outfall," later in this section).

It is impossible to measure critical depth itself, because its exact location is not known and because the water surface is often wavy at that location. For these reasons the gauge is located a short distance upstream from the throat, as shown in Fig. 8.8. The nearer the gauge is located to the throat, the smaller will be the energy loss between the gauge and the location of critical depth. The energy loss may be estimated on the basis of the material presented under "Contractions and Enlargements" later in this section. Equation (8.49) is the Bernoulli equation, written from the gauge location, point 1, to the location of critical depth, point c,

$$D_1 + \frac{V_1^2}{2g} = z + D_c + \frac{V_c^2}{2g} + h_l \tag{8.49}$$

Replacing $(D_c + V_c^2/2g)$ with H_m, and h_l with its value from $0.1\,(V_c^2/2g - V_1^2/2g)$, Eq. (8.49) may be written

$$D_1 + \frac{V_1^2}{2g} = z + H_m + 0.1\left(\frac{V_c^2}{2g} - \frac{V_1^2}{2g}\right) \tag{8.50}$$

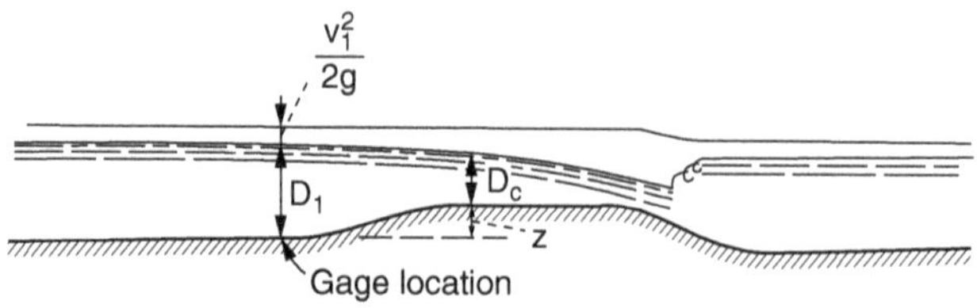

FIGURE 8.8 Critical-depth meter.

For rectangular channels, Eq. (8.50) may be simplified by making use of the relation $V_c^2/2g = H_m/3$ to give the expression

$$D_1 + 1.1\frac{V_1^2}{2g} = z + 1.03H_m \tag{8.51}$$

Because z is known and D_1 is measured by the gauge, the only unknowns in Eq. (8.51) are $V_1^2/2g$ and H_m. The value of $V_1^2/2g$ is usually small in comparison with the other terms. Therefore, as a first approximation, H_m may be determined from Eq. (8.51) by letting $V_1 = 0$. The approximate discharge can now be obtained from Eq. (8.30), and the correct discharge is then determined by successive approximations.

For channels other than rectangular, the transformation from Eq. (8.50) to an equation similar to Eq. (8.51) will require the use of the appropriate relationship for that particular type of channel. The corresponding expression for discharge in terms of H_m must be used in place of Eq. (8.30).

In many cases the energy loss is so small that it may be neglected. If a critical-depth meter is operating properly, a hydraulic jump will occur below the meter when the original depth is greater than D_c, and the jump will occur upstream from the meter if the original depth is less than critical.

Critical Slope

Uniform flow at critical depth will occur when the grade or slope of the channel is just equal to the loss of head per unit length resulting from flow at this depth. In any channel for any given discharge there is one grade that will maintain uniform flow at critical depth. This is termed the "critical slope." For any grade flatter or steeper than the critical slope the depth of flow will be, respectively, greater or less than the critical depth.

In cases where nonuniform (accelerated) flow passes through the critical stage, critical depth will occur at the section at which the energy gradient has critical slope. Examples of flow passing through the critical stage follow.

From the Chezy formula (Sec. 6), the discharge at critical depth is $Q = ac\sqrt{rs_c}$, where a being the cross-sectional area, r the hydraulic radius, s_c the critical slope, and c the Chezy coefficient. Equating this value of Q to the value in the critical depth criterion [Eq. (8.14)] and reducing,

$$s_c = \frac{gD_m}{c^2 r} \tag{8.52}$$

In this formula, D_m is the area divided by the top width or the mean depth. In a channel that is relatively wide for its depth, $D_m = r$ (approximately), and Eq. (8.52) reduces to

$$s_c = \frac{g}{c^2} \tag{8.53}$$

Substituting $c = (1.486/n)r^{1/6}$, as given by the Manning formula [Eq. (6.25)], and writing $g = 32.16$ ft/s^2, Eqs. (8.52) and (8.53) become

$$s_c = \frac{14.56n^2 D_m}{r^{4/3}} \tag{8.54a}$$

and

$$s_c = \frac{14.46n^2}{D_m^{1/3}} \tag{8.55a}$$

In metric units substituting $c = (1.0/n)r^{1/6}$, as given by the Manning formula [Eq. (6.25)], and writing $g = 9.8$ m/s^2, Eqs. (8.52) and (8.53) become

$$s_c = \frac{9.8n^2 D_m}{r^{4/3}} \tag{8.54b}$$

and

$$s_c = \frac{9.8n^2}{D_m^{1/3}} \tag{8.55b}$$

Values of the coefficient of roughness n are contained in Table 7.1. Since s_c varies as the square of this coefficient, errors made in estimating its value are magnified, and slopes computed by the preceding formulas are apt to show considerable variance from actual conditions.

When flow is at or near the critical stage, considerable change in depth may occur without material change in the energy content of the stream. Flow in this region is therefore quite unstable, and a slight disturbance will frequently produce excessive wave action or set up pronounced oscillations of the water surface.

Channel Entrance

The entrance to a channel must be designed so that the size of the channel at the entrance will produce the required discharge in accordance with the limitations on head in the reservoir and the desired depth in the channel. Writing the Bernoulli equation from the

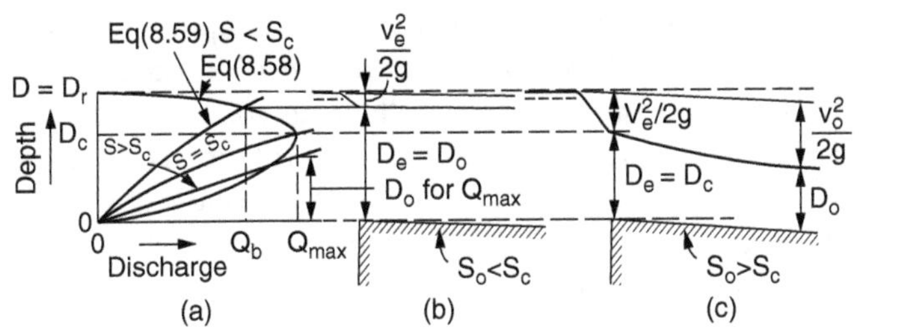

Figure 8.9 Channel entrance.

reservoir to the channel gives the following expression, the symbols being defined in Fig. 8.9*b*:

$$D_r + \frac{V_r^2}{2g} = D_e + \frac{V_e^2}{2g} + h_e \tag{8.56}$$

The kinetic-energy term for the reservoir $V_r^2/2g$ is usually negligible. If the minor loss at the entrance is also neglected, Eq. (8.56) can be written as

$$D_r = D_e + \frac{V_e^2}{2g} \tag{8.57}$$

This is the relationship expressed by Eq. (8.2), D_r being the specific energy with which the water enters the channel. The relationship can also be written in the form of Eq. (8.4),

$$D_r = D_e + \frac{Q^2}{2ga^2} \tag{8.58}$$

Since a is a function of D, this equation can be plotted using Q and D as the variables, as illustrated in Fig. 8.9*a*.

For a channel having a subcritical slope ($s_0 < s_c$), water must enter the channel at uniform flow depth ($D_e = D_0$) because, as shown under "Equations of Gradually Varied Flow and Generalized Profiles" later in this section, for subcritical flow no water surface can exist which approaches the uniform flow depth from a smaller or larger depth. Consequently, for a given shape of channel and a given D_r, the discharge into the channel and the depth at the entrance must satisfy Eq. (8.58) as well as the Manning equation [Eq. (8.59)],

$$Q = \frac{K'b^{8/3}s^{1/2}}{n} \tag{8.59}$$

In each equation the only unknowns are depth and discharge, depth being one of the variables controlling K' in Eq. (8.59). For a given value of b, n, and s_0, the simultaneous solution of Eqs. (8.58) and (8.59) would yield D_e or D_0 and Q. This solution is accomplished graphically in Fig. 8.9*a* by plotting Eq. (8.59) and noting where it crosses the graph of Eq. (8.58). The corresponding discharge is noted as Q_b, and the corresponding value of D_0 is plotted in Fig. 8.9*b*. As the channel slope is increased, lines resulting from Eq. (8.59) will cut the curve of Eq. (8.58) at larger discharges until $s_0 = s_c$, when the crossing of the two curves will be at Q_{max}, as shown in Fig. 8.9*a*.

When the channel has a supercritical slope ($s_0 > s_c$) the discharge is equal to Q_{max} and the intersection of the two curves no longer has any physical meaning. However, the value of D from the curve of Eq. (8.59) at Q_{max} gives uniform flow depth, as illustrated in Fig. 8.9*a* and *c*. The water surface in the channel (Fig. 8.9*c*) is concave upward, the depth varying from D_c at the entrance to D_0. This is known as an S_2 curve (see Fig. 8.20).

When an increase in slope occurs in a channel in which flow is subcritical, the analysis follows the same pattern as that for a channel entrance. When the downstream slope is increased to a value less than s_c, the water enters the steeper channel at D_0 for that channel. If the slope in the steeper channel is increased to a value greater than s_c, water will enter the steeper channel at D_c.

Free Outfall

The reach of the channel illustrated in Fig. 8.10 has a grade less than the critical slope and terminates in a free outfall which discharges freely into the air. Critical depth D_c occurs at section b, a short-distance upstream from the rim a, where the depth is D_1. The depth gradually increases upstream from b, and between b and the rim there is a pronounced drop in water surface, or a *drop-down curve*. The fall possesses interesting hydraulic characteristics, which will be discussed briefly.

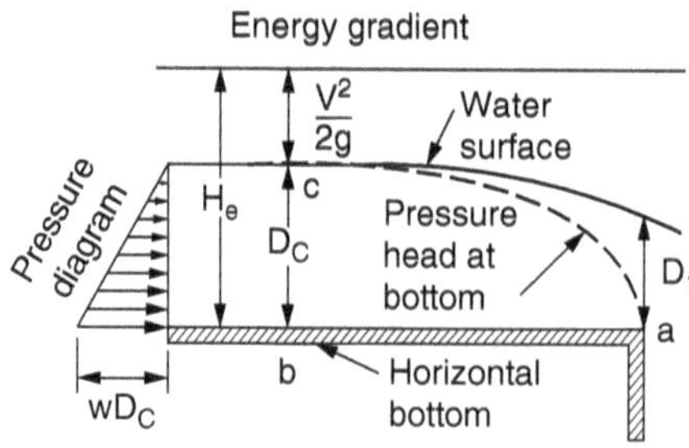

Figure 8.10 Pressures upstream from a free outfall.

In an experimental investigation by O'Brien[1] of a free outfall at the end of a channel of rectangular cross section having a horizontal bottom, the form assumed by the drop-down curve was determined, and hydrostatic pressures were measured within the nappe and on the bottom of the channel. Since contraction continues a short distance beyond the end of the channel, there is, theoretically, a slight hydrostatic pressure within the nappe, continuing as far as the vena contracta. Actually, however, as proved by the measurements, the pressures at all depths in section *a* are very nearly atmospheric. O'Brien found pressure heads on the bottom of the channel to vary from zero at the end section to the full depth of water approximately $3D_1$ (or $2D_c$) upstream from the rim, about as indicated by line *ac* in Fig. 8.10.

Considering the pressure head at section *a* to be zero, only the velocity head and the head of elevation are left. The head due to pressure and part of the elevation head at *b* have been converted into the velocity head at *a*. D_1 must therefore be less than D_c. The depth at *b*, where full hydrostatic pressure exists, cannot in theory be less than the critical depth since this is the depth of minimum energy, and any decrease below this depth would require the addition of energy from an outside source. Because of the instability of flow at this stage, however, depths less than critical may extend some distance back from the drop-down curve, and if waves form on the surface, there may be three or more sections where critical depth occurs. O'Brien found critical depth to be approximately $11.6D_c$ upstream from the rim of the fall. Any increase in the grade of the channel (between zero and critical slope) will cause a corresponding increase in the slope of the energy gradient, and thus it tends to move the position of critical depth farther back from the rim of the fall.

An estimate of the relation between D_c and D_1 can be made for a rectangular channel with a horizontal bottom by writing the Bernoulli equation from *b* to *a*. Because of the absence of hydrostatic pressure, the potential energy term is taken as $D_1/2$ at point *a*. Then, by making use of Eq. (8.25), the following relation is obtained:

$$D_1 = 0.655D_c \tag{8.60}$$

Assuming atmospheric pressure and equal velocities throughout section *a* (Fig. 8.10) from the momentum equation between sections *b* and *a*, after making proper substitutions and reductions, $D_1 = \frac{2}{3}D_c$ is obtained. From his experiments, O'Brien obtained the relation $D_1 = 0.643D_c$.

Hydraulic Jump

When conditions are such that the depth in an open channel must increase from a depth less than D_c to one greater than D_c, the water must pass through a hydraulic jump unless special provisions are made (see "Transition through Critical Depth without Jump"). The location of the jump must be such that the rate of change in momentum is equal to the sum of the forces in the direction of flow. The problem of the jump will be presented in its most general form, after which simplifying assumptions will be made. With reference to Fig. 8.11, the momentum equation is

$$P_2 - W\sin\theta - P_1 + F_{\tau b} + F_{\tau a} = \frac{Qw}{g}(\beta_1 V_1 - \beta_2 V_2) \tag{8.61}$$

where β is the momentum correction coefficient (see Sec. 3), W is the weight of the water in the jump, $F_{\tau b}$ and $F_{\tau a}$ are shear forces due to the channel walls and the air, respectively, and other symbols are identified in Fig. 8.11. The values of P_1 and P_2 are

$$P_1 = wa_1\bar{y}_1\cos\theta \tag{8.62}$$

$$P_2 = wa_2\bar{y}_2\cos\theta \tag{8.63}$$

The cos θ terms in these expressions are inserted on the basis of the assumption that the pressures throughout the fluid will be reduced to correspond with the component of the weight acting on the bottom.

Experimental observations indicate that the shear forces on the bottom and top must be of little importance compared with the other terms. Omitting the shear-force terms, assuming the β factors are

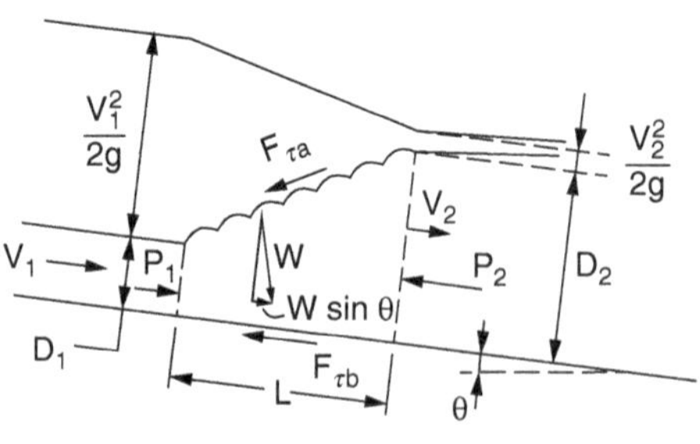

FIGURE 8.11 Hydraulic jump.

unity, and including values of P_1 and P_2 from Eqs. (8.62) and (8.63), Eq. (8.61) becomes

$$wa_2\bar{y}_2 \cos\theta - W \sin\theta - wa_1\bar{y}_1 \cos\theta = \frac{Qw}{g}(V_1 - V_2) \tag{8.64}$$

Hydraulic Jump for Small Slopes

When the slope is small, $\cos\theta \approx 1$, $\sin\theta \approx 0$, and Eq. (8.64) becomes

$$wa_2\bar{y}_2 - wa_1\bar{y}_1 = \frac{Qw}{g}(V_1 - V_2) \tag{8.65}$$

Equation (8.65) can be arranged in the form

$$\frac{QwV_1}{g} + wa_1\bar{y}_1 = \frac{QwV_2}{g} + wa_2\bar{y}_2 \tag{8.66}$$

Dividing through by w and substituting a_1V_1 for Q and a_1V_1/a_2 for V_2, there follows, after algebraic transformation,

$$V_1^2 = g\frac{a_2\bar{y}_2 - a_1\bar{y}_1}{a_1(1 - a_1/a_2)} \tag{8.67}$$

or, expressed in terms of discharge, Eq. (8.67) becomes

$$Q^2 = g\frac{a_2\bar{y}_2 - a_1\bar{y}_1}{1/a_1 - 1/a_2} \tag{8.68}$$

Within the limits of error introduced by ignoring all external forces except P_1 and P_2, a hydraulic jump must occur in conformity to the law expressed by the foregoing equations. Equations (8.66) to (8.68) differ from each other only in symbols and arrangement. They are the basis of all the more specialized hydraulic-jump formulas developed later.

Force Equation

An examination of Eq. (8.66) indicates that for a given discharge in any channel, if water flows at a given depth, there will always be another depth such that the sum of the force due to velocity plus the hydrostatic pressure at the respective cross sections will be the same. If the sum of these forces is designated by the symbol F_m, Eq. (8.66) can be expressed in the general form

$$\frac{F_m}{w} = \frac{Q}{g}V + a\bar{y} \tag{8.69a}$$

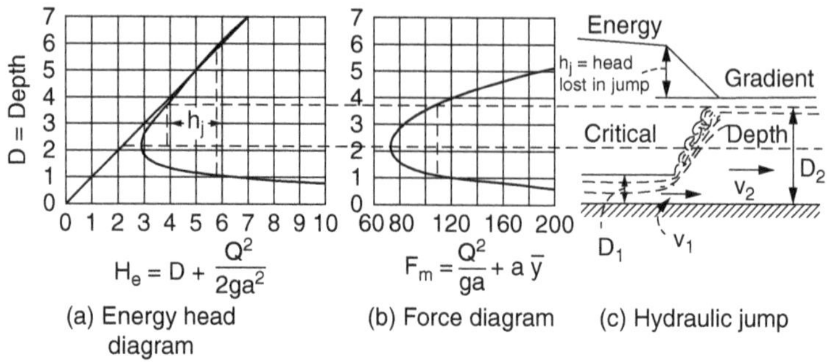

FIGURE 8.12 Conditions producing hydraulic jump.

where Q, V, a, and $\bar{y}$ are, respectively, discharge, mean velocity, area of cross section, and depth to the center of gravity of the cross section. Since $V = Q/a$, Eq. (8.69*a*) can also be written

$$\frac{F_m}{w} = \frac{Q^2}{ga} + a\bar{y} \tag{8.69b}$$

The curve in Fig. 8.12*b* is a graph of Eq. (8.69*a*) or (8.69*b*) for a trapezoidal channel when Q has a constant value. Figure 8.12*a* is a graph of the energy equation, Eq. (8.1) or (8.2), for the same section and discharge. The curves in Fig. 8.12*a* and *b* are similar in that they have conjugate depths for corresponding values of H_c and F_m/w, respectively; also, each curve has a minimum point. If a hydraulic jump occurs in this channel, it is from the lesser to the greater of the conjugate depths indicated in Fig. 8.12*b*.

Mathematical proof that the critical depth is the depth of minimum force (Fig. 8.12*b*) follows. Since a and $a\bar{y}$ are functions of D, Eq. (8.69*b*) can be written

$$\frac{F_m}{w} = \frac{1}{f(D)}\frac{Q^2}{g} + \phi(D) \tag{8.70}$$

An expression for minimum F_m is obtained by differentiating Eq. (8.70) with respect to D and equating to zero. Then

$$d\left(\frac{F_m}{w}\right) = \frac{f'(D)}{f^2(D)}\frac{Q^2}{g} + \phi'(D) = 0 \tag{8.71}$$

In Eq. (8.71), $f(D) = a$ and, as in Eq. (8.10), $f'(D) = TdD$. The value of the term $\phi' D$, which is the first derivative of the static moment $a\bar{y}$, can be determined by examining Fig. 8.4. If the depth D is increased by the increment ΔD, there follows:

$$\Delta(a\bar{y}) = (\bar{y} + \Delta D)a + T\frac{\overline{\Delta D}^2}{2} - a\bar{y} \tag{8.72}$$

or, in the limit, after dropping the infinitesimal of the second order and reducing,

$$d(a\bar{y}) = \phi'(D) = adD \tag{8.73}$$

Equation (8.72) can thus be reduced to

$$\frac{a^3}{T} = \frac{Q^2}{g} \tag{8.14}$$

which is also the criterion for critical depth for the energy curve in Fig. 8.12*a*.

Hydraulic-Jump Computations

The drawings in Fig. 8.12 are to scale and based on a trapezoidal channel having a bottom width of 10 ft, side slopes of 1½ horizontal to 1 vertical, and carrying a discharge of 200 ft^3/s. A jump from a depth of 1 ft is illustrated in Fig. 8.12*c*. The depth after the jump is approximately 3.6 ft, as shown by the curve in Fig. 8.12*b* at the point vertically above the 1-ft (0.3-m) depth. The head lost in the jump is about 2 ft, as indicated by h_j. The jump is always to a depth of lesser energy and always passes through the critical depth.

In problems involving a hydraulic jump, the variable quantities are the discharge Q and the depths before and after the jump, D_1 and D_2, respectively. If Q is the quantity sought, it is given by a direct solution of Eq. (8.68). If either D_1 or D_2 is required, it can be scaled from a force diagram similar to Fig. 8.12*b*. The use of this diagram is particularly advantageous when a number of different depths at the same discharge are required. Values of D_1 and D_2 can also be obtained by trial solutions of Eq. (8.68).

Hydraulic Jump in Trapezoidal Channels

For trapezoidal channels with slopes sufficiently small so that Eq. (8.66) will apply, it is possible to arrange the equation into a dimensionless form, which permits solution by means of a table. The following dimensionless ratios and equations are used:

$$z = \frac{e}{D} \tag{8.74}$$

$$x_1 = \frac{D_1}{H_1} \tag{8.75}$$

$$x_2 = \frac{D_2}{H_1} \tag{8.76}$$

Reference should be made to Figs. 8.6 and 7.2*a*. Then

$$a_{1=bD_1} + zD_1^2 = bx_1H_1\left(1+\frac{zH_1x_1}{b}\right) \tag{8.77}$$

$$a_2 = bD_2 + zD_2^2 = bx_2H_1\left(1+\frac{zH_1x_2}{b}\right) \tag{8.78}$$

The values of $\bar{y}$ become

$$\bar{y}_1 = x_1H_1\frac{\tfrac{1}{2}+\tfrac{1}{3}(zH_1/b)x_1}{1+(zH_1/b)x_1} \tag{8.79}$$

$$\bar{y}_2 = x_2H_1\frac{\tfrac{1}{2}+\tfrac{1}{3}(zH_1/b)x_2}{1+(zH_1/b)x_2} \tag{8.80}$$

Then, these equations along with Eq. (8.68) can be used to compute the depth after a hydraulic jump.

The use of these equations will be illustrated by application to the numerical example given in Fig. 8.12. If D_1 = 1.0 ft and z = 1.5, b = 10 ft and Q = 200 ft³/s, then

$$V_1 = \frac{Q}{a_1} = \frac{200}{11.5} = 17.4 \text{ ft/s}$$

and $$H_1 = D_1 + \frac{V_1^2}{2g} = 1 + 4.7 = 5.7 \text{ ft}$$

Then $$\frac{z_1H_1}{b} = \frac{1.5\times 5.7}{10} = 0.855$$

The value of x_1 is

$$x_1 = \frac{D_1}{H_1} = \frac{1}{5.7} = 0.175$$

and, using Eqs. (8.68), (8.79), and (8.80), D_2 can be found by successive approximations:

D_2	a_2	x_2	y_1	y_2	Q^2	Q
6	114	1.053	0.48	2.53	116,342	341.09
4	64	0.702	0.48	1.75	48,074	219.26
3	43.5	0.527	0.48	1.34	26,678	163.33
3.66	56.74	0.643	0.48	1.61	40,000	200.00

Hydraulic Jump in Rectangular Channels

For rectangular channels of small slope the jump equations can be greatly simplified. A derivation similar to the one for the general case, but using $q = Q/b$ in place of Q and noting that $P \equiv wD^2/2$, yields the equations

$$D_2 = \frac{-D_1}{2} + \sqrt{\frac{2V_1^2 D_1}{g} + \frac{D_1^2}{4}} \tag{8.81}$$

$$D_1 = \frac{-D_2}{2} + \sqrt{\frac{2V_2^2 D_2}{g} + \frac{D_2^2}{4}} \tag{8.82}$$

By dividing Eq. (8.81) by D_2 and introducing the Froude number ($F_1 = V_1/\sqrt{gD_1}$), Eq. (8.81) can be changed to the form

$$\frac{D_2}{D_1} = -\frac{1}{2} + \sqrt{2F_1^2 + \tfrac{1}{4}} \tag{8.83}$$

Similar substitutions in Eq. (8.82) yield

$$\frac{D_1}{D_2} = -\frac{1}{2} + \sqrt{2F_2^2 + \tfrac{1}{4}} \tag{8.84}$$

Hydraulic Jump in Sloping Channels

When the slope is steep, the simplifying assumptions made in Eqs. (8.65) to (8.84) do not apply, and Eq. (8.64) should be used. If Eq. (8.64) is divided by the specific weight of the fluid w, the weight term becomes $W \sin\theta/w$. Noting that W/w is the volume of the liquid in the jump, it can be expressed as

$$\frac{W \sin\theta}{w} = C\frac{a_1 + a_2}{2} L \sin\theta \tag{8.85}$$

where L is the length of the jump and C is a factor which takes into consideration the fact that the water surface may not be a plane surface. Then Eq. (8.64) becomes

$$a_2\bar{y}_2 \cos\theta - C\frac{a_1 + a_2}{2} L \sin\theta - a_1\bar{y}_1 \cos\theta = \frac{Q}{g}(V_1 - V_1) \qquad (8.86)$$

Letting $V_1 = Q/a_1$ and $V_2 = Q/a_2$, Eq. (8.86) can be changed to

$$Q^2 = \frac{g\left[\left(a_2\bar{y}_2 - a_1\bar{y}_1\right)\cos\theta - C\frac{a_1 + a_2}{2} L \sin\theta\right]}{1/a_1 - 1/a_2} \qquad (8.87)$$

or, for rectangular channels,

$$Q^2 = \frac{g\left[\frac{1}{2}\left(D_2^2 - D_1^2\right)\cos\theta - \frac{1}{2}C\left(D_1 + D_2\right) L \sin\theta\right]}{1/D_1 - 1/D_2} \qquad (8.88)$$

The U.S. Bureau of Reclamation[2] has published information on the depth after the jump in sloping channels ($D_{2\theta}$) based on their own tests together with those of other investigators.[3,4] They found that the downstream depth $D_{2\theta}$ exceeded that computed for channels of small slope D_2, as obtained from Eq. (8.81) or (8.83), by an amount that varied with the slope of the bottom, tan θ. This curve is reproduced in Fig. 8.13. In order to use this curve, D_2 would be computed from Eq. (8.81) or (8.83) and then the downstream depth for a particular value of tan θ would be computed by using the value of $D_{2\theta}/D_2$ from Fig. 8.13. This method applies only for cases in which the entire jump

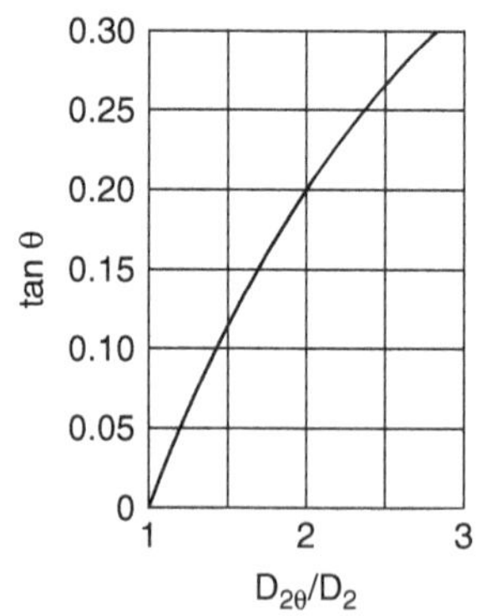

FIGURE 8.13 Depth after jump in sloping channels.

occurs on one bottom slope. It is probable that Fig. 8.13 would also serve as a rough approximation for channels other than rectangular.

Length of Jump

The U.S. Bureau of Reclamation[2] has published curves showing the length of the jump in terms of the depth after the jump $L/D_{2\theta}$ as a function of F_1 for various values of bottom slope tan θ. The maximum length of the jump occurs when F_1 is approximately 6.0. Maximum lengths for each value of tan θ are given in the following tabulation:

tan θ	$L/D_{2\theta}$
0	6.15
0.05	5.20
0.10	4.40
0.15	3.85
0.20	3.40
0.25	3.00

For practical purposes these values may be used for $F_1 \geq 4$. The length decreases when $F < 4$. For example, when $F_1 = 3$, values of $L/D_{2\theta}$ are approximately 10 percent smaller than the tabulated values.

Position of Jump

Hydraulic jumps can occur only when water flowing below the critical stage enters a channel in which flow is normally above the critical stage and where all the requirements expressed by the force equation [Eq. (8.66)] and illustrated in Fig. 8.12 can be fulfilled. In passing through the jump, the flow may change abruptly from nonuniform to uniform or from uniform to nonuniform, or it may be nonuniform both before and after the jump. The method of determining the position of the jump for each of these three conditions will now be discussed.

A case where nonuniform flow becomes uniform after the jump is illustrated in Fig. 8.14. A given discharge Q enters a canal of uniform cross section through gate G at less than the critical stage, the grade of the canal being less than the critical slope, and the uniform flow depth D_2 being less than the upper conjugate depth corresponding to that at which water enters the canal. As the canal grade is not sufficient to compensate for loss of head due to friction, the water increases in depth in an M_3 curve as it passes downstream from G, and its surface slopes upward. A jump will occur at the section where the depth D_1 becomes the lower conjugate depth

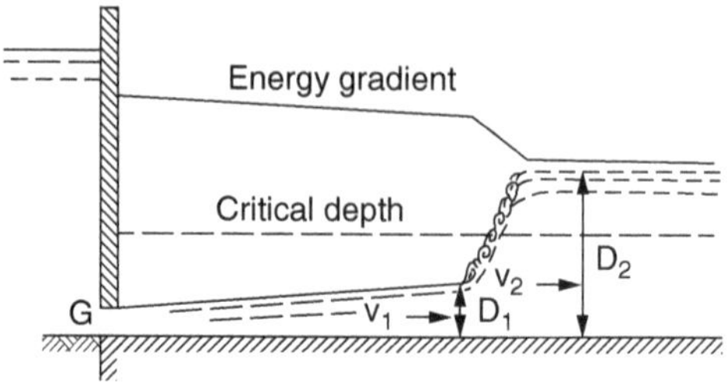

FIGURE 8.14 Hydraulic jump downstream from gate.

corresponding to D_2. Since D_2 and Q are known, D_1 can be computed from Eq. (8.82) for rectangular channels and from Eq. (8.68) for other channels. With the velocity and depth at G known, the distance downstream from the gate to the section of depth D_1, where the jump occurs, can be obtained from Eq. (8.105*a*). If the velocity change is not too great, computations may be made for the single reach; otherwise the distance should be divided into two or more reaches. For the case where the uniform flow depth D_2 is greater than the upper conjugate depth, the jump will drown out the jet and move up to the gate. The solution of the problem under such conditions is described under gates in Sec. 4. Another example of change from nonuniform to uniform flow is that of a jump on the apron of an overflow dam when the apron is approximately parallel to the surface of the tail water (see Fig. 8.25).

Uniform flow changes abruptly in a hydraulic jump to nonuniform flow when a channel having a grade steeper than critical slope encounters backwater (see Fig. 8.20). In this case, by a method corresponding to that described earlier for computing D_1, if D_2 and Q are known, D_2 can be computed from Eq. (8.81) for rectangular channels and from Eq. (8.68) for other channels. The distance upstream from the dam to the section of depth D_2, where the jump occurs, can be obtained from Eq. (8.105*a*).

An example of nonuniform flow occurring both before and after a jump is afforded by the overflow dam illustrated in Fig. 8.15. The slope of the apron is insufficient to provide for loss of head due to friction before the jump, but it is more than sufficient after the jump. In both cases, therefore, the depths increase with the distance downstream and flow is nonuniform. To determine the position of the jump, the discharge Q must be known, and data must be available for determining the profile of water surface in the higher stage

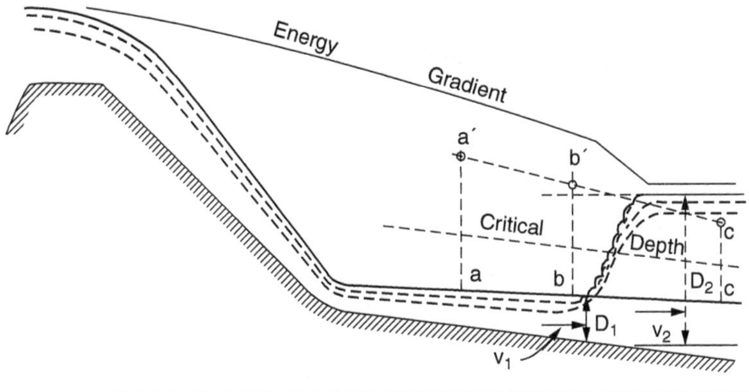

Figure 8.15 Hydraulic jump on sloping apron of dam.

projected some distance upstream from the place where the jump will occur. The water surface in the lower stage must be computed downstream from a known depth, as from the toe of the dam, to a place beyond the position of the jump. Trial values of D_2 should be computed for three sections *a*, *b*, and *c*, which will locate three points *a*′, *b*′, and *c*′ to which the water must jump at the respective sections. One of the points should be on the opposite side of the jump from the other two, and, preferably, the middle point should be near the jump. The intersection of the line *a*′*b*′*c*′ with the higher water-surface profile gives the position of the jump. See Fig. 8.23*d* for another example of nonuniform flow occurring both before and after the jump.

When several determinations of depth after jump in a channel are required for the same discharge, as in the preceding case, it may be less laborious to obtain the values from a graph of the momentum-pressure diagram in Fig. 8.12 than from a solution of Eq. (8.68).

Minor Losses

The losses caused by rapid local changes in magnitude or direction of velocity are called *minor* losses. (For minor losses in pipes, see Sec. 6.) Such losses would occur at bends, contractions, enlargements, or obstructions in channels.

Channel Bends

When a fluid flows around a bend, the centrifugal force tends to develop a water surface which is higher at the outside of the channel. If the velocity in the channel were everywhere equal to the average velocity *V*, the amount that the water surface would rise at the outside wall and the amount that it would fall at the inner wall would be given approximately for *subcritical flow* by

$$D = \frac{V^2 b}{2 g r_c} \tag{8.89}$$

where V is the average velocity, b the width of the channel, and r_c the radius of curvature of the centerline. This equation is derived by noting that the water surface will be perpendicular to the resultant of the radial and gravitational forces on a particle of fluid. However, because the radial force is proportional to the square of the velocity, this force will be greatest on the high-velocity water near the center of the channel, thus developing crosscurrents, eddies, and spiral motion. Also, there may be a tendency toward separation along the inner wall. Furthermore, for supercritical flow, a standing-wave pattern complicates the flow pattern. Supercritical flow in bends is discussed in Sec. 9.

Information on losses at bends in rectangular channels has been presented by various investigators. Yen and Howe[5] reported that K_b in the following expression was 0.38 for a 90° bend having a radius of curvature of 5 ft (1.5 m) and a width of 11 in. (28 cm),

$$h_b = K_b \frac{V^2}{2g} \tag{8.90}$$

Shukry[6] reported test results in which the variables were the angle θ through which the water was turned, the ratio of radius of curvature to width r_c/b, the ratio of depth to width D/b, and the Reynolds number. He used the Reynolds number in the following form:

$$R = \frac{rV}{\nu} \tag{8.91}$$

where r is the hydraulic radius. These tests indicate that K_b is affected very little by D/b except when r_c/b is very small. When $\theta \leq 45°$, the loss is negligible. The loss increases as θ is increased from 45° to 90°, and for values of θ ranging from 90° to 180° the loss is about constant. When $r_c/b \geq 3.0$, losses were found to be negligible. (It should be noted that this does not agree with the results of Yen and Howe reported previously.) For values of $r_c/b < 3$, values of K_b are given in the following tabulation for $R = 31{,}500$. Also shown are experimental values of K_b for various values of R, with $r_c/b = 1.0$.

Bend loss coefficients* ***R* = 31,500**		***r*****$_c$/*****b***** = 1.0**	
r_c/b	K_b	R	K_b
2.5	0.02	1×10^4	0.59
2.0	0.07	3×10^4	0.27
1.5	0.12	5×10^4	0.25
1.0	0.25	7×10^4	0.35

*From Shukry.[6]

Tests on large canals[7] showed that losses due to bends could be estimated from the following equation in which $(\Sigma\Delta°)$ is the summation of deflection angles in the reach,

$$h_b = 0.001\,(\Sigma\Delta^0)\frac{V^2}{2g}$$

In sinuous natural rivers, the bend losses are included in the friction losses.

Contractions and Enlargements

The energy losses for contractions have been expressed by Hinds[8] in terms of the difference in kinetic energy at the two ends,

$$h_c = K_c\left(\frac{V_2^2}{2g} - \frac{V_1^2}{2g}\right) \tag{8.92}$$

and for enlargements,

$$he = K_e\left(\frac{V_1^2}{2g} - \frac{V_2^2}{2g}\right) \tag{8.93}$$

Values of K_e and K_e are given in the following table:

Form of transition	K_c	K_e
Sudden change in area, sharp corners	0.5	1.0
"Well designed":		
Best	0.05	0.10
Design value	0.10	0.20

Additional information on entrance losses is given in Sec. 4. A "well-designed" transition is one in which all plane surfaces are connected by tangent curves and a straight line connecting flow lines at the two ends does not make an angle greater than 12½° with the axis of the channel.

Contracting and enlarging sections are used at channel entrances or to form *transitions* between channels of different size. Hinds has summarized the art of designing transitions for subcritical flow as practiced by the U.S. Bureau of Reclamation as follows (transitions with supercritical flow are discussed in Sec. 9):

1. Sufficient fall must be allowed at all inlet structures to accelerate the flow and to overcome frictional and entrance losses.
2. The theoretical recovery at an outlet structure is reduced by frictional and outlet losses.
3. At open-channel outlets a small factor of safety may be obtained by setting the transition for less than its maximum recovering capacity, but erosion below the structure may be slightly increased.
4. At siphon outlets a small factor of safety may be obtained and erosion avoided by setting the transition for more than its assumed recovering capacity.
5. Simple designs may be prepared by adapting the details of previous designs known to be satisfactory, if proper allowance is made for loss of head.
6. Important structures, where velocities are high, must be carefully designed to conform to a smooth theoretical water surface. Sharp angles must be avoided.
7. Horizontal curvature in the conduit before an outlet appears to reduce its efficiency and to produce objectionable cutting velocities in the canal beyond.
8. K_c [Eq. (8.92)] for a well-designed inlet is likely to be less than 0.05. A value of 0.1 is safe for use in design.
9. K_e [Eq. (8.93)] for a well-designed outlet is likely to be less than 0.2, unless the conduit before the structure is curved. A value of 0.2 is safe for use in design.
10. No definite data as to the best form of water-surface profile, best form of structure, or most efficient length of transition are available.
11. Special care is required where critical depth is approached or where hydraulic jump is involved.

12. The disturbances often observed in long, uncontrolled siphons, at part capacity, are not caused by entrained air but by the hydraulic jump in the pipe.

Losses at Obstructions

Water passing through a constriction in an open channel at *subcritical velocity* decreases in depth, as shown in Fig. 8.3. The depth downstream from the constriction must be the uniform flow depth or normal depth for this discharge because no other water-surface profile can exist (see "Equations of Gradually Varied Flow and Generalized Profiles"). The Bernoulli equation, written from a point just upstream from the obstruction to a point just downstream, is

$$z_u + D_u + \alpha_u \frac{V_u^2}{2g} = z_d + D_d + \alpha_d \frac{V_d^2}{2g} + h \qquad (8.94)$$

The symbols used in Eq. (8.94) are defined in Fig. 8.16. The amount of backwater caused by the obstruction D can then be obtained from Eq. (8.94), using Fig. 8.16 as a reference,

$$\Delta D = (z_u + D_u) - (z_d + D_d) = \alpha_d \frac{V_d^2}{2g} - \alpha_u \frac{V_u^2}{2g} + h_l \qquad (8.95)$$

The losses at obstructions in open channels consist of the loss due to a constriction and an enlargement and, if the obstruction has considerable length in the direction of flow, of a friction loss. Usually the principal loss is that due to the enlargement at the downstream end are decreased than when flow is speeded up. This is illustrated by the coefficients for losses in the previous subsection, the coefficients for enlargements being twice those for contractions under similar conditions. Energy losses at piers can be reduced to a minimum by

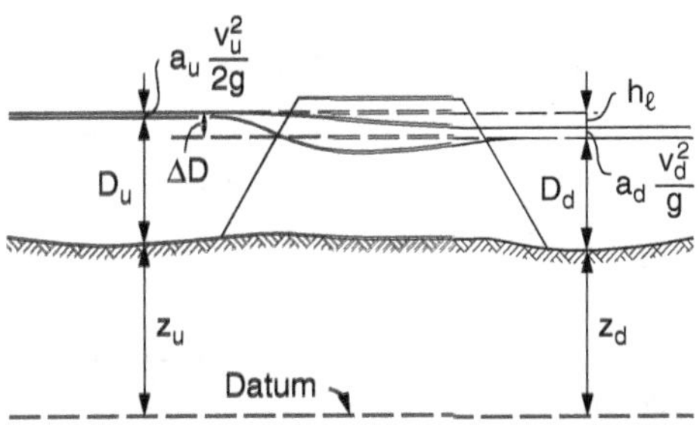

FIGURE 8.16 Flow past obstructions.

rounding the upstream corners and tapering, or "streamlining," the downstream end. The losses could be estimated by treating them as combinations of constriction and enlargements and using the coefficients given in the previous subsection.

Flow through bridge openings has been investigated by means of model studies.[9] The results are presented in a series of curves which are useful in designing bridge openings. The procedure for expressing the losses is based on the equation

$$h_l = K_b \frac{V_n^2}{2g} \tag{8.96}$$

where h_l is the total loss, K_b the loss coefficient, and V_n the average velocity that would occur in the bridge opening if the entire discharge were to pass through the bridge opening at the normal depth in the river for this discharge. Values of K_b are related to the bridge-opening ratio M. The value of M is obtained by dividing the portion of discharge that would normally flow through the bridge opening if no piers were present by the total discharge. Figure 8.17 shows two curves relating K_b to M. One curve applies to abutments with vertical walls and 90° corners, as well as to abutments with sloped embankments on the upstream and downstream sides held in place at the ends by wing walls making an angle of 90° with the piers, as illustrated in Fig. 8.17*b*. For wing walls having angles other than 90°, as shown in Fig. 8.17*c*, the values of K_b are smaller than those shown in the graph, the reduction being, on the average, about 12 percent for a wing-wall angle of 30° and approximately 30 percent for angles of 45° and 60°.

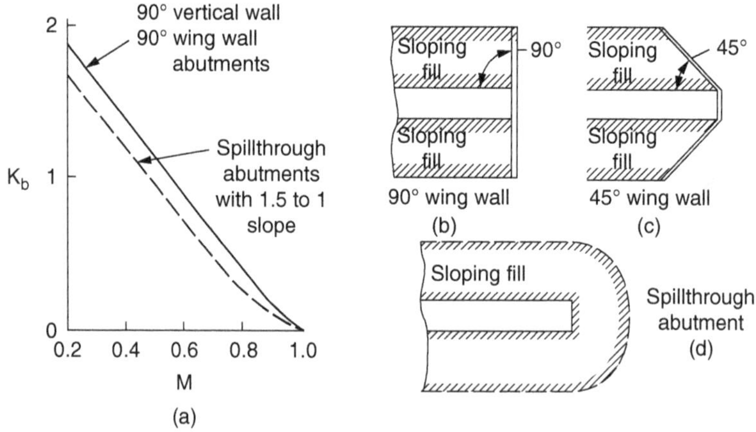

FIGURE 8.17 Losses at bridge piers.

The second curve applies to piers, referred to as spill-through abutments, which have the sloped embankment extending around the ends of the piers, as illustrated in Fig. 8.17*d*. The curve shown is for an embankment slope of 1.5:1, horizontal to vertical. Values of K_b for an embankment slope of 2:1 are 5 to 10 percent larger than those shown, and for a 1:1 slope, the values are 4 to 9 percent lower than those shown by the curve.

A U.S. Bureau of Public Roads publication[9] also provided coefficients ΔK to be added to K_b to take care of minor effects on the losses at the bridge opening. One such coefficient takes care of the increase in loss which occurs when the bridge opening is not in the center of the river. Another includes the additional losses caused by obstructions in the opening. A third one introduces the effect of having the bridge cross the river at an angle differing from 90°.

Transition through Critical Depth without Jump

If water flowing at less than critical depth enters a channel having less than critical slope, change to a higher stage will normally occur in a jump (see "Hydraulic Jump") unless special means are provided for making velocity changes gradually. A transition designed to prevent a jump, for the specific data indicated, is illustrated in Fig. 8.18. The raised bottom has a smooth surface, the elevation at the crest *C* being such that the minimum energy gradient is tangent to the energy gradient of the stream. For this condition a jump is impossible. A similar design could be prepared for channels having other sectional forms.

In Fig. 8.18 lower-stage flow is indicated up to section *C*, where critical depth occurs and then follows higher-stage flow. On both sides

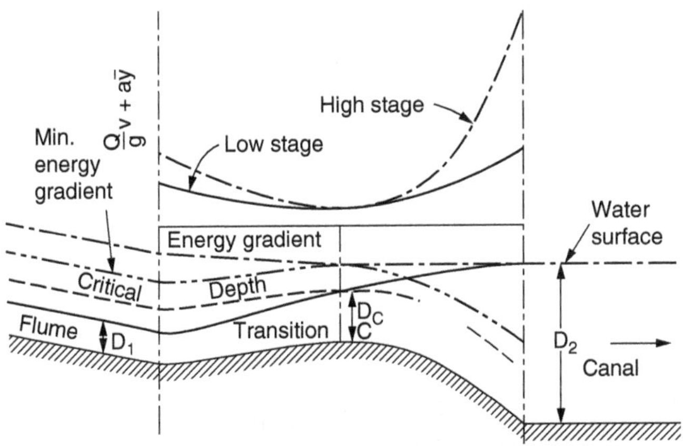

FIGURE 8.18 Transition through critical depth to higher stage without jump.

of C the other stages which could be computed are not shown. The force curves $\left(QV/g + a\bar{y}\right)$ [see Eq. (8.69a)] for the two stages of flow are tangent to each other. If the crest C is lower than that indicated in the figure, the curves will intersect to the right of C at the section where a jump will occur. If the crest is higher than that indicated, backwater will be produced, and there will be a jump to the left of C. The water-surface profile in the transition can be computed in short reaches by the method described under "Constant-Discharge Relations." Energy losses may be estimated using Eqs. (8.92) and (8.93).

Gradually Varied Flow

This phase of nonuniform flow deals with the case where the area of the stream cross section changes so slowly that the energy losses can be computed for various reaches in the same manner as for uniform flow.

Equations of Gradually Varied Flow and Generalized Profiles

Before discussing the methods of computing water-surface profiles for specific conditions, the differential equation giving the rate of change of depth with respect to distance along the channel will be derived. This equation is useful in developing an understanding of the various types of profiles that may occur. With reference to Fig. 8.19, the total energy of fluid is

$$h = z + D + \frac{V^2}{2g} \tag{8.97}$$

If this equation is differentiated with respect to x, the distance along the channel, the following expression is obtained:

$$\frac{dH}{dx} = \frac{dz}{dx} + \frac{dD}{dx} + \frac{d(V^2/2g)}{dx} \tag{8.98}$$

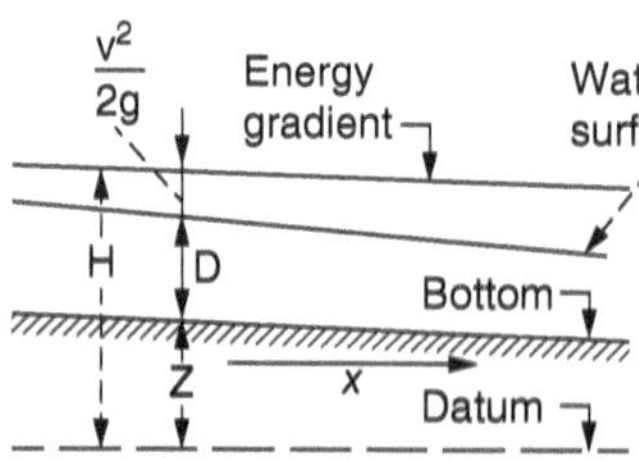

FIGURE 8.19 Energy relationships for open-channel flow.

For the particular case of a rectangular channel the differentiation of the last term may be carried out as follows:

$$\frac{d(V^2/2g)}{dx} = \frac{d(q^2 2gD^2)}{dx} = \frac{-2q^2}{2gD^3}\frac{dD}{dx} = \frac{-V^2}{gD}\frac{dD}{dx}$$

Also, by designating the slope of the bottom tan θ as $-s_0$, θ being the angle between the bottom and the horizontal,

$$\frac{dz}{dx} = \sin\theta = -s_0\cos\theta$$

Then, if the energy loss per unit of length dh/dx is designated as $-s$ and if cos θ is taken as unity, Eq. (8.98) can be written

$$-s = -s_0 + \frac{dD}{dx} - \frac{V^2}{gD}\frac{dD}{dx} \tag{8.99}$$

In Eq. (8.99), s is the slope of the energy gradient and s_0 is the slope of the bottom. The value of s is always positive, and s_0 is positive when the channel slopes down in the direction of flow and negative for the opposite condition. Solving for dD/dx from Eq. (8.99),

$$\frac{dD}{dx} = \frac{s_0 - s}{1 - V^2/gD} \tag{8.100}$$

Equation (8.100) gives the rate of change of depth along a rectangular channel under all conditions. Numerical values may be obtained by solving for s from the Manning formula. When dD/dx is positive, the depth is increasing in the downstream direction, and when dD/dx is negative, the depth is decreasing. When flow is uniform, $s = s_0$ and dD/dx becomes zero. A careful study of Eq. (8.100) will show that there are 12 possible regimes of flow, depending on the relation of the depth to the uniform flow depth D_0 and to the critical depth D_c. For each regime only one type of water-surface profile is possible. These are summarized in Fig. 8.20. The scheme of identifying these curves is based on that used by Bakhmeteff.[10] These curves provide a check on profile computations, because in any regime the computed profile must be concave up or down and the depth must increase or decrease, as shown by the curves.

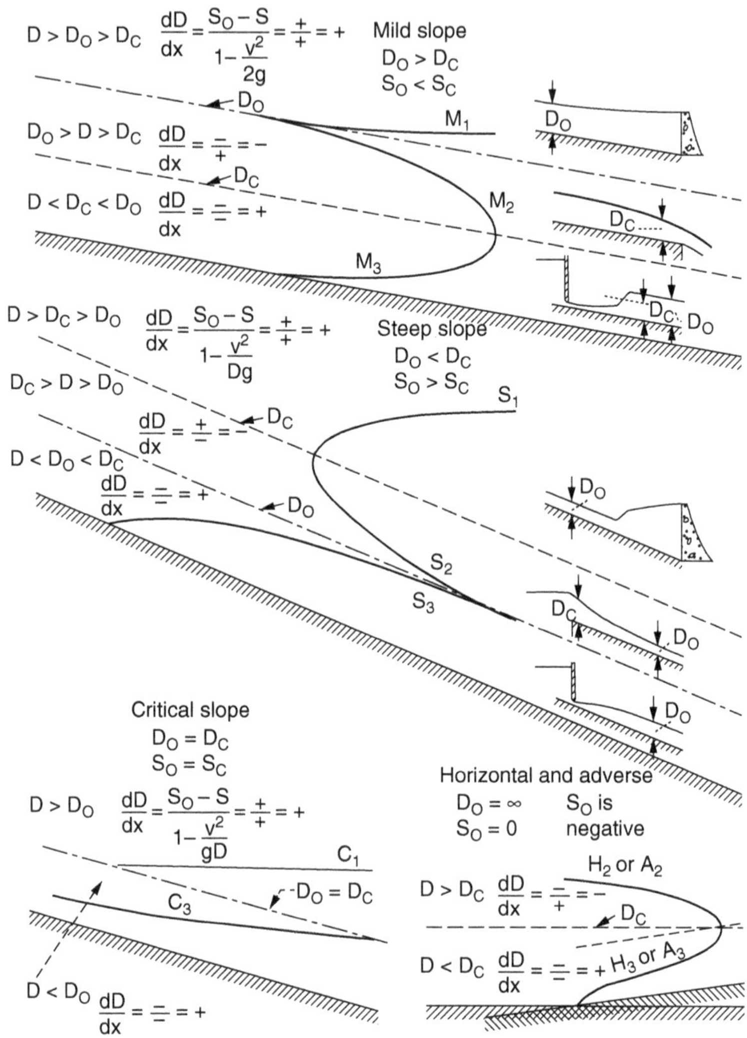

FIGURE 8.20 Water-surface profiles for gradually varied flow in open channels.

If a derivation similar to that leading to Eq. (8.100) is carried out for the general case, the following equation is obtained:

$$\frac{dD}{dx} = \frac{s_0 - s}{1 - Q^2T/ga^3} \tag{8.101}$$

In Eq. (8.101) T is the top width, a is the area, and the other quantities are defined as for Eq. (8.100). Equation (8.101) may be used to determine the rate of change of depth with respect to distance along the bottom for channels of any cross section.

If the width of the channel varies at a known rate dT/dx, a similar derivation will yield Eq. (8.102) for rectangular channels and Eq. (8.103) for trapezoidal channels,

$$\frac{dD}{dx}=\frac{s_0-s+\dfrac{V^2}{gT}\dfrac{dT}{dx}}{1-\dfrac{V^2}{gD}} \tag{8.102}$$

$$\frac{dD}{dx}=\frac{s_0-s+\dfrac{Q^2D}{ga^3}\dfrac{dT}{dx}}{1-\dfrac{Q^2T}{ga^3}} \tag{8.103}$$

As for the case of dD/dx, dT/dx is positive when the width of the channel is increasing in the downstream direction. Equations (8.102) and (8.103) will be found useful in determining whether the depth is increasing or decreasing in channels of varying width.

Methods of Computing Water-Surface Profiles

Retarded and accelerated flows are illustrated in Fig. 8.21. The same analysis applies to both. It is customary to divide the channel into reaches and proceed consecutively with computations for adjoining reaches, either upstream or downstream. The length of reach between sections 1 and 2 is Δl, and the slope of the bottom of the channel is s_0. The loss of head in the reach is H_l, the drop in the energy gradient, and the average loss of head per unit length is $H_l/\Delta l = s_{av}$.

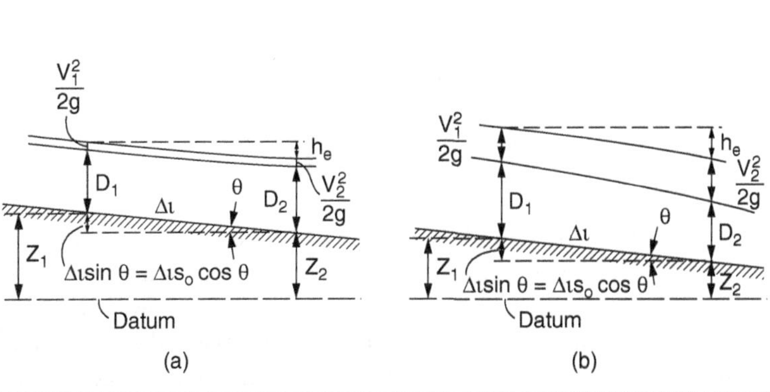

Figure 8.21 Retarded and accelerated flow.

Velocities at the upstream and downstream ends of the reach are V_1 and V_2, and the corresponding depths are D_1 and D_2, respectively. If the datum is the bottom of the channel at the downstream section, from Bernoulli's theorem,

$$\frac{V_1^2}{2g} + D_1 + S_0\,\Delta l \operatorname{Cos}\theta = \frac{V_2^2}{2g} + D_2 + h_l \tag{8.104}$$

or writing $h_l = s_{av}\,\Delta l$, letting cos θ be unity, and transposing,

$$\Delta l = \frac{V_2^2/2g + D_2 - V_1^2/2g - D_1}{s_0 - s_{av}} \tag{8.105a}$$

From Eq. (8.105*a*) it appears that the length of reach is equal to the difference in energy heads at the two sections divided by the difference in slope between the channel bottom and the energy gradient. Expressed in symbols,

$$\Delta l = \frac{H_2 - H_1}{s_0 - s_{av}} \tag{8.105b}$$

The numerator and denominator of this fraction will always be of the same sign.

The value of s_{av} in Eq. (8.105*a*) can be determined by the Manning formula for the average velocity in the reach or for the average depth. Either is an approximation, but the error decreases with the difference in velocities at the two end sections and can be kept within any desired limits by selecting proper lengths of reaches. No general rule can be applied to all conditions, but ordinarily, velocity changes should not exceed 10 to 20 percent. The Manning formula in the general form for computing *s* is

$$s = \frac{n^2V^2}{2.2082r^{4/3}} \quad \text{(English units)}$$

$$s = \frac{n^2V^2}{r^{4/3}} \quad \text{(metric units)} \tag{8.106}$$

where *n* is the coefficient of roughness and *r* the hydraulic radius.

Equation (8.105*a*) provides for the direct determination of the water-surface profile in all cases of nonuniform flow where the channel has a constant cross section. In such cases, if s_0, *n*, *Q*, and the depth at either end of the reach are known, the distance *l* to any other assumed depth upstream or downstream for either accelerated or retarded flow can be computed. For rectangular or trapezoidal channels Eq. (8.105) should be used for determining s, assuming $D = (D_1 + D_2)/2$. If the channel does not have a constant cross section and in all cases where the length of reach is specified, trial solutions of Eq. (8.104) or (8.105) will be necessary.

Example 8.1 The depth at the upstream side of a gate in a trapezoidal concrete channel is 8 m, as shown in Fig. 8.22. The channel has a bottom width of 8 m and side slopes of 1:1. The bottom slope is 0.03. Flow in the channel is uniform upstream from the influence of the gate. The discharge is 100 m³/s. Determine the water-surface profile upstream from the gate.

The uniform flow depth is first computed, using the Manning equation in the following form, assuming n is 0.011

$$Q = \frac{K' b^{8/3} s^{1/2}}{n}$$

or

$$K' = \frac{Qn}{b^{8/3} s^{1/2}} = \frac{100 \times 0.011}{8^{8/3} \times 0.03^{1/2}} = 0.025$$

For a trapezoidal channel,

$$K' = \frac{\left(z + \dfrac{1}{x}\right)^{5/3} \times (x)^{8/3}}{\left[\dfrac{1}{x} + 2\left(z^2 + 1\right)^{1/2}\right]^{2/3}}$$

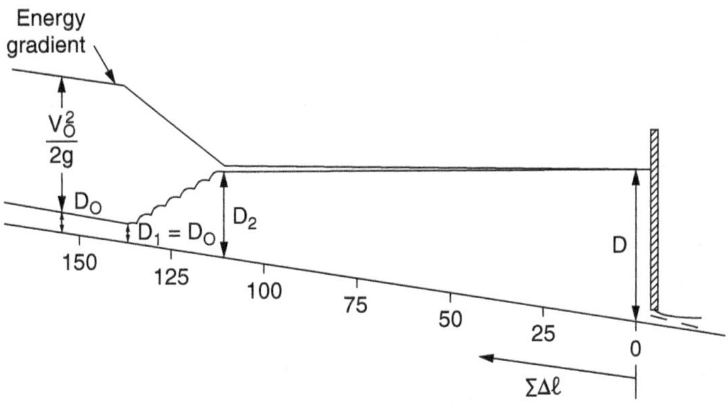

FIGURE 8.22 Gradually varied flow profile, Example 8.1.

where z = the side slope of the channel (H/V) and $x = D/b$. The value of D_0 can then be found by successive approximations.

Example 8.1, Normal Depth Calculation

D_m	b_m	x	z	K'
2	8	0.250	1	0.101
1	8	0.125	1	0.031
0.5	8	0.0625	1	0.010
0.88	8	0.110	1	0.025
$D_0 = 8 \times 0.11 = 0.88$ m				

The value of critical depth is obtained by writing Eq. (8.16) in the form

$$Q = K_c' b^{5/2}$$

from which

$$K_c' = \frac{Q}{b^{5/2}} = \frac{100}{8^{5/2}} = 0.55$$

For a trapezoidal channel

$$K_c' = \frac{(1/y + z)^{3/2}}{(1/y + 2z)^{1/2}} g^{1/2} b^{5/3}$$

where $y = D_c/b$. The value of D_c can then be found by successive approximations.

Example 8.1, Critical Depth Calculation

D_m	B_m	y	z	K_c'
3.0	8	0.375	1	0.88
2.5	8	0.313	1	0.65
2.0	8	0.250	1	0.45
2.28	8	0.285	1	0.55
$D_0 = 8 \times 0.285 = 2.28$ m				

Because $D_c > D_0$, the bottom slope is greater than the critical slope, and uniform flow must change abruptly through a hydraulic jump to an s_1 curve, as illustrated in Fig. 8.20.

The depth after the jump may be determined by using Eqs. (8.68), (8.79), and (8.80).

The specific energy before the jump H_1 is obtained as follows:

$$V_0 = \frac{Q}{a} = \frac{100}{8 \times 0.88 + 0.88^2} = 12.8 \text{ m/s}$$

and

$$H_1 = D_0 + \frac{V_0^2}{2g} = 0.9 + 8.4 = 9.3 \text{ m}$$

Then

$$\frac{zH_1}{b} = \frac{1 \times 9.3}{8} = 1.16$$

and $D_1/H_1 = 0.88/9.3 = 0.095$.

D_2 is solved by successive approximations:

Example 8.1, Hydraulic Jump Calculation

D_2	a_2	x_2	Y_1	y_2	Q^2	Q
8	128	0.867	0.18	3.33	34,562	185.91
6	84	0.650	0.18	2.57	17,975	134.07
5	65	0.542	0.18	2.18	12,054	109.79
4	48	0.434	0.18	1.78	7,509	86.65
4.58	57.7	0.497	0.18	2.01	10,000	100.00
$D_2 = 4.58$ m						

Then gradually varied flow profile may be computed starting from either end. In this case the computations are started at the gate. The arrangement of the computations illustrated in the following table helps to eliminate errors. The water-surface profile is determined by computing the lengths of reaches between depths varying in increments of 1 m until the final reach, where the change is 0.5 m.

A consideration of the terms in Eqs. (8.105*a*) and (8.105*b*),

$$\Delta l = \frac{(D_2 + V_2^2/2g) - (D_1 + V_1^2/2g)}{S_0 - S_{av}} = \frac{H_2 - H_1}{s_0 - s_{av}}$$

along with the column headings in the following table, shows the order of computations for successive values of Δl. The value of D and other values applying to the ends of reaches are aligned horizontally, whereas values of D_{av} and other quantities applying to an entire reach are placed on a horizontal line between the lines that apply to the ends of that reach. Values of s_{av} are computed from the Manning equation in the form of Eq. (8.107),[a]

$$s_{av} = \left(\frac{Qn}{b^{8/3}}\right)^2 \left(\frac{1}{K'}\right)^2 \tag{8.107}$$

and values of K' are obtained from Eq. (8.108):

$$K' = \frac{(z + 1/x)^{5/3}}{\left[1/x + 2(x^2+1)^{1/2}\right]^{2/3}} x^{8/3} \tag{8.108}$$

where z is the side slope of the channel and x is the ratio D/b. All terms on the right side of Eq. (8.107) remain the same for each reach except $(1/K')^2$. Therefore, the equation can be arranged as follows:

$$\begin{aligned} s_{av} &= \left(\frac{Qn}{b^{8/3}}\right)^2 \left(\frac{1}{K'}\right)^2 \\ &= \left(\frac{100 \times 0.011}{8^{8/3}}\right)^2 \left(\frac{1}{K'}\right)^2 \\ &= 0.0000185 \left(\frac{1}{K'}\right)^2 \end{aligned}$$

Example 8.1, Water Surface Profile Calculation

D, m	a, m²	V, m/s	$\frac{V^2}{2g}$, m	H, m	$H_2 - H_1$, m	D_{av}, m	$\frac{D_{av}}{b}$	$\left(\frac{1}{K'}\right)^2$	s_{av}	$s_0 - s_{av}$	Δl, m	$\Sigma \Delta l$, m
8	128	0.78	0.03	8.03								
					0.98	7.5	0.94	0.761	0.000014	0.0300	32.7	
7	105	0.95	0.05	7.05								32.7
					0.98	6.5	0.81	1.37	0.000025	0.0300	32.7	
6	84	1.19	0.07	6.07								65.4
					0.95	5.5	0.69	2.54	0.000047	0.0300	31.7	
5	65	1.54	0.12	5.12								97.1
					0.46	4.8	0.60	4.30	0.000080	0.0299	15.4	
4.5	56	1.79	0.16	4.66								112.5

The computed profile as represented by values of D and Δl in the table is plotted in Fig. 8.22. ▲

[a]See the derivation of this equation under "The Manning Equation," Sec. 7.

Short Channels

Four examples of nonuniform flow in short channels receiving water from a reservoir are illustrated in Fig. 8.23. The first two channels have grades less and the last two have grades greater than the critical slope. Channel *a* terminates in a free outfall and thus has free discharge at the outlet. Channels *b* to *d* discharge into another reservoir, and the depth of submergence at each outlet is D_s.

Critical depth D_c occurs slightly upstream from the outlet of channel *a*, and the relation of the depth at the intake *D* to the discharge is expressed by Eq. (8.58). To determine the discharge of the channel, assume the discharge and compute the corresponding depths of water D_c and *D*. Then, applying the principles of nonuniform flow described in the preceding pages and using Eq. (8.105), determine the length of

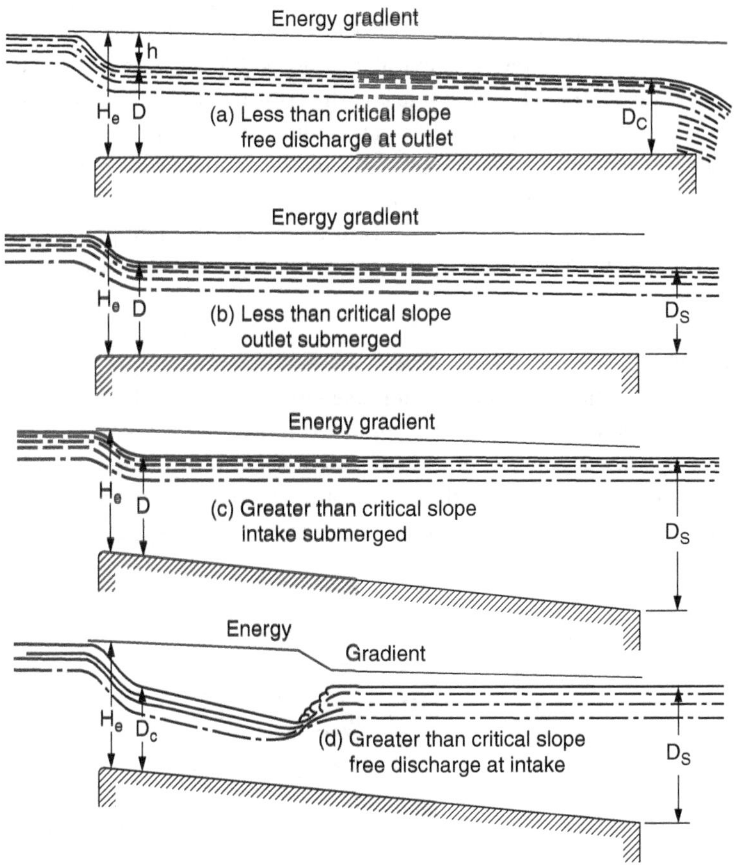

FIGURE 8.23 Short channels.

channel corresponding to these two depths. Continue trial solutions for other assumed discharges. If the assumed discharge is plotted against the computed length, the intersection of the resulting curve with the given length of channel will be the discharge sought.

If the submergence D_s of channel *b* is less than the critical depth, it will have no influence on the flow of the channel, and the discharge will be determined the same as for channel *a*. If the submergence is greater than the critical depth, in making discharge computations, the outlet depth D_s will be constant, and only the intake depth *D* will vary with the discharge. The method of computing the discharge for channel *b* will otherwise be the same as that described for channel *a*.

Examples *c* and *d* represent the same channel with its outlet subjected to different depths of submergence. Since the grade is steeper than the critical slope, the critical depth D_c will occur at the intake if there is no interference from backwater. In channel *c*, because of backwater, the depth at the intake is increased from D_c to *D* and the discharge occurs in accordance with the same principles that apply to channel *b*. In example *d*, backwater is not sufficient to affect conditions at the intake. The discharge is at the critical depth, and a hydraulic jump occurs in the channel. For the condition shown in the figure, the discharge will not be affected by reducing the depth of submergence D_s or by increasing it up to the point that there is still some acceleration beyond the section where the critical depth occurs. The presence of a jump downstream from the intake is evidence that the discharge is occurring at the critical depth.

Chutes

A channel with a steep slope that is used to convey water from a higher to a lower elevation is termed a "chute." As illustrated in Fig. 8.24, water is received by a channel of uniform cross section through a rounded entrance at the critical depth from another channel, the velocity accelerating and gradually approaching uniform flow in the lower reaches.

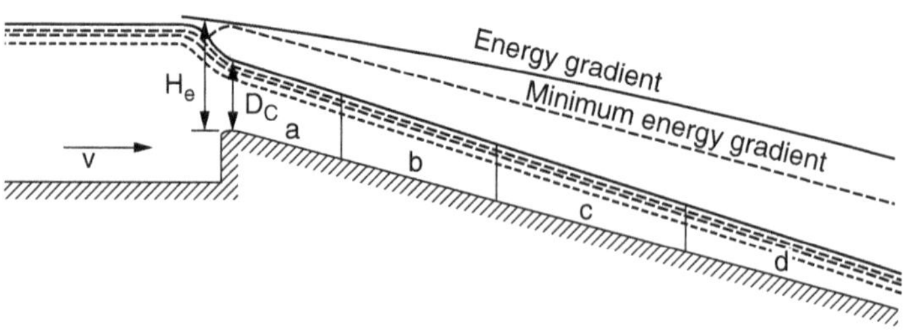

FIGURE 8.24 Chute.

Beginning at the entrance, distances to assumed depths of water at the ends of successive reaches can be computed by Eq. (8.105). The depth at uniform flow can be obtained as in Example 8.1.

In designing a chute, it is usually required to determine the dimensions that will provide for a given discharge. Since flow is accelerated, the cross section of the channel should be gradually reduced to correspond to the reducing cross section of the stream. Before proceeding with computations, it will be necessary to know the form of channel that is to be designed and to assume, at least tentatively, a relation between the depth of water and some linear dimension of the cross section. It may, for example, be decided to give D/b a constant value or to use a constant depth and gradually reduce the width. With the relation of D to b decided upon, the entrance dimensions should be first computed, and then, using Eq. (8.105), the respective distances downstream should be determined to cross sections of successively smaller cross-sectional area.

Flow over Very Steep Inclines

Equation (8.105) cannot be adapted readily to very steep slopes like the one illustrated in Fig. 8.25. This is because heads are measured vertically and the normal cross section, perpendicular to the direction of flow, is inclined. Points at different elevations in the cross section then contain different amounts of energy, and it is not practicable to write Bernoulli's equation in the usual manner. For example, for section b in Fig. 8.25 the energy content of a point m on the bottom is relatively more than that of a point n on the surface by the amount of drop between m' and n', the projections of the respective

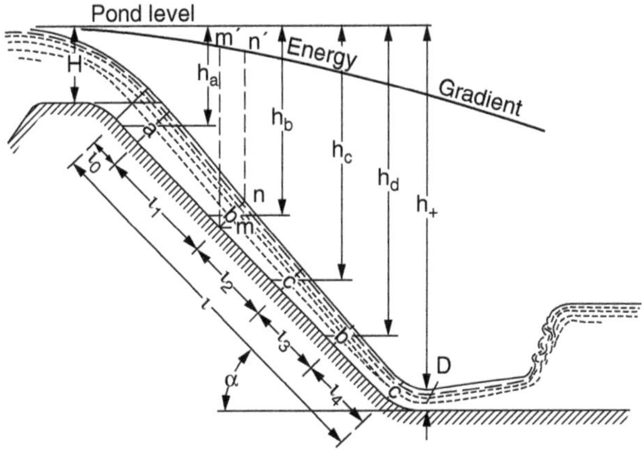

Figure 8.25 Flow over spillway of dam.

points on the energy gradient. The variation in energy of intermediate points in the cross section is indicated by the slope of the energy gradient. The solution for such problems may be obtained by writing the Bernoulli equation from the reservoir to the centers of the selected cross sections, ignoring the hydrostatic pressure. For example, for section c the equation becomes $h_c = V_c^2/2g + \Sigma h_t$, which reduces to $V_c = \sqrt{2g(h_c - \Sigma h_t)}$. The term Σh_l is the sum of the energy losses for the selected reaches from the crest to the selected section determined for individual reaches by means of the Manning equation. The final application of the Bernoulli equation to point D will again include the depth term, and the equation for velocity at D becomes $V_D = \sqrt{2g(h_t - \Sigma h_t)}$.

References

1. M. O. O'Brien, "Analyzing Hydraulic Models for Effects of Distortion," *Eng. News Rec.*, Sept. 15, 1932.
2. "Hydraulic Design of Stilling Basins and Bucket Energy Dissipators," U.S. Bureau of Reclamation, Eng. Monograph 25, 1958.
3. C. E. Kindsvater, "The Hydraulic Jump in Sloping Channels," *Trans. ASCE*, vol. 109, p. 1107, 1944; with discussion by G. H. Hickox.
4. B. A. Bakhmeteff and A. E. Mutzke, "The Hydraulic Jump in Sloped Channels," *Trans. ASME*, vol. 60, p. 60, 1938.
5. C. H. Yen and J. W. Howe, "Effects of Channel Shape on Losses in a Canal Bend," *Civil Eng. (N.Y.)*, p. 28, Jan. 1942.
6. A. Shukry, "Flow around Bends in an Open Flume," *Trans. ASCE*, vol. 115, p. 751, 1950.
7. P. J. Tilp and M. W. Scrivner, "Analyses and Descriptions of Capacity Tests on Large Concrete Lined Canals," U.S. Bureau of Reclamation Tech. Memo 661, 1964.
8. J. Hinds, "The Hydraulic Design of Flume and Siphon Transitions," *Trans. ASCE*, vol. 92, p. 1423, 1928.
9. J. N. Bradley, "Hydraulics of Bridge Waterways, "U.S. Bureau of Public Roads, Div. of Hydraulic Res., Hydraulic Design Ser. 1, 1960.
10. B. A. Bakhmeteff, *Hydraulics of Open Channels*, McGraw-Hill, New York, 1932.

SECTION 9

High-Velocity Transitions

When water flows at supercritical velocities ($D < D_c$), a change in alignment of the walls of a channel creates standing-wave patterns, which must be taken into consideration in the design of a channel. When the change in alignment turns the water toward the center of the channel, as in a constriction or at the outside wall of a bend, the waves create depths which are considerably in excess of those that would be expected under subcritical conditions. Changes in alignment which permit the water to turn away from the center of a channel create negative waves, or depressions. The principles involved, as well as the results of laboratory verifications, have been set forth in a symposium.[1]

Straight-Walled Constrictions

The analytical and experimental study of channel constrictions indicated that straight-walled constrictions are more satisfactory than the smoothly curved transitions used for subcritical flow (see "Minor Losses," Sec. 8).

The effect of a change in wall alignment toward the centerline of the original channel is illustrated in Fig. 9.1. The original velocity, depth, and Froude number are V_1, D_1, and F_1, respectively. The wall is turned through an angle θ, which causes the fluid to turn through the same angle and flow at a new velocity V_2 with a depth D_2 and a Froude number F_2. The changes in depth and velocity occur along a wavefront *bd* oriented at the angle β with respect to the original direction of flow. A change in momentum normal to the wave occurs at the wave, the velocity components normal to the wave being reduced

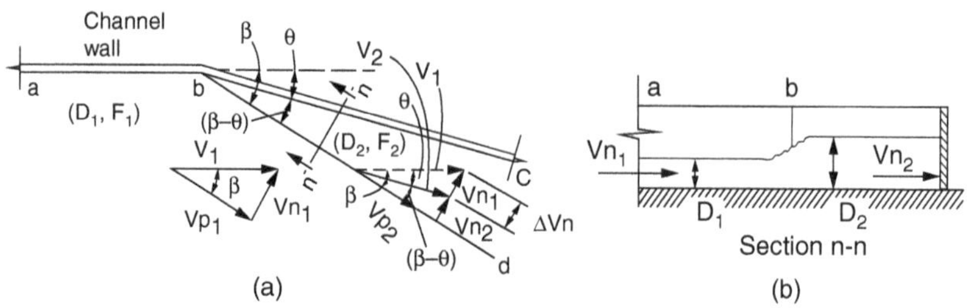

Figure 9.1 Straight-walled constrictions, definition sketch.

from V_{n1} to V_{n2}. It is assumed that the velocity components parallel to the wave V_p are not changed, so that

$$V_{p1} = V_{p2} \tag{9.1}$$

Continuity requires that

$$q_{n1} = q_{n2} \tag{9.2}$$

or

$$D_1 V_{n1} = D_2 V_{n2} \tag{9.3}$$

Assuming that the applied force is due only to the difference in hydrostatic pressures, then a derivation identical with that for the hydraulic jump (see Sec. 8) yields the following expression:

$$V_{n1} = \sqrt{gD_1}\sqrt{\frac{D_2}{2D_1}\left(\frac{D_2}{D_1}+1\right)} \tag{9.4}$$

From Fig. 9.1*a* it may be seen that

$$\sin\beta = \frac{V_{n1}}{V_1} \tag{9.5}$$

By inserting the value of V_{n1} from Eq. (9.4) and replacing $V_1/\sqrt{gD_1}$ with F_1, the following relationship is obtained:

$$\sin\beta = \frac{1}{F_1}\sqrt{\frac{D_2}{2D_1}\left(\frac{D_2}{D_1}+1\right)} \tag{9.6}$$

For small waves in which $D_2 \to D_1$, Eq. (9.6) reduces to the following approximate relation between β and F_1:

$$\sin\beta = \frac{1}{F_1} \tag{9.7}$$

Equation (9.7) is useful for determining the location of the disturbance line for very small channel irregularities, but its principal value is for curved walls or for enlarging sections, as will be shown later.

Equation (9.6) may be solved for D_2/D_1 to obtain the following expression:

$$\frac{D_2}{D_1} = \frac{\sqrt{1+8F_1^2 \sin^2 \beta} - 1}{2} \tag{9.8}$$

Values of D_2/D_1 obtained from this equation are plotted against β for various values of F_1 in the upper right quadrant of Fig. 9.2.

Trigonometric relationships for the vector triangles shown in Fig. 9.1*a* lead to the following expression for θ:

$$\tan\theta = \frac{(1 - D_1/D_2)\tan\beta}{1 + D_1/D_2 \tan^2\beta} \tag{9.9}$$

If the value of D_2/D_1 from Eq. (9.8) is substituted into Eq. (9.9), the following expression involving β, F_1, and θ is obtained:

$$\tan\theta = \frac{\left(\sqrt{1+8F_1^2 \sin^2\beta} - 3\right)\tan\beta}{2\tan^2\beta + \sqrt{1+8F_1^2 \sin^2\beta} - 1} \tag{9.10}$$

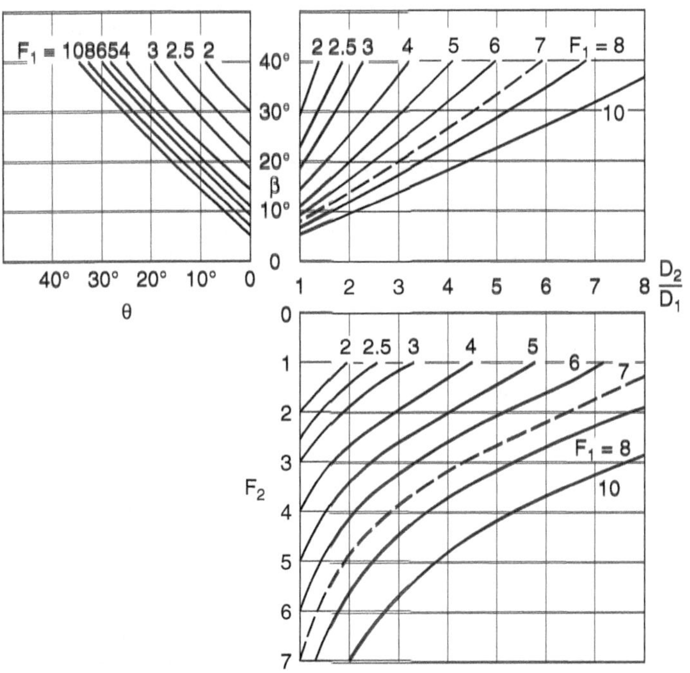

FIGURE 9.2 Straight-walled constrictions, design charts.

This equation can be solved graphically by plotting θ against β for various values of F_1, as shown in the upper left quadrant of Fig. 9.2.

In order to determine the effects at a second change in direction it is necessary to compute the value of F_2. Solution for F_2 may be accomplished by means of the following relationship derived from the velocity vector triangles of Fig. 9.1*a*:

$$V_{p1}^2 = V_1^2 - V_{n1}^2 = V_{p2}^2 = V_2^2 - V_{n2}^2$$

Then, making use of Eqs. (9.3) and (9.4) and inserting the Froude numbers, the following expression is obtained:

$$F_2^2 = \frac{D_1}{D_2}\left[F_1^2 - \frac{1}{2}\frac{D_1}{D_2}\left(\frac{D_2}{D_1} - 1\right)\left(\frac{D_2}{D_1} + 1\right)^2\right] \tag{9.11}$$

This equation was used to derive values for the curves of F_2 versus D_2/D_1 shown in the lower right quadrant of Fig. 9.2. The curves are presented in the manner used by Ippen.[2]

The application of these equations will be illustrated by means of a numerical problem.

Example 9.1 Water is flowing at a depth of 0.3 ft and a velocity of 20 ft/s in a rectangular channel 5-ft wide. Determine the wave pattern and water depths if this channel is constricted by symmetrical straight walls to a width of 2.0 ft with θ taken as 10° (Fig. 9.3).

$$F_1 = \frac{V_1}{\sqrt{gD_1}} = 6.44$$

From Fig. 9.2, $\beta_2 = 18°$, $D_2/D_1 = 2.35$, $F_2 = 4.05$, and $D_2 = 0.72$ ft.

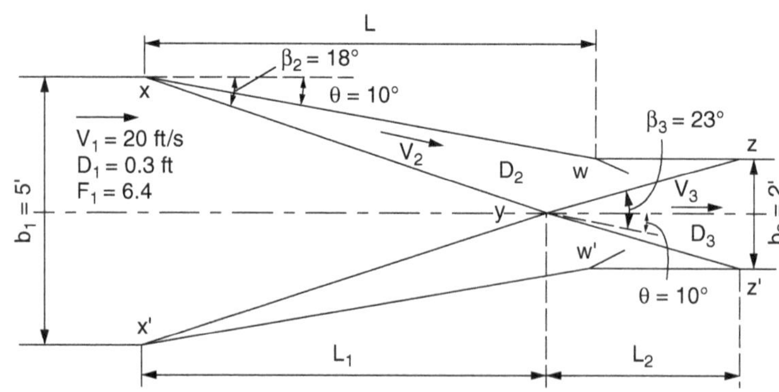

FIGURE 9.3 Examples 9.1 and 9.2.

As the water moves in the regions where the depth is D_2, it approaches the center from each side at an angle of 10°. At the centerline it may be assumed that the velocity is again deflected through 10° ($\theta = 10°$) and that the water again flows parallel to the centerline. Then new standing waves *yz* and *yz′* are developed. The angle β_3 and the depth D_3 can be obtained from Fig. 9.2 by letting D_2 be D_1 and D_3 be D_2. Then $\beta_3 = 23°$, $F_3 = 2.85$, $D_3/D_2 = 1.80$, and $D_3 = 1.30$ ft. Following this second change in direction and depth, no further reflection will occur if $L_1 + L_2 > L$. Investigation of this phase of the example will be continued after the following derivations.

A consideration of the geometry of Fig. 9.3 yields the following relationship involving dimensions of the plan of the constriction and the wave pattern:

$$L = \frac{b_1 - b_2}{2 \tan \theta} \tag{9.12}$$

$$L_1 = \frac{b_1}{2 \tan \beta_2} \tag{9.13}$$

$$L_2 = \frac{b_2}{2 \tan (\beta_3 - \theta)} \tag{9.14}$$

If $L_2 + L_1 < L$, the point *z* would be upstream from the end of the transition *w*, and another increase in depth would be created. Such a design is usually avoided. It should be noted that the equations developed in this section would not be valid if the downstream conditions were such that subcritical flow would occur ($F < 1$). Under such conditions flow in the constriction would be affected by a downstream control.

Example 9.1 (Continued) From Eq. (9.12),

$$L = \frac{5 - 2}{2 \times 0.176} = 8.5 \text{ ft}$$

From Eq. (9.13),

$$L_1 = \frac{5}{2 \times 0.325} = 7.7 \text{ ft}$$

From Eq. (9.14),

$$L_2 = \frac{2}{2 \times 0.231} = 4.3 \text{ ft}$$

Thus $L_1 + L_2 = 12$, which is greater than L, and therefore D_3 would be the highest depth encountered. The depth D_3 will occur only in the region near the centerline (Fig. 9.3), because negative waves will emanate from w and w', causing a decrease in depth. Negative waves will be discussed next. ▲

Enlargements and Curved-Wall Constrictions

In channel enlargements and in curved-wall constrictions, the change in direction takes place gradually instead of in one relatively steep standing wave (as for straight-wall constrictions), and the resulting wave configuration may be determined by considering the total change in fluid direction θ to be made up of many small angles $\Delta\theta$. The following equations are based on Ippen's work.[1] For very small directional changes it has already been shown that Eq. (9.7) may be used to present the relationship between β and F_1,

$$\sin\beta = \frac{1}{F_1} \tag{9.7}$$

Referring to the velocity vector triangles of Fig. 9.1*a* and assuming that θ is such a small angle $d\theta$ that

$$\sin(90 + \beta - d\,\theta) = \sin(90 + \beta)$$

application of the law of sines yields the relationship

$$dV_n = \frac{-V d\theta}{\cos\beta} \tag{9.15}$$

Application of the momentum relationship for very small changes in depth dD yields

$$-dV_n = \frac{g\,d\,D}{V_n} \tag{9.16}$$

Then, from Eqs. (9.15) and (9.16),

$$\frac{dD}{d\theta} = \frac{V^2}{g}\tan\beta \tag{9.17}$$

The specific energy H is defined as

$$H = D + \frac{V^2}{2g} \tag{9.18}$$

from which

$$V = \sqrt{2g(H - D)} \tag{9.19}$$

or
$$\frac{H}{D}=1+\frac{F^2}{2} \tag{9.20}$$

In the case of very small angles, Eq. (9.4) reduces to

$$V_n=\sqrt{gD} \tag{9.21}$$

The vector triangles in Fig. 9.1*a* yield the relationship

$$V_p=\sqrt{V^2-V_n^2} \tag{9.22}$$

and
$$\tan\beta=\frac{V_n}{V_p} \tag{9.23}$$

Assuming that the specific energy H will be constant, Eqs. (9.17) through (9.23) can be combined to yield the expression

$$\frac{dD}{d\theta}=\frac{2H\sqrt{(D/2H)}(1-D/H)}{\sqrt{1-\frac{3}{2}D/H}} \tag{9.24}$$

and upon integration,

$$\theta+\theta_1=\sqrt{3}\tan^{-1}\sqrt{\frac{D/\frac{2}{3}H}{1-D\frac{2}{3}H}} - \tan^{-1}\frac{1}{\sqrt{3}}\sqrt{\frac{D/\frac{2}{3}H}{1-D/\frac{2}{3}H}} \tag{9.25}$$

Equation (9.25) can also be stated in terms of F, using Eq. (9.20),

$$\theta+\theta_1=\sqrt{3}\tan^{-1}\sqrt{\frac{3}{F^2-1}}-\tan-1\sqrt{\frac{1}{F^2-1}} \tag{9.26}$$

In these equations θ may be considered as $\Sigma(\Delta\theta)$ for any location. θ_1 may be evaluated by means of Eq. (9.25) or (9.26) for the original conditions, in which case $\theta=0$, $V=V_1$, and $F=F_1=V_1/\sqrt{gD_1}$. Values of $\theta+\theta_1$ for various values of D/H as obtained from Eq. (9.25) are plotted in Fig. 9.4. Also shown in Fig. 9.4 is a curve of F versus D/H, determined from Eq. (9.20).

Equations (9.25) and (9.26) are not intended to be used for abrupt straight-walled constrictions, because by neglecting energy losses, the results cannot be expected to be as accurate as when the curves of Fig. 9.2 are used. Yet for angles as small as 10°, the differences in computed values of D_2 and F_2 are negligible and the straight-walled

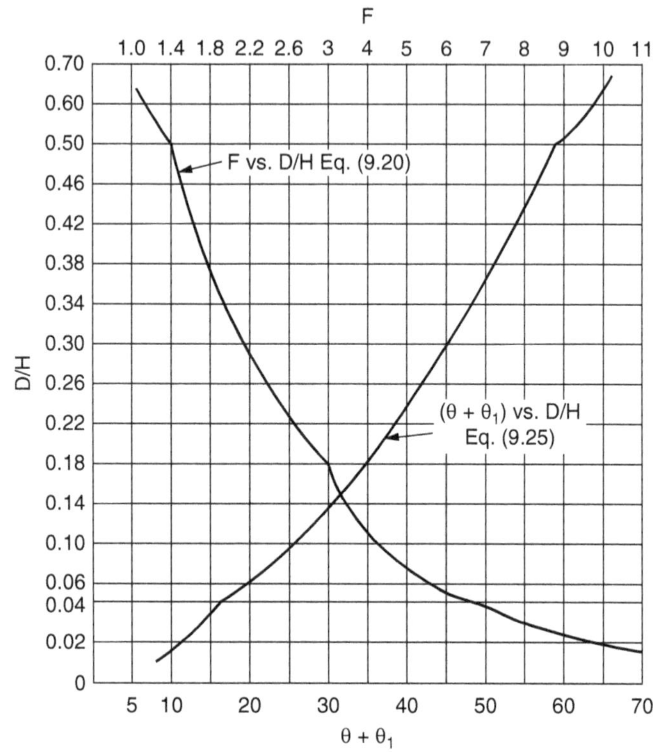

FIGURE 9.4 *D/H* versus *F* and $\theta + \theta_1$ from Eqs. (9.20) and (9.25).

constriction shown in Fig. 9.3 will be used in Example 9.2 as the first illustration of the application of Eqs. (9.25) and (9.26). Example 9.3 illustrates the use of Eqs. (9.25) and (9.26) in the solution of a non-symmetrical straight-walled constriction.

Example 9.2 Data are the same as in Example 9.1.

$$F_1 = 6.44 \qquad D_1 = 0.3\ \text{ft} \qquad \theta = 10°$$

From Fig. 9.4, θ_1 is determined by entering at $F_1 = 6.44$, locating $D_1/H = 0.045$ and $\theta + \theta_1 = 17.2°$. Then, when $\theta = 10°$, $\theta + \theta_1 = 27.2°$. From the curves $D_2/H = 0.11$ and $F_2 = 4.0$,

$$D_2 = \frac{0.11}{0.045} \times 0.30 = 0.73\ \text{ft}$$

As the water deflects through another 10° at the centerline, $\theta + \theta_1 =$ 37.2°, and from the curves $D_3/H = 0.20$,

$$D_3 = \frac{0.20}{0.11} \times 0.73 = 1.33 \text{ ft}$$

Although these values agree closely with those obtained from Fig. 9.2 in Example 9.1, it should be noted that β must still be obtained from Eq. (9.8) or Fig. 9.2. ▲

Example 9.3 Figure 9.5*b* shows a rectangular channel in which $F_1 = 4.0$, $V_1 = 7.4$ m/s, and $D_1 = 0.35$ m. The left side, looking downstream, converges at an angle $\Delta\theta_{L1}$ of 4°, and the right side at an angle $\Delta\theta_{R1}$ of 2°. Determine the locations of the standing waves *LC*, *RC*, *CL′*, and *CR′* and the water depths and velocities.

Conditions after the first changes in direction, in the regions *LCL′* and *RCR′*, could be determined from Fig. 9.2 as in Example 9.1; however, Eq. (9.25) and Fig. 9.4 will be used in this example.

Entering Fig. 9.4 with $F_1 = 4.0$, then $D_1/H = 0.111$ and $\theta_1 = 27.2°$. Then $\Delta\theta_{L1} + \theta_1 = 31.2°$ and $\Delta\theta_{R1} + \theta_1 = 29.2°$. The corresponding values of the parameters as obtained from Fig. 9.4 are

$$\frac{D_{L2}}{H} = 0.144 \qquad \frac{D_{R2}}{H} = 0.127$$

$$F_{L2} = 3.41 \qquad F_{R2} = 3.70$$

and

$$D_{L2} = \frac{0.144}{0.111} \times 0.35 \qquad D_{R2} = \frac{0.127}{0.111} \times 0.35$$

$$= 0.45 \text{ m} \qquad = 0.40 \text{ m}$$

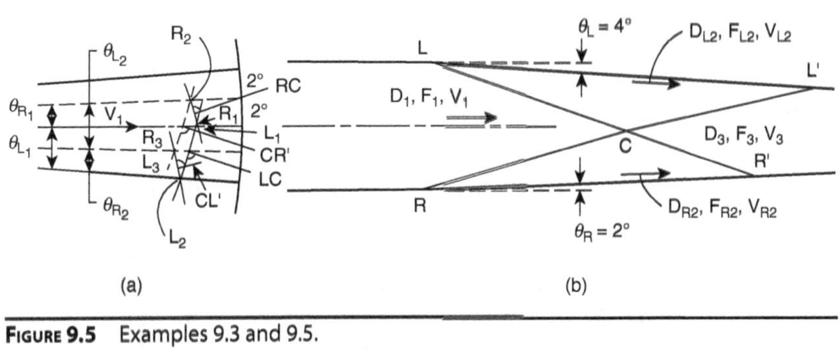

Figure 9.5 Examples 9.3 and 9.5.

After the waters from the two sides converge in the region downstream from $L'CR'$, a common velocity V_3, depth D_3, and Froude number F_3 must be attained. Since V_{L2} and V_{R2} differ in direction by $\Delta\theta_{L1}+\Delta\theta_{R1}=6°$, it is apparent that $\Delta\theta_{L2}+\Delta\theta_{R2}$ must also be 6°. Knowing this, the solution for D_3 and V_3 could be obtained from Fig. 9.2 by successive approximation. However, a consideration of Eq. (9.25), as plotted in Fig. 9.4, shows that a common depth and Froude number can be achieved only if

$$\Delta\theta_{L1}+\Delta\theta_{L2}+\theta_1=\Delta\theta_{R2}+\theta_1$$

and therefore if $\Delta\theta_{L1}+\Delta\theta_{L2}=\Delta\theta_{R1}+\Delta\theta_{R2}$. Since $\Delta\theta_{L1}$ and $\Delta\theta_{R1}$ are 4° and 2°, respectively, it follows that $\Delta\theta_{L2}=2°$ and $\Delta\theta_{R2}=4°$. Then, knowing that $\Delta\theta_{L1}+\Delta\theta_{L2}+\theta_1=33.2°$, the following values may be obtained from Fig. 9.4:

$$\frac{D_3}{H}=0.163 \qquad D_3=\frac{0.163}{0.111}\times 0.35=0.51\text{ m}$$

and $$F_3=3.2$$

V_3 can be obtained as follows:

$$F_3=\frac{V_3}{\sqrt{gD_3}}=\frac{V_3}{\sqrt{9.8\times 0.51}}=3.2$$

and $$V_3=7.16\text{ m/s}$$

One of the most important applications of Eqs. (9.25) and (9.26) is in the computation of depths in an *enlarging section*. The method will be illustrated for the enlargement shown in Fig. 9.6. This example is one of those tested and reported by Ippen and Harleman.[2] ▲

Example 9.4 A channel section is enlarged by a straight wall at an angle of 15°; $F_1=2.94$. Determine lines of equal depth, and thus establish the water-surface form.

The assumption is made that the velocity will ultimately turn through the total 15° and that flow near the wall will turn without separation. The computations, which are shown in Table 9.1, are carried forward in five uniform angular increments $\Delta\theta$ of 3°. The value of θ_1 is found to be 35.7° from Fig. 9.4. F_n is also taken from Fig. 9.4. β_n is computed from Eq. 9.7. D_n/H is taken from Fig. 9.4. D_n/D_1 is then the ratio of D_n/H and the initial value of D_n/H, where $n=1$.

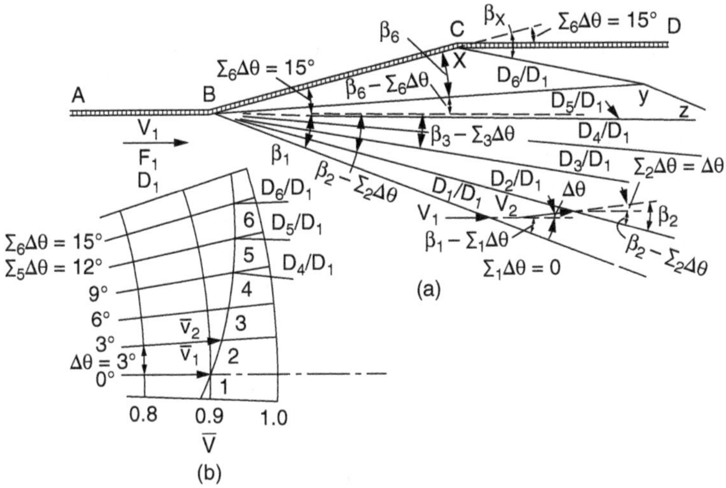

Figure 9.6 Straight-walled enlargement, Examples 9.4 and 9.6.

The water crosses the line $D_n/D_1 = 1.0$ with a velocity V_1. In the space between $D_n/D_1 = 1.0$ and $D_n/D_1 = D_2/D_1 = 0.84$, the velocity changes its magnitude to V_2 and its direction through $\Delta\theta = 3°$. At each line of D_n/D_1, the summation of angular increments is denoted by $\Sigma_n \Delta\theta$. The subscripts n are given in column 1 of Table 9.1. Values of F_n and D_n/H (columns 3 and 8) are read from Fig. 9.4, and values of β are computed from Eq. (9.7). Each β_n is the angle between the imaginary small wave, or line of constant D_n/D_1, and the velocity vector V_n approaching that line, and since the angle between V_n and the original velocity V_1 is $\Sigma_n \Delta\theta$, the angle between lines of D_n/D_1 and V_1 is $\beta_n - \Sigma_n \Delta\theta$. This is illustrated in Fig. 9.6. Values of $\beta_n - \Sigma_n \Delta\theta$ are listed in column 7 of Table 9.1.

Table 9.1 Example 9.4

(1) n	(2) $\theta + \theta_1$	(3) F_n	(4) $\frac{1}{F_n}$	(5) β_n	(6) $\Sigma_n \Delta\theta$	(7) $\beta_n - \Sigma_n \Delta\theta$	(8) $\frac{D_n}{H}$	(9) $\frac{D_n}{D_1}$
1	35.7°	2.94	0.340	19°50′	0°	19°50′	0.187	1.00
2	32.7°	3.30	0.303	17°40′	3°	14°40′	0.158	0.84
3	29.7°	3.65	0.274	15°50′	6°	09°50′	0.130	0.70
4	26.7°	4.15	0.241	14°00′	9°	05°00′	0.106	0.57
5	23.7°	4.70	0.213	12°20′	12°	00°20′	0.084	0.45
6	20.7°	5.38	0.186	10°40′	15°	−04°20′	0.065	0.35

These values provide a convenient method of locating the lines of equal depth. It follows also from Fig. 9.6 that the angle between the new wall direction BC and any constant depth line D_n/D_1 is $\Sigma_6 \Delta\theta + (\beta_n - \Sigma_n \Delta\theta)$. This is of particular interest for the final line (D_6/D_1 in this case), beyond which no further changes in depth or velocity occur, because its angle with the new wall direction is then simply equal to β_6. Therefore, the values of β_6, D_6, and F_6 can be computed without determining the intermediate conditions if this is all the information needed. The values of β and D_n/D_1 computed in this example differ by approximately 10 percent from measured values.[2]

In this example, no consideration has so far been given to possible effects from the opposite wall. If an enlargement also occurred on the opposite side of the channel, the disturbance lines would cross, or if the other side continued in its original direction, some of the disturbance lines shown in Fig. 9.6 would reflect from this opposite wall and cross other disturbance lines. This situation is much more complex than the case of a single symmetrical crossing of an abrupt wave which was solved in Example 9.2. A method of solving such complex problems using characteristic curves will be presented in the following section. At point C, where the wall resumes its original direction, an abrupt positive wave xy is formed in the same manner as discussed previously. ▲

Method of Characteristics

Another method of solving problems in which the disturbances may be treated as a series of small waves, which is particularly useful for complex wave systems, is known as the method of characteristics. The equations are derived as follows. Equation (9.18) may be rearranged in the form.

$$1 = \frac{D}{H} + \frac{V^2}{2gH} \tag{9.27}$$

If $V/\sqrt{2gH}$ is replaced by $\bar{V}$, Eq. (9.27) can be written as

$$\frac{D}{H} = 1 - \bar{V}^2 \tag{9.28}$$

or

$$\bar{V} = \sqrt{1 - \frac{D}{H}} \tag{9.29}$$

Substitution of the value for D/H from Eq. (9.28) into Eq. (9.25) yields the following equation for θ:

$$\theta + \theta_1 = \sqrt{3}\tan^{-1}\sqrt{\frac{1-\overline{V}^2}{\overline{V}^2 - 1/3}} - \tan^{-1}\frac{1}{\sqrt{3}}\sqrt{\frac{1-\overline{V}^2}{\overline{V}^2 - 1/3}} \tag{9.30}$$

Corresponding values of $\overline{V}$ and D/H obtained from Eq. (9.28) or (9.29) are plotted in Fig. 9.7. Shown in the same figure is a curve of $\overline{V}$ versus $\theta + \theta_1$ as obtained from Eq. (9.30). When $\overline{V}$ and $\theta + \theta_1$ are plotted on polar coordinates, Eq. (9.30) represents a family of curves such as those shown in Fig. 9.8. These curves are called characteristic curves, perhaps because $\overline{V}$ has the same characteristics as velocity in that values of $\overline{V}$ plotted in Fig. 9.8 have the same direction as the actual velocity and a length that is proportional to the length of the velocity vector. The use of Eq. (9.30) in this graphic form provides a convenient tool for solving problems in which many small changes in direction are involved. As an illustration, the problem shown in Fig. 9.5, previously solved in Example 9.3 by other methods, will now be solved by this method.

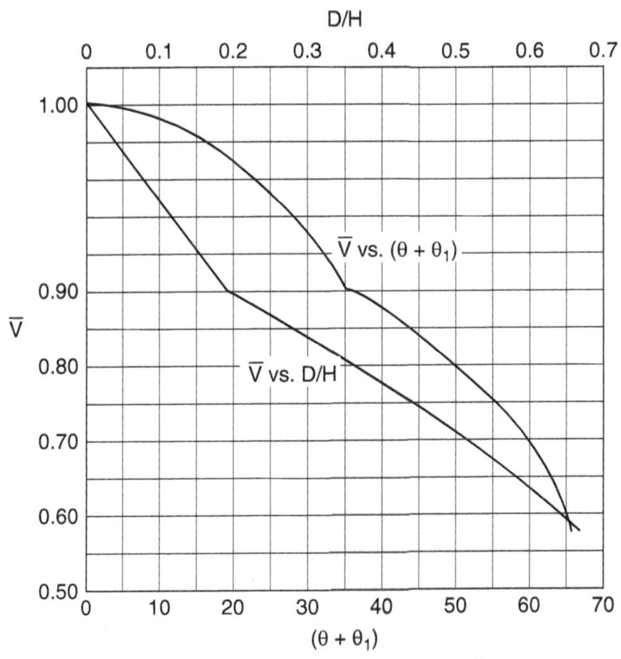

FIGURE 9.7 $\overline{V}$ versus D/H and $\theta + \theta_1$ from Eqs. (9.28) and (9.30). (Note change in vertical scale.)

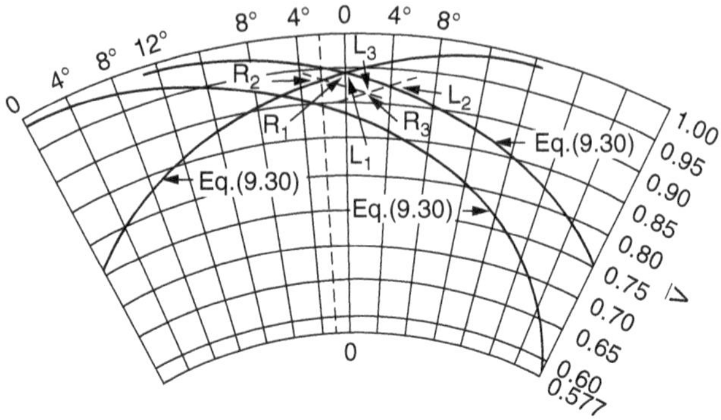

FIGURE 9.8 $\bar{V}$ versus $\theta + \theta_1$ from Eq. (9.30) using polar coordinates.

Example 9.5 Determine the depths, velocities, and wave pattern for the unsymmetrical straight-walled constriction shown in Fig. 9.5*b*. The initial conditions are as follows:

$$V_1 = 7.4 \text{ m/s} \qquad F = 4.0 \qquad D_1 = 0.35 \text{ m}$$

D_1/H may be read from Fig. 9.4, knowing θ_1 from Example 9.3, or computed as follows:

$$\frac{V_1^2}{2g} = 2.79 \text{ m}$$

$$H_1 = D_1 + \frac{V_1^2}{2g} = 0.35 + 2.79 = 3.14 \text{ m}$$

$$\frac{D_1}{H} = \frac{0.35}{3.14} = 0.111$$

From Fig. 9.7,

$$\bar{V}_1 = 0.943$$

The direction of the initial velocity V_1, or $\bar{V}_1$, is taken as the 0° line in Fig. 9.8, and the end of the vector $\bar{V}$ is located at the point designated as R_1 and L_1, because this point represents the initial condition for both the left and right deflections. Noting that Eq. (9.30) must be satisfied for the initial conditions as well as after the first deflections from each side, curves of Eq. (9.30) are plotted in both directions through point L_1R_1. The curve L_1L_2 gives the change in magnitude and direction of $\bar{V}$ as the water is deflected 4° at the left side. After the deflection, the direction of flow is along the ray through L_2, and the value of $\bar{V}_{L2}$, read at L_2, is 0.925. Then, from Fig. 9.7, D_{L2}/H = 0.148 and D_{L2} = 0.148 × 3.14 = 0.465 m, and from Fig. 9.4, F_{L2} = 3.4. Similarly, R_2 represents conditions after the deflection of 2° from the right wall. The corresponding values for the right side are $\bar{V}_{R2}$ = 0.935, D_{R2}/H = 0.126, D_{R2} = 0.40 m, and F_{R2} = 3.7.

The orientation of the standing waves on the left side, *LC* in Fig. 9.5*b*, must be perpendicular to the segment of the characteristic curve L_1L_2. This is because L_1L_2 has the same direction as the vector difference between V_1 and V_{L2}, and therefore this is the component of velocity normal to the standing wave, which has been designated as V_n in the derivation of the equations. A portion of Fig. 9.8 has been reproduced in Fig. 9.5*a*, in which the line *LC* is drawn perpendicular to L_1L_2 and serves to establish the direction of *LC* in Fig. 9.5*b*. Correspondingly, the line *RC*, perpendicular to R_1R_2 in Fig. 9.5*a*, establishes the direction of the standing wave *RC*.

Until this point the use of the characteristic curves has provided little advantage over the methods used previously to solve this problem. However, in determining the conditions downstream from *LCR* there is an advantage in this graphic procedure in that a direct solution (without successive approximation) can be obtained. This is done in Figs. 9.8 and 9.5*a* by noting that, after the waters from the two sides converge beyond *LCR*, there will be a common value of $\bar{V}$, $\bar{V}_3$, and that the relation between the new angle $\theta + \theta_1$ and $\bar{V}_3$ must be in accordance with Eq. (9.30). Consequently, curves of Eq. (9.30) (L_2L_3 and R_2R_3) are plotted through L_2 and R_2, respectively. These curves cross at point L_3R_3. This point gives the value of $\bar{V}_3$ as well as the angle θ_{L2} through which the waters from the left side are deflected to the left and the angle θ_{R2} through which the waters from the right are deflected back to the right. Furthermore, the orientation of the standing waves *CL′* and *CR′* can be determined from lines perpendicular to L_2L_3 and R_2R_3, respectively, as shown in Fig. 9.5*a*. Numerical values of D_3, V_3, and F_3 are obtained as follows. $\bar{V}_3$, from Fig. 9.8 or 9.5*a*, is 0.916. Then, from Fig. 9.7,

$$\frac{D_3}{H} = 0.16$$

or

$$D_3 = 0.16 \times 3.14 = 0.50$$

From Fig. 9.4,

$$F_3 = 3.24$$

Finally,

$$\overline{V}_3 = \frac{V_3}{\sqrt{2gh}} = 0.916 = \frac{V_3}{\sqrt{19.6 \times 3.14}} = \frac{V_3}{7.84}$$

and $$V_3 = 7.19 \text{ m/s}$$

V_3 might also be obtained from F_3. Additional reflections would be handled in the same manner. ▲

Example 9.6 Another illustration of the use of this method is presented for the expanding channel (Example 9.4) shown in Fig. 9.6. The problem here is to determine the flow directions and lines of uniform depth. Figure 9.6*b* shows a graph of $\overline{V}$ versus $(\theta + \theta_1)$ as plotted from Eq. (9.30). The graph was laid out so that the ray for 0° is parallel to the centerline of the original channel, and rays were laid out at 3° angles so that a total deflection of 15° may be illustrated. Knowing $F_1 = 2.94$, $\overline{V}$ and D_1/H were determined from Figs. 9.7 and 9.4 as 0.902 and 0.187, respectively. The curve of Eq. (9.30) was so located in Fig. 9.6*b* that, at 0°, $\overline{V}$ has the value of 0.902. Then the values of $\overline{V}$ may be read for all deflection angles, and the corresponding values of D/H may be obtained from Fig. 9.7. Values of any D can then be computed, knowing D_1/H and D_1. The computations are not shown because they are the same as those in Table 9.1. The direction of the various lines of constant D/H can be taken from the graph, as illustrated in the figure. When the channel wall abruptly resumes its original direction, a positive wave *xyz* is developed, as shown in Fig. 9.6*a*. The angles β_x and β_y were determined from Fig. 9.2. The depth downstream from *xyz* is equal to the original depth D_1. ▲

High-Velocity Flow at Channel Bends

As water enters a bend in a channel at supercritical velocity, it undergoes a continuous change in direction, which can be analyzed in the manner described in the previous section. At the outer wall a standing positive wave is developed along *AM*, whereas a decrease in depth occurs along *A′M′* (Fig. 9.9*a*). The growth of the waves is interrupted at *M* and *M′* because at these locations waves have crossed the channel. Downstream from *M* the wave height decreases because of the negative wave emanating from *A′*. This decrease continues to *N*, where the reflected positive wave again begins to build up the wave height. At *N* the water depth is

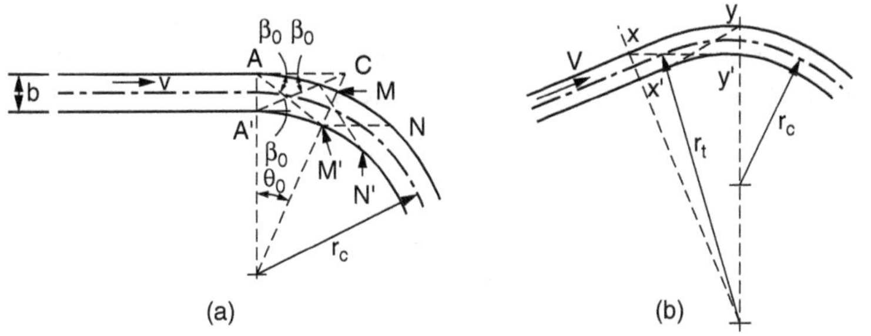

FIGURE 9.9 Channel bend, Example 9.7.

approximately the same as at A. The distance from A to N is one wavelength L, and the distance from A to the crest of the wave at M is one-half wavelength. The elevation of the water surface at M is a critical design criterion.

If it is assumed that the initial wave orientation, given by β_0, extends across the channel from A' to C as a straight line, then the distance AC is approximately equal to AM. The angle θ_0 is the angular change required to develop the maximum water-surface elevation. The geometry of Fig. 9.9*a* leads to the following approximate relationships, using symbols defined in the figure:

$$\tan \beta_0 = \frac{b}{AC} \tag{9.31}$$

or

$$\frac{L}{2} = AC = \frac{b}{\tan \beta_0} \tag{9.32}$$

and

$$\tan \theta_0 = \frac{AC}{r_c + b/2} \tag{9.33}$$

Inserting the value of AC from Eq. (9.32) into Eq. (9.33) yields the following expression for $\tan \theta_0$:

$$\tan \theta_0 = \frac{b}{(r_c + b/2)\tan \beta_0} \tag{9.34}$$

With θ_0 determined from Eq. (9.34), the depth at the wave crest M can be readily evaluated by means of Eqs. (9.20) and (9.25), using Fig. 9.4 for convenience.

It was observed by Knapp[1] that the increase in elevation of the water surface above the uniform flow depth D_0 at wave crests such as M was approximately given by the expression

$$\Delta D_M = \frac{V^2 b}{g r_c} \tag{9.35}$$

where V is the average velocity in the channel. It may be noted that this is twice the increase in depth caused by centrifugal force, as developed under "Channel Bends" in Sec. 8. Thus the water surface along the outer wall during supercritical flow may be visualized as a series of undulations above and below the surface that would occur because of centrifugal force alone, the depth at the troughs being D_0 and at crests

$$D_M = D_0 + \frac{V^2 b}{g r_c} \tag{9.36}$$

The determination of the depth at a wave crest such as M will be illustrated by the following example, which is one for which observed values were reported by Knapp.

Example 9.7 The discharge in a rectangular channel 0.45 m wide is 0.10 m^3/s. The uniform flow depth is 0.06 m, and the average velocity is 3.7 m/s. The radius of the bend is 15 m. Determine the location of the first wave crest and the depth at that location.

$$F_1 \frac{V_1}{\sqrt{gD_1}} = 4.8$$

From Eq. (9.7),

$$\sin \beta_0 = \frac{1}{F_1} = 0.208$$

and

$$\beta_0 = 12^\circ$$

Then, from Eq. (9.34),

$$\tan \theta_0 = \frac{0.45}{(15 + 0.22)0.2125} = 0.139$$

and

$$\theta_0 = 7.9^\circ$$

The observed value[1] of θ_0 was 7.75°.

The distance from the beginning of the bend to the wave crest (AM in Fig. 9.9) is, from Eq. (9.32),

$$AM \approx AC = 0.45/0.2125 = 2.12\ \text{m}$$

The depth at M will first be computed by means of Eqs. (9.20) and (9.25), using Fig. 9.4. Values of D_1/H and θ_1 are determined to be 0.08 and 23.0°, respectively. Knowing $\theta_0 = 8°$, $(\theta + \theta_1)$ is 31.0°, and from Fig. 9.4, $D_M/H = 0.143$. Then

$$D_M = \frac{D_M/H}{D_1/H} D_1 = \frac{0.143}{0.08} \times 0.06 = 0.107\ \text{m}$$

Using Eq. (9.36),

$$\begin{aligned} D_M &= 0.06 + \frac{3.7^2 \times 0.45}{9.8 \times 15.0} \\ &= 0.06 + 0.042 \\ &= 0.102\ \text{m} \end{aligned}$$

The observed value[1] of D_M for this example was 0.098 m. ▲

Several methods of reducing the height of the wave crests at the outside of a bend have been suggested. One method is to begin the bend with a transition section, xy in Fig. 9.9*b*, in which the radius of curvature is larger than that of the bend and the length is such that at the end of the transition yy' the negative wave from x' will begin to reduce the depth on the outside at the same time that the increased curvature, beyond yy', would tend to increase the depth. If the increase or decrease in depth due to the transition is made just half as much as that due to the bend itself, the two effects will tend to nullify each other beyond yy', and the total increase in depth will be reduced by approximately half.

This can be accomplished by either a spiral or a circular transition. The use of a circular transition will be described here. It may be seen from Eq. (9.35) that the maximum increase in depth is inversely proportional to the radius of curvature. Therefore, in order to cause an increase in depth at y of one-half of this amount, the radius of the transition section r_t should be $2r_c$. The distance xy can then be computed from Eq. (9.32) and the angle θ_t from Eq. (9.34), using r_t in place of r_c.

Other methods of reducing the height of the waves are banking the channel and the use of diagonal sills. Sills are subject to possible erosion or sedimentation and may cause cavitation at very high velocities. Nevertheless, under laboratory conditions, satisfactory results have been obtained by the use of sills. If the channel is banked at the angle that the water surface would assume because of the centrifugal force, there will be no increase or decrease in depth along the walls. However, it is necessary to begin the curvature of the bend gradually in a transition so that the maximum curvature of the bend is reached at the same time that the curvature of the channel attains its full banking angle.

References

1. "High Velocity Flow in Open Channels: A Symposium," *Trans. ASCE*, vol. 116, p. 265, 1951.
2. A. T. Ippen and R. F. Harleman, "Verification of Theory for Oblique Standing Waves," *Trans. ASCE*, vol. 121, p. 678, 1956.

SECTION 10

Wave Motion and Forces

This section is devoted to a discussion of surface-water waves. The first part will deal with oscillatory waves, the second with translatory waves. The propagation of a change in surface configuration, which is the visible manifestation of water-wave motion, is accompanied by the transmission of energy, pressure changes, and velocity changes. The forward motion of the water is very small for oscillatory waves, particularly in deep water. However, in shallow water a considerable volume of water is transported, with the result that currents are created as the waves approach a shoreline. As the name implies, translatory waves are characterized by a transverse displacement of water. The characteristics of the wave motion and currents near shore establish the design criteria for harbors and, along with the wind, constitute the dominant factors in all shore processes. The forces resulting from wave motion must be evaluated for the design of marine structures.

Water waves may be generated by any disturbance which affects the local velocity, surface elevation, or pressure. In nature, waves are most commonly generated by the action of the wind. Other natural phenomena which create waves are tides, seismic disturbances, and barometric-pressure fluctuations. Waves may be generated artificially by the opening or closing of a gate, the breaking of a dam, or, in the laboratory, by means of plungers designed for the purpose.

Once wave motion has been started, it is controlled primarily by gravitational forces, although short waves of low amplitude are influenced by surface tension. The principal role of viscosity lies in the gradual damping of waves. Elasticity plays no part in water-wave motion.

Oscillatory Waves

Oscillatory waves are so called because water particles oscillate in circular or elliptical paths, the forward motion visible to the eye being primarily that of the waveform.

Surface Form

The surface form of oscillatory waves is that of the simple sine curve, as described by the equation

$$y = \frac{H}{2}\cos\theta \tag{10.1}$$

where y is the vertical coordinate of the water surface with respect to the still-water level (SWL), H is the wave height, and the phase angle θ is defined by

$$\theta = 2\pi\left(\frac{x}{L} - \frac{t}{T}\right) \tag{10.2}$$

In Eq. (10.2), x is the horizontal coordinate of the water surface measured from a selected wave crest, L is the wavelength, T is the wave period, and t is the time measured from the instant that a wave crest is located at $x = 0$. All these terms are illustrated in Fig. 10.1.

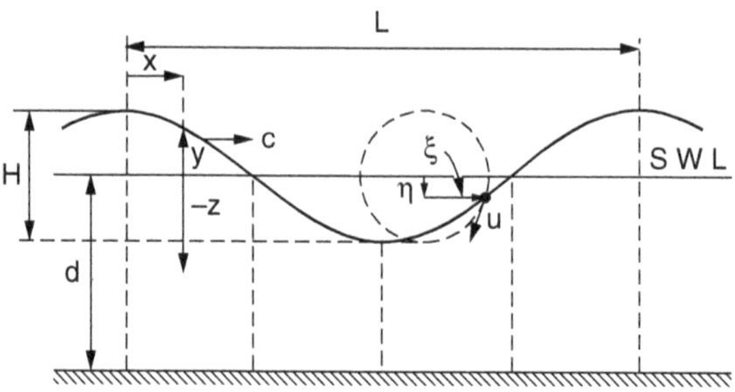

$\frac{x}{L} = 0$	$\frac{1}{4}$	$\frac{1}{2}$	$\frac{3}{4}$	1
$\frac{t}{T} = 0$	$-\frac{1}{4}$	$-\frac{1}{2}$	$-\frac{3}{4}$	−1
$\theta = 0°$	90°	180°	270°	360°
Cos θ, η, u_x = max +	0	max −	0	max +
Sin θ, u_z, $\frac{\partial u_x}{\partial t}$ = 0	max +	0	max −	0
−Cos θ, $\frac{\partial u_z}{\partial t}$ = max −	0	max +	0	max −
− Sin θ, ξ = 0	max −	0	max +	0

FIGURE 10.1 Definition of oscillatory waves.

Laboratory measurements confirm that the waveform is approximately that of a sine curve for deep water and for small wave heights, whereas for shallow water and relatively large wave heights the peaks become somewhat sharper and higher and the trough somewhat flatter and higher than would be indicated by a sine curve. In nature, waves are rarely found which do not consist of several wave trains of differing periods and lengths. Because the vertical coordinates of oscillatory waves are additive, natural waves usually appear as a series of large waves followed by smaller ones, often with much smaller waves superimposed on the entire complex system. Furthermore, the longer-period waves move much faster and thus overtake the shorter-period waves, with the result that the wave pattern is constantly changing.

The effect of both the depth and the wave height on the form of waves, as well as any other wave characteristics, is dependent on the wavelength. For example, water having depths such that $d/L < 0.5$ is designated as shallow water because it is only in this range that depth affects the wave characteristics. Similarly, the effect of wave height is usually related to H/L, the *wave steepness*, rather than to the depth alone.

Wave Celerity

The velocity of the waveform in relation to the body of fluid through which the waves are being propagated is the *celerity* C. The celerity is related to wavelength and wave period by definition, as shown by the equation

$$C = \frac{L}{T} \tag{10.3}$$

The expression for celerity is

$$C = \sqrt{\frac{gL}{2\pi}} \tanh \frac{2\pi d}{L} \tag{10.4}$$

In deep water, where $d/L \geq 0.5$, the value of $tanh(2\pi d/L)$ approaches 1.0 (Fig. 10.2), and Eq. (10.4) is reduced to the subscript 0 being introduced to indicate deep-water conditions:

$$C_0 = \sqrt{\frac{gL_0}{2\pi}} \tag{10.5}$$

The variation of C with d/L in accordance with Eq. (10.4) is illustrated in Fig. 10.3 for three values of deep-water wavelength L_0. The value of C_0 obtained from Eq. (10.5) for $L_0 = 100$ ft is also illustrated in Fig. 10.3.

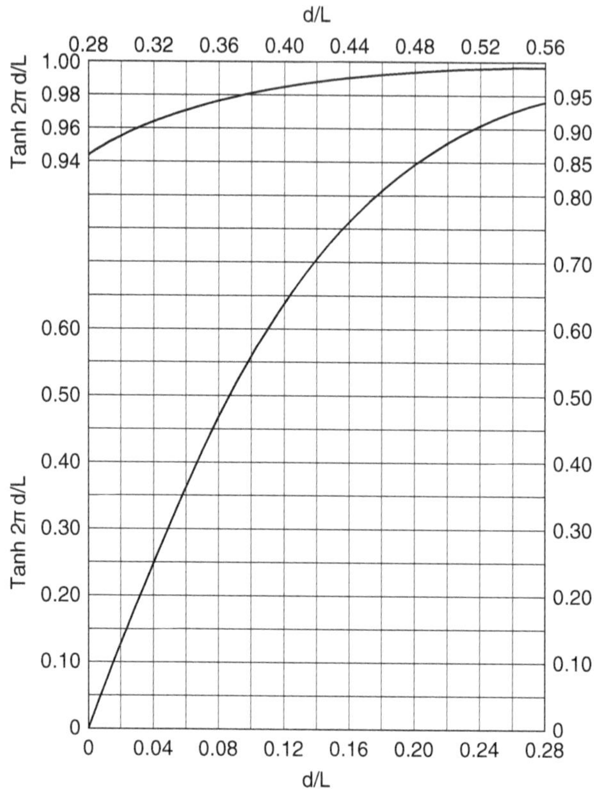

FIGURE 10.2 Tanh ($2\pi d/L$) versus d/L.

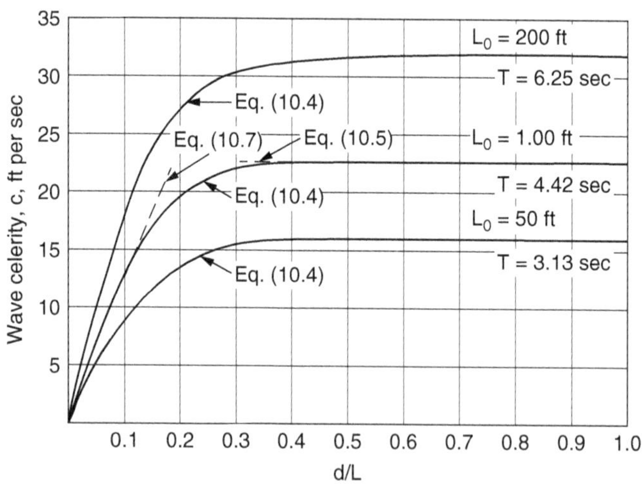

FIGURE 10.3 *C* versus d/L from Eqs. (10.4), (10.5), and (10.7).

Computations for C by means of Eq. (10.4) are facilitated by the use of Fig. 10.2 or by published tables.[1,2] The relation $C_0 = L_0/T$, from Eq. (10.3), may be introduced into Eq. (10.5) to obtain the following useful relationship:

$$L_0 = \frac{gT^2}{2\pi} \tag{10.6}$$

The values of T shown in Fig. 10.3 were computed from Eq. (10.6). Because the wave period remains nearly constant as waves enter shallow water, the subscript for deep water is not used with T. Because the wave period does not vary with depth, the deep-water wave-length may be computed by means of Eq. (10.6) from values of T observed near shore. However, it should be noted that the waves reaching shore may be complex combinations of two or more wave trains, thus making it difficult to detect the principal wave period.

When waves reach *very shallow* water, $d/L \leq 0.1$, the value of tanh $2\pi d/L$ approaches the magnitude of $2\pi d/L$ (Fig. 10.2), and Eq. (10.4) may be reduced to the following approximate relationship:

$$C = \sqrt{gd} \tag{10.7}$$

A segment of a curve obtained from Eq. (10.7) for $L_0 = 100$ ft is shown in Fig. 10.4 to illustrate the convergence of Eqs. (10.7) and (10.4) for small values of d/L.

Wavelength

Because the wave characteristics depend on the value of d/L, it is necessary to derive the equations that permit the evaluation of L for any depth. The determination of L requires that either the deep-water wavelength L_0 or the wave period T be measured or selected arbitrarily. Knowing either L_0 or T, the other can be obtained by means of Eq. (10.6). Equation (10.8) is the functional relationship between d/L and d/L_0.

$$\frac{d}{L_0} = \frac{d}{L} \tanh \frac{2\pi d}{L} \tag{10.8}$$

The solution of Eq. (10.8) is facilitated by the use of Fig. 10.2, or values may be read directly from Fig. 10.4 or from published tables.[1,2] It may be noted that as d/L increases in magnitude, it becomes more nearly equal to d/L_0, and when d/L becomes 0.5, it differs from d/L_0 by only about 0.4 percent.

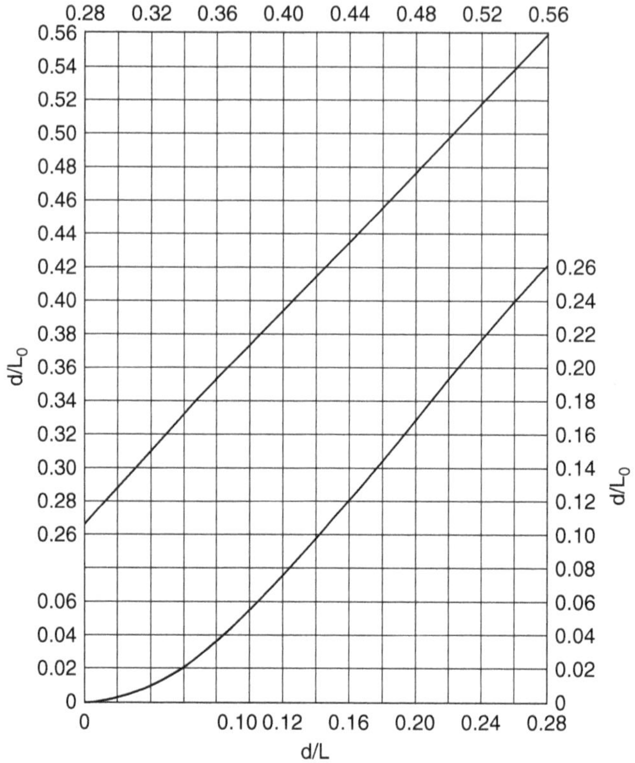

FIGURE 10.4 *d/L* versus d/L_0 from Eq. (10.10).

As an example of the use of Eq. (10.8) it will be shown how points on the curve in Fig. 10.3 for $L_0 = 100$ ft were obtained. After selecting $L_0 = 100$ ft, the period was found to be 4.42 s by means of Eq. (10.6). Then, for selected values of depth, d/L_0 is computed and d/L is determined from Eq. (10.8) or Fig. 10.4. The corresponding values of C are determined from Eq. (10.4), thus providing points on the curve.

Orbital Motion

The equations for orbital displacements, velocities, and accelerations are needed for problems dealing with wave energy and wave forces. The Airy equations[3] for horizontal displacement ξ and vertical displacement η are given by

$$\xi = -\frac{H}{2}\frac{\cosh[2\pi(d+z)/L]}{\sinh(2\pi d/L)}\sin\theta \tag{10.9}$$

$$\eta = \frac{H}{2}\frac{\sinh[2\pi(d+z)/L]}{\sinh(2\pi d/L)}\cos\theta \tag{10.10}$$

In these equations, z is the distance from the water surface, upward being positive, to the point where the displacement is computed. The phase angle θ varies with x or t, or both, as defined by Eq. (10.2). It should be noted that z is measured from the same horizontal plane as y in Eq. (10.1). However, y is used in place of z for those points which fall on the water surface. The displacements of a point originally on the SWL ($z = 0$) are illustrated in Fig. 10.1 for the case of a circular orbit.

The equations for orbital velocity and acceleration may be determined from Eqs. (10.9) and (10.10) by successive differentiations with respect to t. The equations for the horizontal and vertical components of the orbital velocity are

$$\frac{\partial \xi}{\partial t} = u_x = \frac{\pi H}{T} \frac{\cosh[2\pi(d+z)/L]}{\sinh(2\pi d/L)} \cos\theta \tag{10.11}$$

$$\frac{\partial \eta}{\partial t} = u_z = \frac{\pi H}{T} \frac{\sinh[2\pi(d+z)/L]}{\sinh(2\pi d/L)} \sin\theta \tag{10.12}$$

The corresponding expressions for orbital acceleration are

$$\frac{\partial u_x}{\partial t} = \frac{2\pi^2 H}{T^2} \frac{\cosh[2\pi(d+z)/L]}{\sinh(2\pi d/L)} \sin\theta \tag{10.13}$$

$$\frac{\partial u_z}{\partial t} = \frac{2\pi^2 H}{T^2} \frac{\sinh[2\pi(d+z)/L]}{\sinh(2\pi d/L)} \cos\theta \tag{10.14}$$

The manner in which the quantities defined by Eqs. (10.9) to (10.14) vary with θ, x, or t is illustrated in the tabulation at the bottom of Fig. 10.1. The manner in which the horizontal components of the orbital parameters vary with z can be illustrated by arranging all three in the following dimensionless form, assuming that the phase angle in each case is such that maximum positive values occur:

$$\frac{\xi}{H/2} = \frac{u_x}{\pi H/T} = \frac{\partial u_x/\partial t}{2\pi^2 H/T^2} = \frac{\cosh[2\pi(d+z)/L]}{\sinh(2\pi d/L)} \tag{10.15a}$$

It may be seen that, for given values of H and T, the horizontal displacement, velocity, and acceleration vary with z and d/L in accordance with the right-hand term of Eq. (10.15a). The variations of this term with z for shallow water ($d/L = 0.2$) and for deep water ($d/L = 0.6$) are shown in Fig. 10.5. A corresponding dimensionless

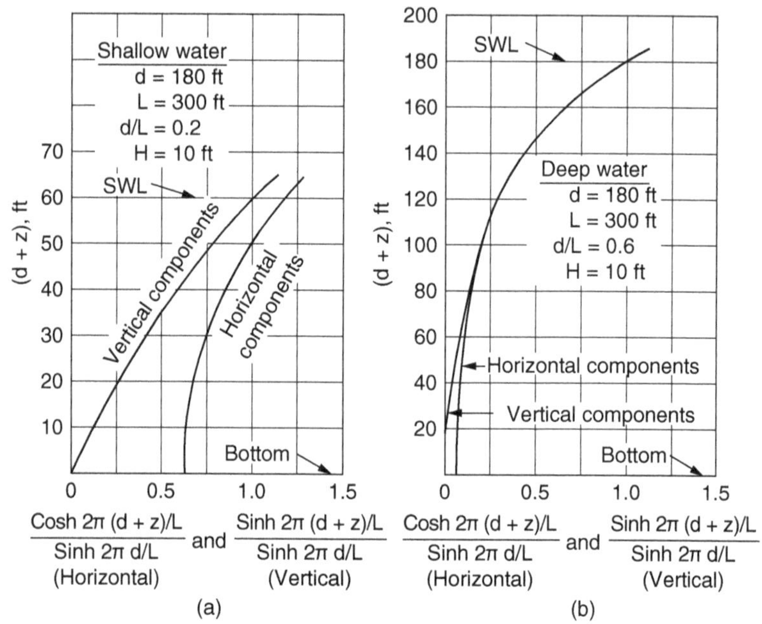

FIGURE 10.5 Variation of oscillatory wave parameters with depth.

arrangement of the vertical components of the orbital parameters results in the equation

$$\frac{\eta}{H/2} = \frac{u_z}{\pi H/T} = \frac{\partial u_z/\partial t}{\pi^2 H/T^2} = \frac{\sinh[2\pi(d+z)/L]}{\sinh(2\pi d/L)} \tag{10.15b}$$

The variations of the vertical components with z for two values of d/L, in accordance with the right-hand term of Eq. (10.15*b*), are also shown in Fig. 10.5. Computations of the orbital components are facilitated by the use of Fig. 10.6 or of published tables.[1,2] Figure 10.5*a* shows, for shallow water, that the orbits are elliptical in shape, that the horizontal components have relatively large magnitudes near the bottom, and that the vertical components all become zero at the bottom. Thus at the bottom, the motion of the water is simply a horizontal oscillation. Figure 10.5*b* shows, for deep water, that the orbits are circular near the water surface and become very small and elliptical near the bottom.

Group Velocity and Energy Transmission

If a group of waves is generated, as by a limited number of strokes of a wave generator in a laboratory or by the motion of a boat, the velocity

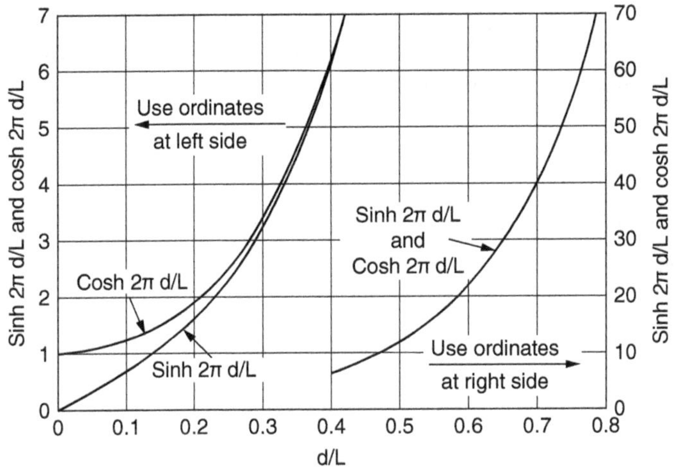

Figure 10.6 Sinh ($2\pi d/L$) and cosh ($2\pi d/L$) versus d/L.

of the center of the group of waves is less than the celerity of the individual waves. This may be explained by the fact that the energy of oscillatory waves is nearly equally divided between kinetic and potential energies and that the kinetic energy is retained at any location while the potential energy is transferred forward at the same speed as the celerity of the individual waves. Then, if a newly generated group of, say, five waves is considered, half of the energy of each wave will move forward one wavelength during one wave period. This will create a new (sixth) wave at the front of the group and leave the rear wave of the original group, each with half as much energy as the original waves, whereas the four waves in between will retain their original energy content. If now it is assumed that the front and rear waves will divide their half-energies equally between kinetic and potential, the next wave period will result in seven waves, the front and rear waves having one-fourth of the original energy, the second waves at front and rear having three-fourths of their original energy, and the three waves in the middle still having their original energy content. The center of the group will now have progressed forward one full wavelength, whereas individual waves will have moved two wavelengths. Thus for the assumptions made, which are approximately correct for waves of small amplitude in deep water, the group velocity is one-half the wave celerity.

An analytical approach based on the assumption that two trains of waves of slightly different wavelengths are traveling in the same direction leads to the following expression for the group velocity[a]:

$$U_G = C - L\frac{dC}{dL} \tag{10.16}$$

Inserting the expression for C given by Eq. (10.4) into Eq. (10.16), the group velocity becomes

$$U_G = C\left[\frac{1}{2} + \frac{2\pi d/L}{\sinh(4\pi d/L)}\right] \tag{10.17}$$

or

$$U_G = nC \tag{10.18}$$

where

$$n = \frac{1}{2} + \frac{2\pi d/L}{\sinh(4\pi d/L)} \tag{10.19}$$

For deep water n becomes approximately ½, which agrees with the previous discussion, whereas for very small values of d/L, n approaches unity, as shown in Fig. 10.7. Values of n may be read from Fig. 10.7 or from published tables.[1,2]

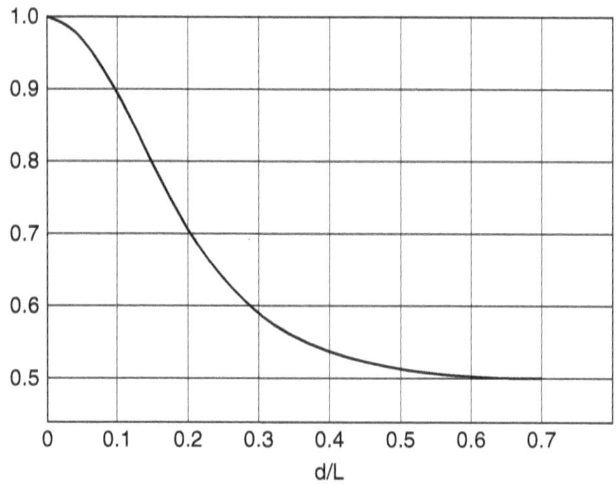

Figure 10.7 *n* versus *d/L*.

[a]This derivation is presented by Ramsey,[4] who credits Stokes, Hamilton, and Rayleigh with original work on this topic.

As was indicated in the preceding discussion, group velocity is related to the rate of transmission of energy. If the instantaneous rate of doing work is taken as

$$W_i = \int_o^d p\,dz\,u_x \tag{10.20}$$

where W_i is the energy per second per unit length of wave crest and p is the pressure, then the approximate expression for the average rate of transmission of energy is[4]

$$W = \int_o^L \frac{W_i\,dx}{L} = \frac{wH^2C}{8}\left[\frac{1}{2} + \frac{2\pi d/L}{\sinh(4\pi d/L)}\right] \tag{10.21}$$

Substitution of n from Eq. (10.19) yields

$$W = \frac{wH^2C}{8}n \tag{10.22}$$

The group velocity as defined in Eq. (10.18) may now be introduced to obtain the following expression for the transmission of energy:

$$W = \frac{wH^2}{8}U_G \tag{10.23}$$

Because $wH^2/8$ is the total energy [Eq. (10.22)], it may be seen that the energy is transmitted at the group velocity or that the potential portion of the energy is transmitted at twice the group velocity.

Refraction

Investigation of Eqs. (10.3), (10.4), and (10.8) will show that as waves enter shallow water ($d/L < 0.5$), the wavelength and the wave celerity decrease. When the waves approach the shore in such a manner that the angle α between the wave crests and the bottom contours is greater than zero, the depth will vary from point to point along each wave crest. Therefore, the portion of the wave that encounters shallow water first will lag behind that portion of the crest which is in deep water. The wave crests then become curved in plan, and the angle between waves and bottom contours becomes smaller as the waves approach the shoreline. This is illustrated in Fig. 10.8, in which the bottom contours are assumed to be parallel straight lines and the value of α in deep water α_0 is taken as 63.5°. Such a refraction diagram is constructed by projecting orthogonals forward one or more wavelengths at a time at selected locations along the wave crests. In Fig. 10.8, the orthogonals were projected three wavelengths at a time; therefore only every third wave crest appears on the diagram.

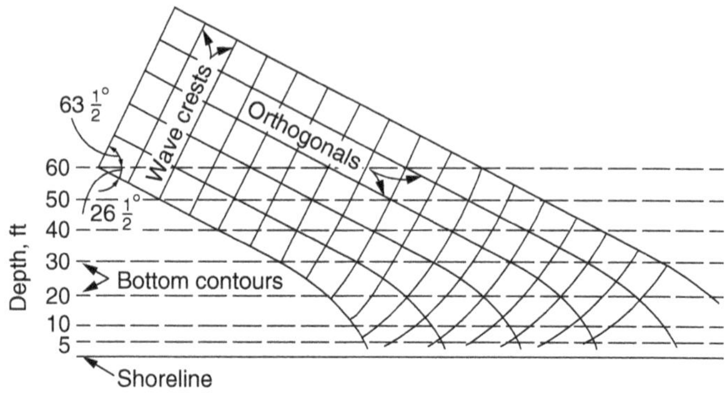

FIGURE 10.8 Refraction diagram.

In order to project an orthogonal one step (three wavelengths) toward shore, the average depth over the projected distance is first estimated from the graph. Then the wavelength is computed from Eq. (10.8), and the orthogonal is carried forward the required distance perpendicular to the previous crest. When this has been done for all the orthogonals, the new crest is sketched in by connecting the ends of the projections. The average depth in each projection must then be checked to ensure that the first estimate was correct. Further adjustments may then be made to provide for the change in direction of the orthogonals.

The essential characteristics of the refraction diagram can also be determined analytically for simple topography. The analysis is similar to that used in optics. Referring to Fig. 10.9, two parallel orthogonals r and s are shown crossing contours 1 and 2, for which the depths, celerities, and wavelengths are indicated by the corresponding subscripts. It is assumed that the change in depth occurs abruptly at the midpoint between contours, that the velocity will change from C_1 to C_2, and that the value of α will change from α_1 to α_2 at that location. A wave crest in the position

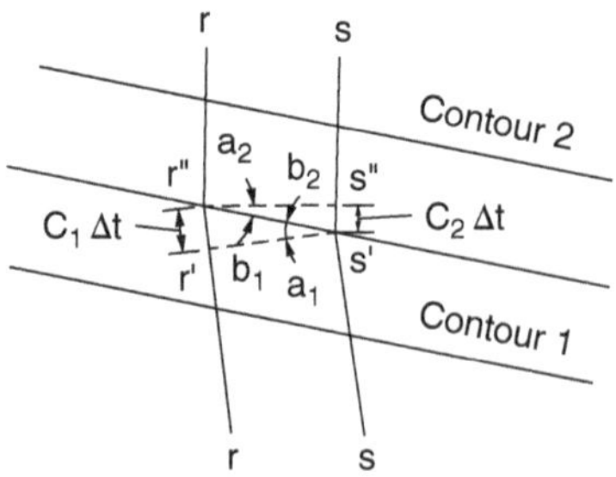

FIGURE 10.9 Refraction at bottom contour.

shown as b_1 will then continue to travel along the orthogonal r at the velocity C_1 over the distance $C_1\,\Delta t$ until the midcontour is reached. During the same time interval, the orthogonal s will progress a smaller distance $C_2\,\Delta t$. At the end of the time Δt, the wave crest will extend from point r'' to s'' in such a manner that it will be perpendicular to the orthogonal s and so that s'' is the distance $C_2\,\Delta t$ from s'. Thus the direction of wave motion across contour 2 is established by $s's''$, and the new distance between contours becomes b_2. The following equation, expressing α_2 in terms of α_1, may be derived from the construction shown in Fig. 10.9:

$$\sin\alpha_2 = \frac{C_2}{C_1}\sin\alpha_1 \tag{10.24}$$

This equation may also be written in terms of wavelength by making use of Eq. (10.3),

$$\sin\alpha_2 = \frac{L_2}{L_1}\sin\alpha_1 \tag{10.25}$$

These equations may be generalized by letting the subscript 1 refer to the original conditions in deep water and removing the subscript 2 entirely, so that they give the orientation of the waves at any depth d, where the celerity is C and the wavelength is L. Equations (10.24) and (10.25) may then be rewritten as

$$\sin\alpha = \frac{C}{C_0}\sin\alpha_0 = \frac{L}{L_0}\sin\alpha_0 \tag{10.26}$$

Again referring to Fig. 10.9, the ratio of the distances between orthogonals may be expressed as

$$\frac{b_1}{b_2} = \frac{\cos\alpha_1}{\cos\alpha_2} \tag{10.27}$$

Here again the equation may be expressed in the following more general form:

$$\frac{b_0}{b} = \frac{\cos\alpha_0}{\cos\alpha} \tag{10.28}$$

where b and α refer to conditions at any depth d and the subscript 0 refers to deep-water conditions. The square root of b_0/b is denoted by K and is called *refraction coefficient*. Then

$$K = \sqrt{\frac{\cos\alpha_0}{\cos\alpha}} \tag{10.29}$$

The refraction coefficient is used in computing wave heights in shallow water.

The application of Eqs. (10.24) to (10.28) is illustrated by the example given in Fig. 10.8. At the 20 ft contour the value of L computed from Eq. (10.10) is 106 ft for $L_0 = 128$ ft. Then the value of α_{20} from Eq. (10.33) is 48°. This agrees very well with values of α_{20} read from Fig. 10.8. The value of K computed from Eq. (10.29) for a depth of 20 ft is 0.82, whereas K computed from values of b and b_0 scaled from Fig. 10.8 has a value of approximately 0.83. After a number of computations of this type, an entire system of orthogonals and wave crests could be drawn. A number of techniques for constructing refraction diagrams based on similar analytical procedures have been devised.[5–7]

Variation of Wave Height with Depth

One of the principal applications of a refraction diagram is in the determination of the height and orientation of the waves at various locations in shallow water for the purpose of designing structures and studying shore processes. The assumption is made in this derivation that no energy will cross the orthogonals and that no energy is lost. Dissipation of energy causes a reduction in wave height over that computed from the refraction diagram. This effect is considered in this section.

The rate at which energy progresses in the direction of wave motion was given by Eq. (10.22), which is repeated here,

$$W = \frac{wH^2C}{8}n \tag{10.22}$$

The total energy transmission between two orthogonals separated by the distance b (Fig. 10.8) is

$$bW = \frac{wbH^2C}{8}n \tag{10.30}$$

For deep water this expression reduces approximately to

$$b_0W_0 = \frac{wb_0H_0^2C_0}{16} \tag{10.31}$$

If it is assumed that no energy is lost either by dissipation or by flowing across the orthogonals, then

$$bW = b_0W_0 \tag{10.32}$$

and
$$\frac{wbH^2c}{8}n = \frac{wb_0H_0^2C_0}{16} \tag{10.33}$$

which reduces to

$$\frac{H}{H_0} = \sqrt{\frac{b_0}{b}}\sqrt{\frac{C_0/C}{2n}} \tag{10.34}$$

The introduction of K from Eq. (10.29) leads to the expression

$$\frac{H}{H_0} = K\sqrt{\frac{C_0/C}{2n}} \tag{10.35}$$

The term K, which shows the effect of refraction, can be computed from refraction diagrams or from Eq. (10.29), as described previously. When the direction of wave motion is perpendicular to straight parallel bottom contours ($\alpha = 0$), K is unity and Eq. (10.35) may be written as

$$\frac{H}{H_0'} = \sqrt{\frac{C_0/C}{2n}} \tag{10.36}$$

where H_0' is the deep-water wave height when K is unity or the hypothetical deep-water wave height corresponding to the value of H at any location, neglecting refraction. The value of H_0' is required to determine the breaker height and the depth at which waves break. Equation (10.35) can now be written in the form

$$\frac{H}{H_0} = K\frac{H}{H_0'} \tag{10.37}$$

from which it can be seen that $K = H_0'/H_0$.

The value of H/H_0' may be computed from Eq. (10.36) by inserting the value of C_0 from Eq. (10.5), C from Eq. (10.4), and n from Eq. (10.19). Because, for a particular deep-water wave, these quantities depend only on the depth, the refraction diagram is not needed to determine H/H_0'.

A graph showing the variation of H/H_0' with d/L_0 is given in Fig. 10.10. Tables showing the variation of H/H_0' with d/L_0 and d/L_0 are also available.[1,2] For the case of waves traveling directly toward a straight shoreline ($K = 1$), this graph also represents H/H_0. It may be seen that, neglecting refraction, the wave height tends to decrease until near the breaking point, when it increases again. Also shown in Fig. 10.10 is a graph of H/H_0 for the waves of Fig. 10.8, which are approaching from $\alpha_0 = 62.5°$. Computations involved are summarized

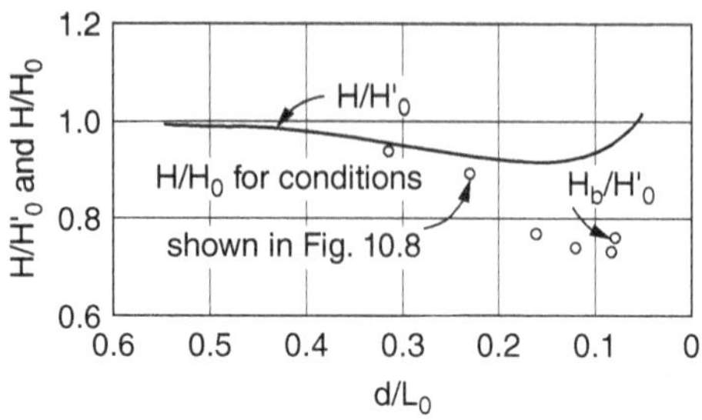

Figure 10.10 H/H_0 and H/H_0' versus d/L_0.

in Table 10.1. Selected values of d are given in column 1 and the corresponding magnitudes of d/L_0 in column 2 ($L_0 = 39$ m). Values of H/H_0' obtained from Fig. 10.10 are shown in column 3. Values of b scaled from Fig. 10.8 and the corresponding values of K are listed in columns 4 and 5. The wave heights shown in column 6 were then determined from Eq. (10.37).

Breaking Waves

The previous subsection dealt with the change in wave height as the wave progressed into shallow water. For the case of no refraction or for converging orthogonals, the wave steepness H/L gradually increases as the waves approach the shore, until eventually the waves become unstable and break. When orthogonals are diverging, the effect of the divergence may cause the steepness to decrease as the waves first enter shallow water, but this trend is eventually reversed, and then the steepness begins to increase rapidly until the breaking

Table 10.1 Example Refraction Calculation

(1) d, m	(2) $\frac{d}{L_0}$	(3) $\frac{H}{H_0'}$	(4) d, m	(5) K	(6) H, m	(7) H_0', m	(8) $\frac{d}{H_0'}$	(9) $\frac{H_0'}{L_0}$
24	0.63	1.0	117	1.0	3.00	3.00		
18	0.47	1.0	117	1.0	3.00	3.00		
15	0.39	0.97	117	1.0	2.96	3.00		
12	0.31	0.95	119	0.99	2.86	3.00		
9	0.23	0.93	129	0.95	2.71	2.93	3.1	0.075
6	0.16	0.91	166	0.84	2.35	2.59	2.4	0.067
4.5	0.12	0.92	184	0.80	2.26	2.47	1.9	0.063
3	0.078	0.96	203	0.76	2.23	2.32	1.3	0.059

$L_0 = 39$ m.

point is reached. Several investigators have concluded that the theoretical limiting case of stability occurs when H/L becomes 0.143. Waves in shallow water usually break before this limiting condition is reached. Laboratory and field observations on the height of waves at breaking H_b and the still-water depth at the breaking point d_b serve as the basis for predicting these parameters under field conditions. The most extensive set of observations, gathered by the Hydrographic Office of the U.S. Navy,[8] are represented by the curves of d_b/H'_0 versus H'_0/L_0 and H_b/H'_0 versus H'_0/L_0, designated as H.O. in Fig. 10.11. The data which established these curves have a maximum scatter of about 25 percent above and below these curves. It should be noted that the deep-water wave height H'_0 used in these correlations is a hypothetical value computed from Eq. (10.37).

The slope of the beach has an important effect of H_b and d_b. This is illustrated by curves, based on laboratory tests, presented by Iversen.[9] His curves for the extreme slopes, in the range of slopes tested, are also reproduced in Fig. 10.11.

Figure 10.11 further includes a curve derived by Munk[10] on the basis of the theory of solitary waves. Included in this theory is the

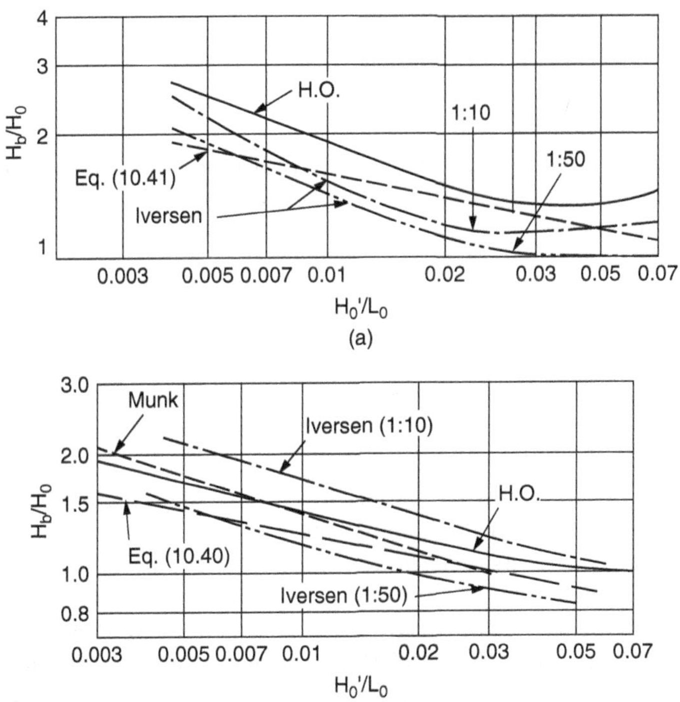

Figure 10.11 Depth at breaking.

concept that the ratio of depth at breaking to breaker height d_b/H_b, is 1.28. This value agrees quite well with the average of all observations. The range is from slightly less than 1.0 to about 1.5. The effect of the steepness of the beach slope may be estimated from Iversen's data,[9] which showed an average value of d_b/H_b of about 1.0 for 1:10 slopes, whereas for a slope of 1:50 the average value was near 1.2.

The Airy wave equations also can be used to make an approximate analytical approach to the determination of these relationships. If Eq. (10.37) is written for the breaker location, it becomes

$$\frac{H_b}{H_0'} = \sqrt{\frac{C_0/C_b}{2n}} \tag{10.38}$$

At the breaker zone shallow-water conditions exist, and therefore n [Eq. (10.19)] may be taken at unity, and from Eq. (10.7) the value of C_b becomes

$$C_b = \sqrt{gd_b} \tag{10.39}$$

If it is assumed that $d_b = 1.28H_b$, then the insertion, of the value of C_0 from Eq. (10.5) permits the transformation of Eq. (10.38) into the following expression:

$$\left(\frac{H_b}{H_0'}\right) = \frac{0.031}{H_0'/L_0} \tag{10.40}$$

This equation is plotted in Fig. 10.11*b*. If now H_b in Eq. (10.40) is replaced by $d_b/1.28$, the following expression for the depth at breaking is obtained:

$$\left(\frac{d_b}{H_0'}\right)^5 = \frac{0.106}{H_0'/L_0} \tag{10.41}$$

This equation is plotted in Fig. 10.11*a*. Here again there is fair agreement with the H.O. curve, particularly for the large values of H_0'/L_0.

These curves are used to determine the depth of breaking and the breaker height for any deep-water wave conditions in a manner that will be illustrated for the example given in Figs. 10.8 and 10.10. The wave heights are first determined for selected depths in the manner previously described and demonstrated in Table 10.1. Values of H_0' are then obtained by dividing H by H/H_0'. These values are shown in column 7. Values of d/H_0' and H_0'/L_0 are then computed for the smaller depths (columns 8 and 9). The depth of breaking is determined by finding the point where the values in columns 8 and 9 fall on the breaker index curves of Fig. 10.11*a*. If the values shown in columns 8 and 9 were plotted on Fig. 10.11, it would be seen that they cross the H.O. curve for a depth slightly larger than 3 m. Therefore for practical purposes, the breaker zone is located at a depth of 3 m. The corresponding breaker height H_b is then determined from Fig. 10.11*b*. Using the H.O. curve, at H_0'/L_0 the value of H_b/H_0' is approximately 1.0, thus establishing H_b as 2.3 m. The breaker location and height are also shown in Fig. 10.10. The lower points in Fig. 10.10 are a dimensionless representation of the variation of H with d for the example given in Fig. 10.8, with the exception that, at the point of breaking, the wave height departs from the curve and becomes H_b. It may be noted that, although the upper curve in Fig. 10.10 is completely general and applies to all conditions, it may also be considered as representing the variation of wave height with depth for a wave traveling perpendicular to a straight shoreline ($a_0 = 0$).

Characteristics of Wind-Generated Waves

As wind moves across a body of water, the shear stress between wind and water sets the upper layers of water into motion. As a result of the large-scale turbulence in the air, slight pressure and velocity differences exist which create small disturbances in the water surface. As soon as waves are formed, the wind produces a positive pressure on the windward sides of the waves and a reduced pressure on the leeward sides. The resultant pressure force combined with the drag force transmits energy to the water and causes the waves to grow. Because of the unsteady nature of the wind with respect to both location and time, a spectrum of wave heights, wave periods, and wavelengths is developed. Usually a dominant pattern is developed upon which waves of different heights and lengths are superimposed. Because the wave celerity varies with the wavelength [Eqs. (10.4) and (10.5)], the various wave patterns pass in and out of phase with the dominant pattern, with the result that the composite pattern is often quite irregular, consisting of alternating groups of high and low waves.

The variation in wave height within a continuous series of waves is illustrated by the following tabulation of ratios:

$$\frac{H_{1/3}}{H_1} = 1.6$$

$$\frac{H_{1/3}}{H_{1/3}} = 1.3$$

$$\frac{H_{1/100}}{H_{1/3}} = 1.7$$

$$\frac{H_{MAX}}{H_{1/3}} = 1.8$$

The numerical values were selected as representative of published results of wave train analyses.[11–13] H_1 is the mean of all wave heights in a series, $H_{1/3}$ is the average of the highest one-third of the waves in the series, $H_{1/10}$ is the average of the highest one-tenth of the waves, and $H_{1/100}$ is the average of the highest 1 percent of the waves. $H_{1/3}$ is called the *significant wave height.* This value has been adopted as a design parameter for many purposes, and it is the wave height used in the following relations between wave height, fetch, and wind.

Correlations of wind characteristics with wave characteristics were made by Sverdrup and Munk[14,15] and revised by Bretschneider[16,17] as additional data became available. The wave characteristics used in the correlations were significant wave height and significant wave period. Knowing the wave period, the wavelength can be computed from Eq. (10.6) for deep water and from Eq. (10.8) for shallow water. The wind characteristics which must enter the correlations are velocity U, duration t, and fetch F. The fetch is the distance over which the wind is in contact with the water. On lakes this is usually the distance from shore to shore, whereas on the oceans the fetch must be scaled from synoptic charts. Wind velocities and durations may be obtained from measurements near the shorelines of lakes or from synoptic charts of pressure gradients for the oceans.[14] The wind velocity used in the correlations was taken about 26 ft (8 m) above the ground or water surface. When records are obtained at other elevations, they must be reduced to this elevation by assuming that a typical velocity gradient existed.[18] Dimensional analysis suggests the arrangement of variables shown in Eq. (10.42) for relating wave characteristics to fetch and wind speed,

$$\frac{gH}{U^2} \text{ and } \frac{gT}{U} = f\left(\frac{gF}{U^2}\right) \tag{10.42}$$

The corresponding parameters when wind duration is included rather than fetch are

$$\frac{gH}{U^2} \text{ and } \frac{gT}{U} = f\left(\frac{gt}{U}\right) \tag{10.43}$$

The data were plotted according to these functional relationships and average curves drawn through rather widely scattered points.[14–17] Based on the most recently revised curves, the chart presented in Fig. 10.12 was prepared by Bretschneider.[16] This convenient chart includes all the variables needed to predict wave characteristics from wind. The proper use of the chart requires the following consideration of wave growth. At a particular location, assuming a certain fetch and a

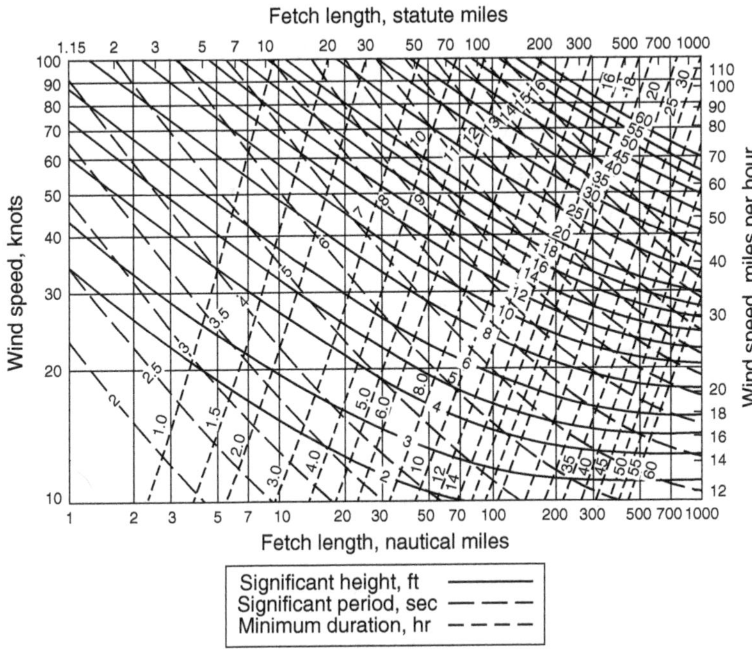

Figure 10.12 Deep-water wave-forecasting curves as a function of wind speed, fetch length, and wind duration.

constant wind speed, the wave height and period continue to grow with the wind duration until steady (maximum) conditions are attained.[b] During this growing period the wave characteristics at this location are a function of wind duration and velocity. However, when steady conditions are reached, the wave characteristics are established by the fetch together with the wind velocity. At locations farther downwind (larger fetch), the steady conditions are reached later (larger minimum duration), and the ultimate wave height and period have correspondingly larger values. For example, suppose that, for a fetch of 100 miles and an average wind velocity of 30 mi/h, the wind duration was 12 h. The chart would be entered at 30 mi/h, and proceeding from left to right, it will be noted that, for a fetch of 100 miles, the minimum duration is approximately 9.5 h. Therefore equilibrium conditions were reached at the end of 9.5 h, and the wave height and period can be read at the intersection of the lines of $F = 100$ and $U = 30$ miles per hour, the values being approximately 8.8 ft and 7.8 s, respectively. If, however, the duration of

[b]The time required for conditions to maximize at a given location is termed minimum duration.

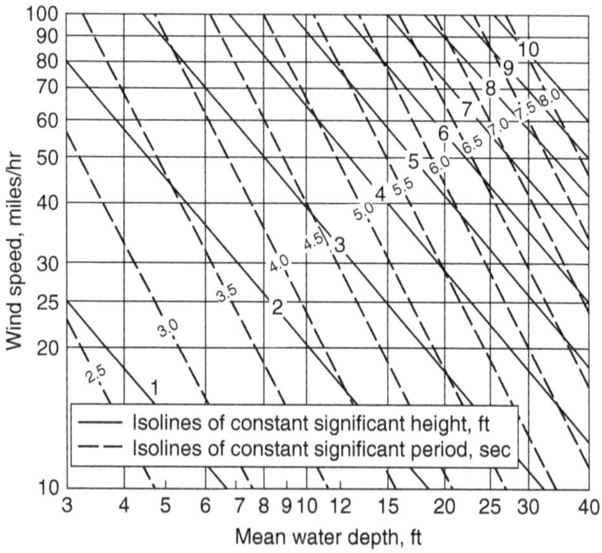

FIGURE 10.13 Wave-forecasting relationships for shallow water of constant depth with unlimited fetch.

the wind were only 6 h, the duration would control and the wave height and period would be determined at the intersection of $U = 30$ mi/h and $t = 6$ h, the resulting values for H and T being 7.1 ft and 6.7 s, respectively.

When water is shallow, $d/L < 0.5$, the wave height and period produced by a given wind are smaller than for deep water because of the energy dissipation due to bottom friction and percolation. Curves showing the probable wave height and wave period have been prepared[16,19,20] for the shallow-water conditions with unlimited wind duration. The curves are based in part on observations and in part on computations assuming the friction factor k to be 0.01 and neglecting percolation losses. Two charts derived from these curves by Bretschneider[16] are shown in Figs. 10.13 and 10.14. These apply to a constant depth. The significant wave height and wave period can be read directly from Fig. 10.13 for an unlimited fetch. When the fetch is short, values can be read from the corresponding gF/U_2 curve of Fig. 10.14, as illustrated in the following example.

Example 10.1 Suppose the fetch is 15 mi, the depth is 30 ft, and the wind velocity is 30 mi/h. Then

$$\frac{gF}{U^2} = \frac{32.2 \times 15 \times 5280}{30^2 \times (5280/3600)^2} = 1310$$

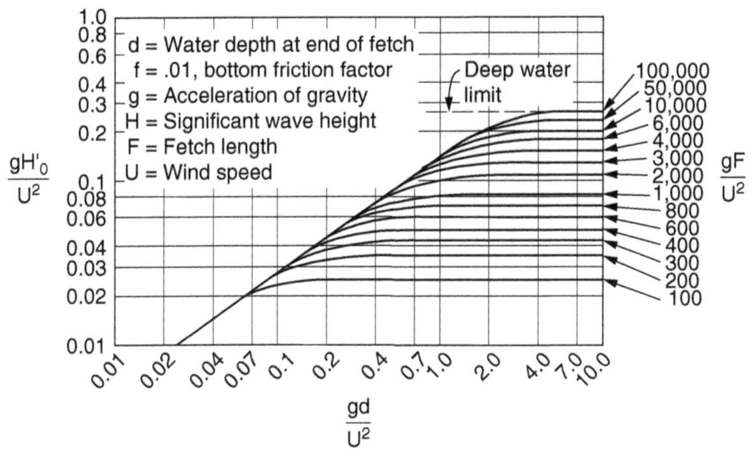

FIGURE 10.14 Generation of wind waves over bottom of constant depth for unlimited duration.

$$\frac{gd}{U^2} = \frac{32.2 \times 30}{30^2 \times (5280/3600)^2} = 0.50$$

Then, from Fig. 10.14, interpolating between the curves for $gF/U^2 =$ 1000 and 2000,

$$\frac{gH}{U^2} = 0.076$$

and

$$H = \frac{0.076 \times 30^2}{32.2} \times \left(\frac{5280}{3600}\right)^2 = 4.6 \text{ ft}$$

The wave period for this example can be estimated from Fig. 10.13 by entering at $U = 30$ m/h and noting that, for $H = 4.6$ ft, $T = 5.9$ s.

If the fetch extends from deep water into relatively shallow water, the resulting wave height will be between that estimated from Figs. 10.13 and 10.14 for shallow water of a constant depth and that determined for deep water from Fig. 10.15. The U.S. Corps of Engineers has presented curves for estimating wave heights under these conditions for three bottom slopes.[20] It is recognized that for narrow bodies of water the effective fetch to be used in estimating wave height may be less than the actual fetch. Field data are not available, but an analytical approach has been developed,[21] which indicates that substantial effects can be expected when the fetch width becomes less than half the fetch. ▲

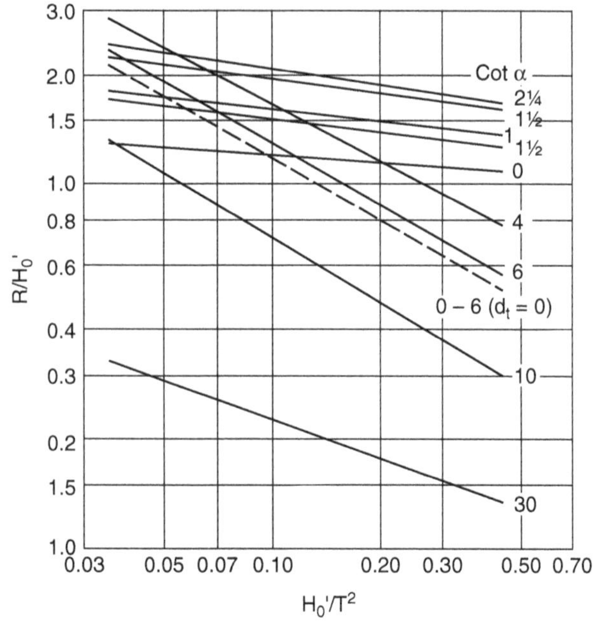

Figure 10.15 Wave run-up, smooth surfaces.

Wave Run-Up and Overtopping

The selection of the crest elevations of sea walls, breakwaters, revetments, and other shore structures requires information regarding the height to which waves will run up the face of the structure or an estimate of the discharge over the wall when overtopping occurs. This section presents a summary of the results of laboratory tests on these phenomena.

Run-up R is defined as the vertical distance above the SWL that a wave will run up the slope. It should be noted that during an onshore wind, the SWL may rise considerably because of wind setup or wind tide (see "Wind Tides"). The run-up is a function of the wave height H, the wave period T, the depth of water at the toe of the structure d_t, the slope of the face of the structure tan a, the roughness and permeability of the structure, and the local wind. It has been shown[22,23] that a satisfactory arrangement of the variables for smooth slopes is as given in Fig. 10.18, where R/H_0' is plotted against H_0'/T^2 for various values of cot α. The effect of variations in d_t is also included in these curves, as will be explained later. R/H_0' is called the relative run-up, H_0' being the wave height in deep water if there were no refraction, as defined more fully under "Variation of Wave Height with Depth." H_0'/T^2 is a convenient method of expressing wave steepness, which is usually written as H/L. The close relationship between T^2 and L is shown by Eq. (10.6). The lines shown in Fig. 10.15 are a summary of values presented in the references given previously for the practical range of H_0'/T^2. The solid lines are for values of d_t such that $d_t > H_0'$. For this condition the wave breaks on the structure, and as can be seen from Fig. 10.15 there is considerable variation with α. For the case of zero depth at the face of the structure ($d_t = 0$), the slope of the structure has little effect on run-up for slopes of 1:6 or steeper, and the dashed line in Fig. 10.15 can be used for all values of cot $\alpha \lesseqqgtr 6$. For slopes of 1:10 and 1:30, the run-up is affected very little by d_t, and the solid curves in Fig. 10.15 may be used for all values of d_t. When cot $\alpha \lesseqqgtr 6$, and for values of d_t between H_0' and zero, values of run-up can be estimated by interpolating between the dashed line and the appropriate solid line.

The effects of roughness and permeability have been determined[24] by measuring run-up with material of a certain size fastened to an impermeable surface and then using the same size material to form the entire wall. Based on the values for 10-mm stone, curves were developed showing run-up on rough impermeable and rough permeable revetments in terms of the comparable run-up on smooth impermeable surfaces.[23,24] These curves, shown in Fig. 10.16, are the averages for the range of H_0'/T^2 from 0.04 to 0.40. Model data have also been derived for run-up on slopes protected with various types of precast armor units.[25]

Several series of laboratory tests on *overtopping*[22,26,27] provide data from which the discharge over a sea wall can be estimated. Some of the data are summarized in Fig. 10.17*a*. Here Q is the discharge per unit length of wall, H is the wave height at the toe of the wall, T the

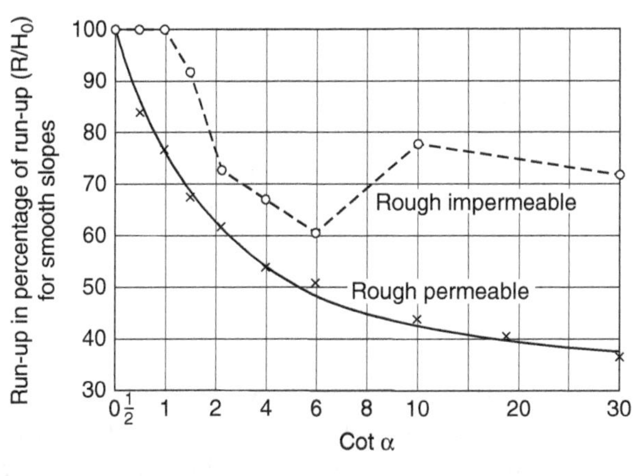

FIGURE 10.16 Wave run-up, rough surfaces.

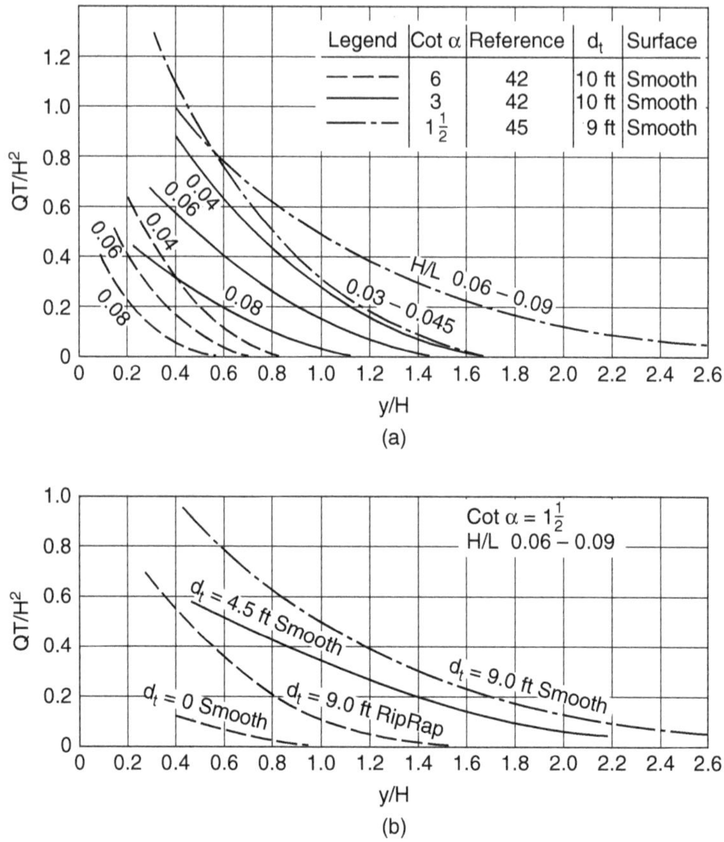

Figure 10.17 Wave overtopping.

wave period, L the wavelength at the toe of the wall, tan α the slope of the face of the wall, and y the height of the wall above the wind-tide level. It may be seen that the group of variables QT/H^2 varies with the height of the structure relative is the wave height y/H, the wave steepness H/L, and cot α. The relative spacing of three sets of curves (each for a particular value of α) shows that overtopping increases as the slope increases (cot α decreases) up to

$$\cot \alpha = 1\frac{1}{2}$$

However, for a vertical wall (not shown in the figure) the values of discharge drop back to magnitudes similar to the ones for cot $\alpha = 3$. The effect of the depth at the toe of the wall d_t is illustrated in Fig. 10.17b, in which curves for cot $\alpha = 1\frac{1}{2}$ and for H/L in the range of 0.06 to 0.09 for smooth slopes are shown for three values of d_t. It may be seen that overtopping increases with d_t. The effect of roughness is also demonstrated in Fig. 10.17*b* by comparing the location of the line for a slope covered with riprap to that of the smooth-surface curve for the same value of d_t.

Laboratory tests made with wind-generated waves[27] indicate that overtopping is increased by the action of the wind at the structure for winds greater than 30 mi/h. This increase may be by as much as 50 percent. It should be noted, however, that most of the values gives in these graphs are for structures having a smooth face. The one curve in Fig. 10.17*b* for a riprap-covered wall indicates that a roughened surface tends to reduce the overtopping and may compensate for the wind effect.

Wave Forces

The topic of wave forces can be divided into two subdivisions, forces on shore structures and forces on submerged offshore structures. The first deals with the forces needed to design breakwaters, retaining walls, jetties, groins, and revetments, located in such a manner that long segments of the waves are reflected or absorbed. The second category deals with forces on free-standing structures located far enough from shore so that waves pass by with no distortion except that caused by the structure itself.

Forces on shore structures located in deep water are caused by the increase in depth resulting from a reflected wave, whereas structures located in shallow water must be designed to withstand the impact of breaking waves and the erosive force resulting from the uprush and return of the waves. Minikin[28] presents a simplified procedure for determining forces on a vertical wall from nonbreaking waves, which is easy to use. Pressure varies from zero at a distance h above the water surface to a maximum at the SWL, as shown in Fig. 10.18. The Minikin method assumes a uniform pressure from the SWL to the bottom. The diagrams in Fig. 10.18 show only the pressures that are produced by the wave reflection, and the hydrostatic pressure resulting from water at a depth d (shown by line *SS* in Fig. 10.18) must be added to obtain the total force on the wall. In most cases this force can be neglected because it is balanced by an equal and opposite pressure on the back of the wall. The equations for the various parameters are as follows:

$$h = 1.66H \tag{10.44}$$

$$p_1 = wH \tag{10.45}$$

$$p_2 = wH \tag{10.46}$$

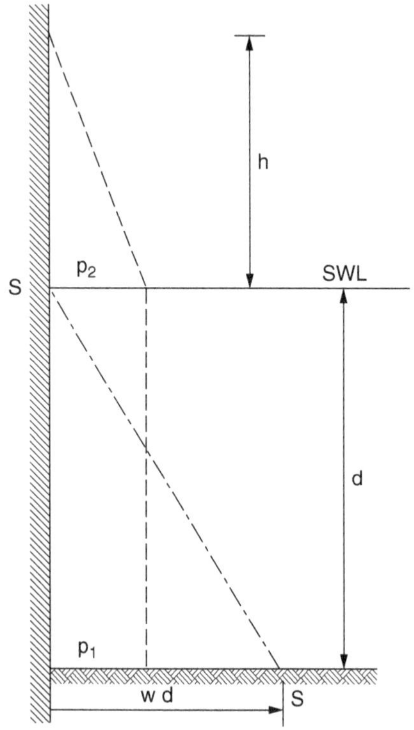

FIGURE 10.18 Wave forces on sea walls, deep water.

The graphs shown in Fig. 10.18 were drawn to represent the following numerical example:

$$H = 3 \text{ m}, \quad L = 150 \text{ m}, \quad d = 8 \text{ m}$$

The numerical values for this problem are as follows:

h, m	5.1
p_1, kN/m^2	29.9
p_2, kN/m^2	29.9
Total force, kN/m	315
Moment at bottom, m·N	483,000

Hudson[29] has reported model test results which show that there is a systematic deviation of the measured overturning moment M_M, depending on variations in d/L. Hudson's curve is reproduced in Fig. 10.19. It will be seen that for the value of d/L in the numerical

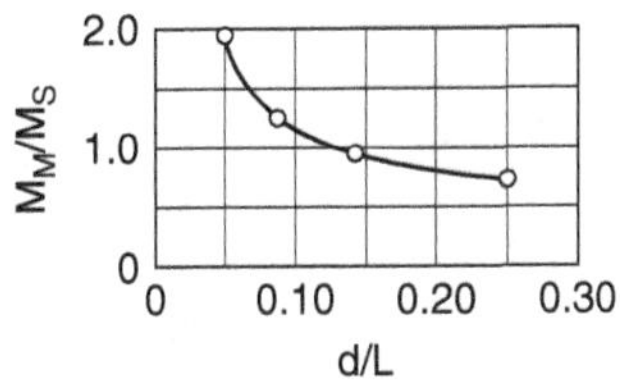

FIGURE 10.19 Wave-force correction factors.

example (0.05), the measured moment would be nearly twice the computed value. A safe design procedure would be to multiply the computed values by the ratio from Fig. 10.19.

Forces due to breaking waves are caused by the impact of the onrushing water. Laboratory measurements show that there is a great variation in the pressure intensity and that some local areas of very high pressure occur for short intervals. The intensity of pressure is also affected by the amount of entrapped air.[30] Some observations of damage from breaking waves have provided information regarding pressures over relatively large areas.[31] These pressures were computed by estimating the force required to move large blocks of stone, taking into consideration the coefficient of friction and the uplift. The maximum pressures were computed by assuming that the horizontal variation in pressure was such that the average pressure was from 0.50 to 0.67 times the maximum. The two cases reported indicated maximum pressures of approximately 3000 and 1800 lb/ft² for wave heights of 13.5 and 12.0 ft, respectively. The same writers presented the following semi empirical relationship:

$$p_{max} = 133.6H \quad \text{lb/ft}^2 \tag{10.47}$$

This equation agrees very well with measurements made by Gaillard[32] in the Atlantic Ocean and Lake Superior, as well as with the two values reported in the foregoing.

Minikin[28] derived the following semiempirical equation for forces due to breaking waves:

$$p_{mi} = 2\pi wg\frac{d}{L}\left(\frac{D+d}{2D}\right)H \quad \text{lb/ft}^2 \tag{10.48}$$

where p_{mi} is the maximum pressure occurring at the SWL with a vertical distribution such that the pressure becomes zero a distance $H/2$ above and below the water surface, and the average pressure over the vertical distance H is $p_{mi}/3$. The other new term introduced in Eq. (10.48) is D, which is defined in Fig. 10.20, Minikin having

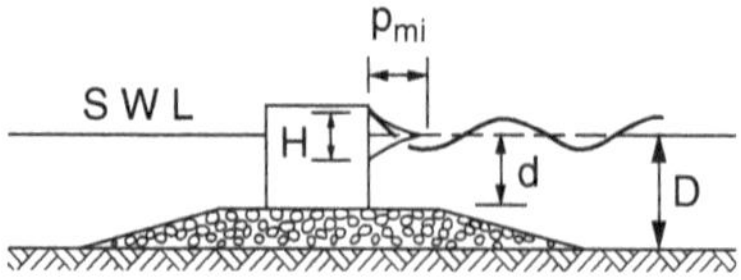

FIGURE 10.20 Wave forces on sea walls, breaking waves.

developed his equation for a composite breakwater such as the one shown. The equation is sometimes applied to simple break-waters (without the rubble base) by assuming that D is the depth one wavelength seaward of the wall.[20] Equation (10.48) gives pressures somewhat higher than Eq. (10.47).

When breaking waves approach a breakwater from a direction such that the angle β between the wave crests and the wall is different from zero, the pressure is reduced from the values given by Eq. (10.47) or (10.48). As illustrated in Fig. 10.21, if the average force per unit area of wall resulting from direct wave impact is P_{av}, then the component perpendicular to the wall is $P_{av} \cos \beta$. This component is reflected from a wall area equal to $1/\cos \beta$ and the average pressure on the wall $p_{w,av}$ may be obtained as

$$p_{w,av} = \frac{P_{av} \cos \beta}{(1/\cos \beta)} = p_{av} \cos^2 \beta \tag{10.49}$$

Rubble-mound breakwaters fail because of the displacement of individual stones. Extensive model tests conducted at the U.S. Waterways Experiment Station[33] have provided information for the design of such structures. The following equation for the weight of stone required for stability was derived by Hudson[33]:

$$W_r = \frac{w_r H^3}{K_\Delta (S_r - 1)^3 \cot \alpha} \tag{10.50}$$

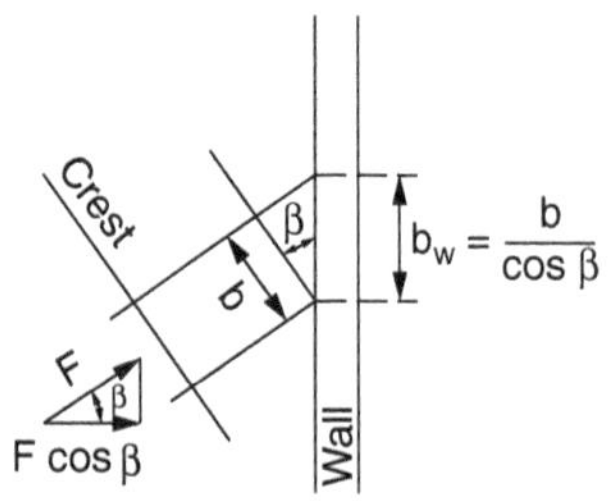

FIGURE 10.21 Wave forces on sea walls.

where W_r is the weight of a stone in newtons, w_r is the specific weight of the stone, in newtons/m^3, H is the wave height in meters, K_Δ is a coefficient depending on the shape of the stone, S_r is the specific gravity of the stone relative to the water in which it is being used, and α is the angle between the breakwater face and the horizontal. Tests were conducted for rock and formed concrete units. Values of K_Δ vary from 4 for randomly placed angular rock to more than 10 for formed units. The tests showed that two layers of armor, or cover stones, provided the optimum amount of protection for the volume of material used. The seaward ends of the breakwater are more vulnerable to wave attack than the body of the breakwater, and the values of K_Δ for nonbreaking waves should be reduced by about 25 percent from those shown. For breaking waves values of K_Δ should be about 12 percent less than for nonbreaking waves.

Some of the formed sections tested in the laboratory were found to be exceedingly stable under wave attack, as indicated by values of K_Δ much larger than those for quarry stones.[20,25,33,34] The question of which type of construction to use in any particular case would depend on the relative costs of quarry stone and formed units.

Wave Forces on Piling and Submerged Structures

The forces produced on submerged structures by oscillatory waves are of two types, the drag force, resulting from orbital velocity, and the inertial force, resulting from orbital acceleration. The drag force can be expressed as

$$F_d = C_d A \frac{\rho u^2}{2} \tag{10.51}$$

where C_d is the drag coefficient, ρ the density, A the cross-sectional area of the structure in a plane normal to the force, and u the component of the orbital velocity in the direction of the force, as given by Eqs. (10.11) and (10.12). The flow pattern referred to is that which would occur if flow were not disturbed by the presence of the submerged body. Because of the nature of the velocity variation [Eq. (10.11)], the horizontal drag force varies from a maximum downwave value under the crest ($x/L = 0$) to a maximum upwave value under a trough ($x/L = ½$), as illustrated in Fig. 10.1. For steady flow the drag coefficient has been evaluated for many different shapes and has been found to be a function of the form and of the Reynolds number. However, for the periodic unsteady flow associated with oscillator wave motion, this quantity may differ considerably from steady-flow values. An additional complication arises for the case of large structures because of the variation of the velocity within the dimensions of the body.

The inertial force is expressed by

$$F_i = C_m \rho \forall \frac{\partial u}{\partial t} \tag{10.52}$$

where C_m is the coefficient of inertial resistance, $\forall$ the displaced volume, and $\partial u/\partial t$ the component of acceleration in the direction of the force. This type of force occurs for two reasons. The acceleration of the fluid is caused by a pressure gradient so that the pressure on one side of the structure is different from that on the other side. The resulting force can be shown to be equal to the product of the acceleration and the mass of the displaced volume. Furthermore, because of the presence of the structure, the accelerations in the flow around the body are increased, thus requiring an additional pressure difference. This force divided by the acceleration is called *virtual mass* and is conveniently expressed by its ratio to the mass of the fluid displaced. Thus C_m is the sum of a term, approximately unity, for the pressure-gradient force and the coefficient of virtual mass. The variation of the horizontal acceleration [Eq. (10.15)] is such that the horizontal inertial force varies from a maximum downwave value at $x/L = ¼$ to a maximum upwave value at $x/L = ¾$, as illustrated in Fig. 10.1. Values of C_m must be determined experimentally.

The total force is the sum of the two forces given by Eqs. (10.51) and (10.52),

$$F = C_d A \frac{\rho u^2}{2} + C_m \rho \forall \frac{\partial u}{\partial t} \tag{10.53}$$

Because the two forces are 90° out of phase, the maximum combined downwave force may occur at any value of x from $x/L = 0$ to $x/L = ¼$, depending on the relative magnitudes of the two components. For any particular value of wave height and wave period and for a particular value of z, F is a function only of the phase angle, and Eq. (10.53) can be expressed as

$$F = C'_d \cos^2 \theta + C'_m \sin \theta \tag{10.54}$$

where

$$C'_d = C_d \rho A \frac{\pi^2 H^2}{2T^2} \left\{ \frac{\cosh[2\pi(d+z)/L]}{\sinh(2\pi d/L)} \right\}^2 \tag{10.55}$$

$$C'_m = C_m \rho \forall \frac{2\pi^2 H}{T^2} \frac{\cosh[2\pi(d+z)/L]}{\sinh(2\pi d/L)} \tag{10.56}$$

The phase angle corresponding to the maximum force is obtained by setting $dF/d\theta = 0$, with the result

$$\sin\theta_{\max} = \frac{C'_m}{2C'_d} \tag{10.57}$$

A similar derivation for the phase angle of the maximum moment on piles has been presented by Morison.[35]

The equations presented here may be applied to structures having dimensions which are small in the direction of wave motion by taking the velocity and acceleration in Eq. (10.53) at the center of the structure. For large structures it is necessary to take into consideration the variation in velocity and acceleration over the dimension of the body. The methods for finding wave forces on large submerged structures are treated later in this section.

The computation of wave forces on submerged structures from the equations presented requires information regarding values of C_d and C_m. Values must be determined from unsteady flow conditions because, as explained previously, steady-flow values do not apply. Model tests on circular cylinders[36] indicate that a value of 2.0 for C_m will give conservative values of the inertial force. However, the same study showed values of C_d varying from 1.2 to 4.4. These tests were made by measuring moments on the model piles. Thus the values of C_m and C_d include the integrated effects of distance below the water surface. The same study indicated that when three piles are placed in a row perpendicular to the direction of wave motion, the moment on the center pile will be greater than that on a single pile by an amount varying from about 7 percent for a pile gap of 1½ pile diameters to approximately 100 percent for a gap of only ½ diameter. For three piles in a row parallel to the direction of wave travel, the moment on the center pile is about 25 percent less than that on a single pile.

Tests were also conducted on piles formed of *H* sections. In this case the volume term $\forall$ in Eqs. (10.52) and (10.53) is taken as the volume of a cylinder having a diameter equal to the width of a projection of the section on a vertical plane perpendicular to the direction of wave motion. It was found that when the sides of the *H* were either perpendicular or parallel to the direction of wave motion, the moments were about twice as large as for a circular pile of the same diameter as the height and width of the *H* section. When the *H* section was oriented so that the sides of the *H* were at a 45° angle with the direction of wave motion, the moment was a maximum and was about 2.75 times greater than for a circular pile of the same size. Measured values of maximum force on horizontal flat plates[37] with the face of the plate perpendicular to the wave motion could be duplicated as to both phase angle [Eq. (10.57)] and magnitude of the force [Eq. (10.53)] by using $C_d = 3.5$ and $C_m = 1.75$, when the volume term in Eq. (10.53) was taken as the volume of the circumscribing cylinder. Tests on vertical plates 2.5 C_m wide[36] indicated that the force was approximately 1.3 times the force on a 2.5-cm pile.

The computation of wave forces on piles is illustrated by the following numerical examples.

Example 10.2 Assume that the depth is 30 m, the wave height 3 m, and the wave period 10 s. Determine the force distribution and the moment at the bottom for a circular column 1 m in diameter. Let $C_d = 2.0$ and $C_m = 2.0$.

From Eq. (10.6), $L_0 = 156$ m. Then $d/L_0 = 0.192$. From Fig. 10.4, $d/L = 0.221$, and $L = 136$ m.

The phase angle for which the maximum force occurs will be determined on the basis of conditions 6 m below the water surface ($z = -6$). Then, from Eq. (10.55), $C'_d = 718$ and from Eq. (10.56), $C'_m = 836$. From Eq. (10.57), $\theta_{max} = 35.6°$. By using Eq. (10.1) it is found that the total depth at $\theta = 35.6°$ is 31.3 m. Then values of u_x and $\partial u_x/\partial t$ may be computed from Eqs. (10.13) and (10.15), respectively. Values F_d, F_i, and F were then computed for 1-m sections at the selected values of z by means of Eqs. (10.51) to (10.53), as shown in columns 7 to 9 of Table 10.2. The moment at any point can then be computed by breaking the load into increments of length, as illustrated in columns 1 to 5 of Table 10.3 for moments at the bottom.

If the diameter of the column had been 2 m instead of 1 m, Eq. (10.57) would have indicated that $\theta_{max} = 90°$. For this phase angle the force is entirely inertial and the depth would be 30 m. The corresponding values of forces and moments are shown in columns 6 to 8 in Table 10.3.

Submerged pipelines make up a special class of structures that are frequently exposed to wave forces. Forces are predominantly inertial, but for the smaller diameters and wave heights the effect of the drag force becomes noticeable. When $D^2/HL \leq 0.02$, the drag force starts to influence the maximum force.

The values of C_m derived from the tests covered a wide range of experimental conditions, but it was found that a high degree of correlation occurred when C_m was related to z/L. When the pipe is above the bottom ($0.25 < -z/d < 0.75$), Eq. (10.58) fits the data very closely,

$$C_m = 1.73 + 5.37\frac{-z}{L} \tag{10.58}$$

TABLE 10.2 Wave Force on Piles Example

(1) z, m	(2) $\frac{d+z}{L}$	(3) $\cosh\frac{2\pi(d+z)}{L}$	(4) $\frac{\cosh[2\pi(d+z)/L]}{\sinh(2\pi d/L)}$	(5) ux, m/s	(6) $\frac{\partial u_x}{\partial t}$, m/s^2	(7) Fd, N/m	(8) Fi, N/m	(9) Fi, N/s
+1.3	0.230	2.25	1.20	0.98	0.38	960	597	1557
0	0.221	2.13	1.13	0.91	0.36	828	565	1393
–6	0.177	1.69	0.90	0.73	0.29	533	455	988
–12	0.133	1.38	0.74	0.61	0.24	372	377	749
–18	0.088	1.18	0.63	0.52	0.20	270	314	584
–24	0.044	1.04	0.55	0.46	0.18	212	283	495
–30	0.000	1.00	0.53	0.43	0.17	185	267	452

Table 10.3 Example Pile Moment Calculation

Diameter of column = 1 m					Diameter of column = 2 m		
(1) z, m	(2) F, N/m	(3) F_{av}, N/m	(4) Moment arm, m	(5) M, kN·m	(6) F, N/m	(7) F_{av}, N/m	(8) M, kN·m
+1.3	1557						
		1475	30.7	59			
0	1393				4260		
		1190	27	193		3775	612
−6	988				3290		
		868	21	109		3005	379
−12	749				2720		
		666	15	60		2530	228
−18	584				2340		
		540	9	29		2195	119
−24	495				2050		
		474	3	9		1895	34
−30	452				1740		
Total				459			1372

When the pipe is at the bottom,

$$C_m = 3.07 + 4.13\frac{-z}{L} \tag{10.59}$$

For a half-buried pipe,

$$C_m = 1.01 + 2.85\frac{-z}{L} \tag{10.60}$$

When the pipe is below the bottom in a trench,

$$C_m = 1.0 + 1.4\frac{-z}{L} \tag{10.61}$$

Some additional refinements can be obtained by referring to the original article.[38] Values of C_d were obtained from measured forces[38] at wave crests and troughs ($x/L = 0$ and ½), where the forces can be expected to be entirely due to drag. The values of C_d show clear trends when plotted against the Reynolds number R. However, values of R only extended to 10,000, and prototype values of R would be much larger. The test results showed C_d dropping to 2 as R increased to 10,000 for pipes at or above the bottom. ▲

Example 10.3 As an example of the application of this information to a practical problem, complete the force per unit length for a 5-m pipe located on the bottom at a depth of 12 m. Waves having a height of 3.5 m and a period of 8 s pass over the pipe in a direction such that the crests are parallel to the pipe.

From Eq. (10.6), $L_0 = 328$ ft and $d/L_0 = 0.122$. From Fig. 10.4, $d/L = 0.160$, and then

$$L = 250 \text{ ft}$$

The value of D^2/HL is then found to be 0.09, which shows that the inertial force predominates and the maximum force can be computed from Eq. (10.52) for a phase angle of 90° or 270°. The value of z for the center of the pipe is −32 ft. The corresponding value of z/L is −0.128, and C_m may be computed from Eq. (10.59),

$$C_m = 3.07 + 4.13\frac{-z}{L} = 3.60$$

Then

$$\begin{aligned} F_i &= C_{m\rho} \forall \frac{\partial u}{\partial t} \\ &= 3.60 \times 201 \times 3.20 = 4490 \text{ lb/ft} \end{aligned} \tag{10.62}$$

Structures which have large horizontal dimensions in the direction of wave motion, for example, on the order of magnitude of a quarter of a wavelength, have negligible friction forces compared with the inertial forces, and the maximum forces occur at $x/L = ¼$ and ¾.[37,38] When computing the inertial force from Eq. (10.61) for such large structures, it is necessary to obtain an average value of acceleration $(\partial u/\partial t)_{av}$. A convenient way to approach this problem is to express the inertial force in terms of the difference in pressure on the two sides of the submerged body. Consider a mass of fluid of rectangular cross section, as shown in Fig. 10.22*a*, centered at $x/L = ¼$ (or ¾) with an area A perpendicular to the direction of wave motion. Then

$$F_i = \forall \left(\frac{\partial u}{\partial t}\right)_{av} = (p_1 - p_2)A \tag{10.63}$$

where p_1 and p_2 are the pressure intensities on the up- and downwave sides of the fluid mass, respectively. If the fluid mass is then replaced by a solid body, Eq. (10.63) must have the coefficient of inertial resistance added, and the expression for the inertial force becomes

$$F_i = C_m \forall \left(\frac{\partial u}{\partial t}\right)_{av} = C_m (p_1 - p_2)A \tag{10.64}$$

The value of $p_1 - p_2$ determined from the equation of motion in the vertical direction[37] for the special case of structures centered at the location of maximum acceleration, $x/L = ¼$, is given by

$$p_1 - p_2 = wH\left(K - 2\sinh^2 \frac{\pi H \cos(2\pi x_1/L)}{2L}\right)\cos\frac{2\pi x_1}{L} \tag{10.65}$$

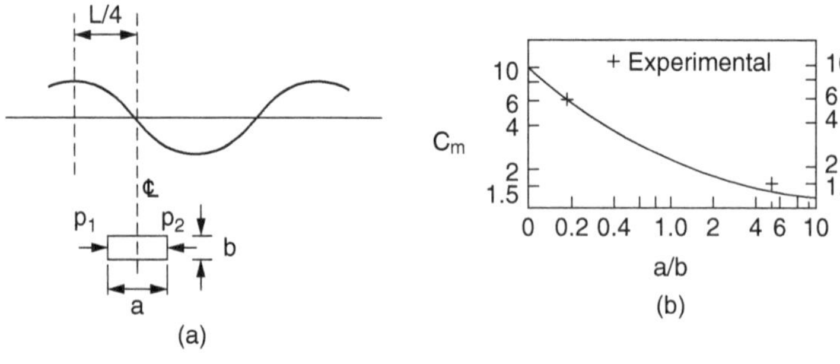

Figure 10.22 Coefficient of inertial resistance, submerged bodies of rectangular cross section.

in which x_1 is the horizontal coordinate of the upwave side of the structure, that is, the location where the pressure is p_1, and K is defined as

$$K = \frac{\cosh[2\pi(d+z)/L]}{\cosh(2\pi d/L)} \tag{10.66}$$

An interesting application of Eq. (10.65) is to the case of the pressure difference between a crest and a trough. Then, x_1/L being zero, Eq. (10.65) can be written

$$p_c - p_t = wH\left(K - 2\sinh^2\frac{\pi H}{2L}\right) \tag{10.67}$$

Solution of this equation for H would give the value for wave height corresponding to pressure differences recorded by a pressure gauge at the bottom of the sea. Measurements made by Folsom[39] indicate that values of H computed from this equation from measured values of $p_c - p_t$ will be somewhat smaller than the actual wave heights.

Measurements of vertical and horizontal forces on large rectangular-shaped structures[37] provided values of C_m which agree reasonably well with values estimated from theory for flow without separation by Riabauchinski.[40] A curve based on his derivation is shown in Fig. 10.22*b* along with two experimental values.[37] The values of C_m vary with H/L, or for a given L they vary with H as well as with the location of the structure with respect to the water surface z. This is illustrated in Fig. 10.23*a*, in which C_m is plotted with respect to H/L for three values of z/d. The model from which the values were determined was a rectangular parallelepiped 2 to 6 in long 10 in wide, and 1⅞ in high. The large surfaces were parallel to the water surface, and the long side was perpendicular to the direction of wave motion.[37] The variation of C_m with H for a constant L is shown in Fig. 10.23*b*, in which force is plotted against z for three different wave heights. The curves were computed by means of Eq. (10.64), using the value of C_m (indicated on the graph) that gave the best fit in each case. It may be noted, however, that variations of C_m are relatively small. ▲

The computation of forces on a large structure is illustrated by the following example.

Example 10.4 A rectangular barge, 200 ft long, 50 ft wide, and 15 ft high, is centered 49 ft below the water surface in a depth of 100 ft. The orientation of the barge would be as shown in Fig. 10.25*a* if *a* were 59 ft and *b* were 15 ft. The wave height is 10 ft, and the period is 10 s, as in the problem illustrating the computation of forces on piling. The horizontal

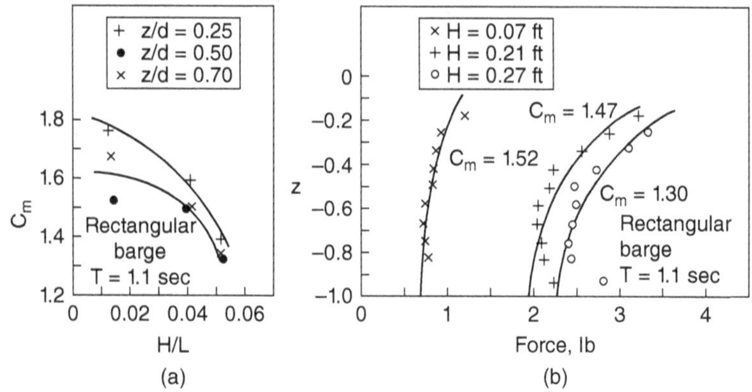

FIGURE 10.23 Wave forces on submerged structures.

force may be computed from Eqs. (10.64) and (10.65). The value of C_m is estimated as 1.5 from Fig. 10.23*b*, knowing that $a/b = 50/15 = 3.3$. $A = 200 \times 15 = 300$ ft². Then K is evaluated using Eq. (10.65) as 0.65. If

$$x_1 = \frac{L}{4} - \frac{50}{2} = \frac{453}{4} - 25 = 88 \text{ ft}$$

and

$$p_1 - p_2 = 62.4 \times 10 \times (0.65 - 0.02) \times 0.342$$
$$= 135 \text{ lb/ft}^2$$

Then the force can be computed from Eq. (10.51).

$$F = 1.5 \times 135 \times 3000 = 600{,}000 \text{ lb}$$

The vertical force may be computed from Eq. (10.52). In this case, $a/b = 0.3$, and from Fig. 10.23, $C_m = 4.1$. From Eq. (10.14), noting that the maximum vertical acceleration occurs at $\theta = 0$,

$$\frac{\partial u_z}{\partial t} = \frac{2\pi^2 H}{T^2} \frac{\sinh[2\pi(d+z)/L]}{\sinh(2\pi d/L)}$$
$$= -\frac{2\pi^2 \times 10}{100} \frac{0.94}{1.88} = 0.99 \text{ ft/s}^2$$

And then from Eq. (10.62)

$$F_2 = -4.1 \times 1.94 \times 150000 \times 0.99 = -1{,}200{,}000 \text{ lb} \quad \blacktriangle$$

Wind Tides

The action of wind on a body of water not only produces waves, but also induces a drag which moves surface water in the direction of the wind. Thus the water becomes deeper on the leeward end of a body of water and shallower on the windward end. The increase in depth at the leeward end is also referred to as wind setup. The difference in pressure at the two ends of a lake induces a return flow along the bottom. Wind tides must be taken into consideration in planning the crest elevations of dams and levees. The SWL referred to under "Wave Run-Up and Overtopping" should be taken as the wind tide level associated with that particular wind. The low tide on the windward end of the lake must be taken into consideration in selecting the elevations of water intakes and the bottoms of navigation channels. Wind tides as large as 2 m above or below the mean lake level have been recorded on Lake Erie.[41]

Sibul[42] has developed the following empirical relationships which represent quite well his laboratory data and also data from larger bodies of water, including Lake Okeechobee and observations in the Baltic Sea:

$$\frac{h}{d} = 2.44 \times 10^{-5} \left(\frac{F}{d}\right)^{1.66} \left(\frac{U^2}{Fg}\right)^{a} \tag{10.68}$$

and

$$a = 2.02 \left(\frac{F}{d}\right)^{-0.0763} \tag{10.69}$$

The equations may be used to estimate wind setup when no records are available. The decline in water-surface elevation at the windward and of a lake is usually slightly less than the positive tide h at the leeward end. The following is an example of the use of Eqs. (10.68) and (10.69).

Example 10.5 Suppose the fetch is 16 km, the wind velocity is 12 m/s, and the average depth is 6 m. Then

$$\frac{U^2}{Fg} = \frac{12^2}{16{,}000 \times 9.8} = 0.00092$$

and

$$\frac{F}{d} = \frac{16{,}000}{6} = 2667$$

Then

$$a = 2.02 \times 2667^{-0.0768} = 1.1$$

and

$$\frac{h}{d} = \frac{2.44}{10^5} \times 2667^{1.66} \times 0.00092^{1.1} = 0.0055$$

from which it is found that $h = 0.033$ m.

The effect of bottom roughness on setup has been studied by means of laboratory tests.[43] This research indicated that for lake bottoms covered by dense weed growth, the setup may be as much as twice the value computed for a smooth bottom when the water depth is equal to, or just a little greater than, the thickness of the roughness. For water depths varying from one to four times the height of the roughness, reasonable results can be obtained by taking the depth of the water as the depth above the roughness. For depths greater than four times the roughness, the roughness has little effect. When the depth becomes less than the roughness, the tides are reduced. For example, for a depth of half the height of the roughness it was found that the tide was only 0.08 that for a smooth bottom.

If the wind crosses the long dimension of a lake at an angle θ, the wind setup may be estimated by applying $\cos\theta$ to h. When the shape of the lake is such as to concentrate the water in a small area on the leeward end, it has been recommended that a form factor be applied.[44] ▲

Translatory Waves

A sudden increase or decrease in the discharge in an open channel creates a translatory wave. As the name indicates, the water is displaced during the passage of the wave. This is in contrast to oscillatory waves, in which the fluid oscillates in a nearby fixed orbit. The wave shown in Fig. 10.24*a* might have been produced by the sudden opening of a gate or the failure of a dam at an upstream location. The wave of Fig. 10.24*b* would result from the sudden closing of a gate at a downstream location. The tidal bore is an example of a wave of translation which is produced on certain rivers by rising tides. A form of translatory wave which has a single crest rising above the SWL is called a *solitary wave*. Waves of this type may be produced on the Great Lakes by the passage of a low-pressure area. The waves discussed so far are all positive waves in that the depth is increased. At the opposite sides of the gates that may have produced the positive waves shown in Fig. 10.24 a sudden decrease in depth, called a *negative wave*, would be projected along the channel.

The conditions upon which the following equations are based are, in all cases, transient ones. After the passage of the initial wave there may be a series of irregular waves, including in many cases reflected waves. The flow will eventually reach a steady condition for which the depth and velocity will be controlled by friction alone.

Positive Waves

The depths and velocities on either side of a translatory wave as well as the velocity C of the wave itself must be such as to satisfy continuity and momentum relationships. If the conditions before the passage of the wave, D_1 and V_1, are known, there remain three unknown quantities, D_2, V_2, and C. These can be determined by applying the continuity and momentum equations together with one known condition resulting from the cause of the wave. Most commonly this latter condition is the new discharge. In the derivations of the equations, velocities from left to right are considered to be positive. Thus in Fig. 10.24*a* all quantities will be positive, whereas in Fig. 10.24*b* the value of C will be negative. The derivation will be carried out for the conditions shown in Fig. 10.24*a* for a unit width of a wide rectangular channel. However, the corresponding equations for Fig. 10.24*b*, as well as for a channel of any cross section, are presented in Table 10.4.

The discharge entering the moving wave through $w_1 - w_1$ is

$$q_{w1} = D_1(C - V_1) \tag{10.70a}$$

and that leaving the wave through $w_2 - w_2$ is

$$q_{w2} = D_2(C - V_2) \tag{10.71a}$$

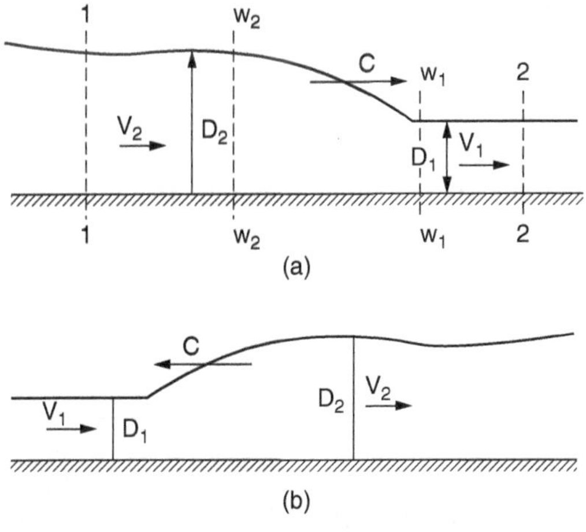

Figure 10.24 Translatory waves.

Table 10.4 Translatory Wave Equations

Wave travels in direction of flow (Fig. 10.24*a*)				Wave travels in opposite to direction of flow (Fig. 10.28*b*)	
Wide rectangular channels	**Eq.**	**General case**	**Eq.**	**Wide rectangular channels**	**Eq.**
$q_{w1} = D_1(C - V_1)$	(10.70*a*)	$Q_{w1} = a_1(C - V_1)$	(10.70*a*)	$q_{w1} = D_1(-C + V_1)$	(10.70*b*)
$q_{w2} = D_2(C - V_2)$	(10.71*a*)	$Q_{w2} = a_2(C - V_2)$	(10.71*aa*)	$q_{w2} = D_2(-C + V_2)$	(10.71*b*)
$V_2 = \frac{D_1}{D_2}(V_1 - C) + C$	(10.72*a*)	$V_2 = \frac{a_1}{a_2}(V_1 - C) + C$	(10.72*aa*)	$V_2 = \frac{D_1}{D_2}(V_1 - C) + C$	(10.72*b*)
$V_1 = \frac{D_2}{D_1}(V_2 - C) + C$	(10.73*a*)	$V_1 = \frac{a_2}{a_1}(V_2 - C) + C$	(10.73*aa*)	$V_2 = \frac{D_2}{D_1}(V_2 - C) + C$	(10.73*b*)
$C = \sqrt{\frac{gD_2}{2D_1}(D_2 + D_1)} + V_1$	(10.74*a*)	$C = \sqrt{\frac{g(a_2\bar{y}_2 - a_1\bar{y}_1)}{a_1(1 - a_1/a_2)}} + V_1$	(10.74*aa*)	$C = -\sqrt{\frac{gD_2}{2D_1}(D_2 + D_1)} + V_1$	(10.74*b*)
$C_s = \sqrt{\frac{gD_2}{2D_1}(D_2 + D_1)}$	(10.75*a*)			$C_s = \sqrt{\frac{gD_2}{2D_1}(D_2 + D_1)}$	(10.75*b*)
$C_s = \sqrt{\frac{gD_1}{2}}\sqrt{\frac{D_2}{D_1}\left(\frac{D_2}{D_1} + 1\right)}$	(10.76*a*)			$C_s = -\sqrt{\frac{gD_1}{2}}\sqrt{\frac{D_2}{D_1}\left(\frac{D_2}{D_1} + 1\right)}$	(10.76*b*)
$C = C_s + V_1$	(10.77*a*)			$C = -C_s + V_1$	(10.77*b*)
$C = \sqrt{gD} + V_1$	(10.78*a*)	$C = \sqrt{gD_m} + V_1$	(10.78*aa*)	$C = -\sqrt{gD_1} + V_1$	(10.78*b*)

If the form of the wave between w_1 and w_2 remains constant, q_{w1} is equal to q_{w2}. Equating these expressions for q_w and rearranging terms results in the following forms of the continuity equations:

$$V_2 = \frac{D_1}{D_2}(V_1 - C) + C \tag{10.72a}$$

$$V_1 = \frac{D_2}{D_1}(V_2 - C) + C \tag{10.73a}$$

The expression for wave velocity in rectangular channels:

$$C = \sqrt{\frac{gD_2}{2D_1}(D_2 + D_1)} + V_1 \tag{10.74a}$$

The general form of this equation, which applies to channels of any cross-sectional form follows:

$$C = \sqrt{\frac{g(a_2\bar{y}_2 - a_1\bar{y}_1)}{a_1(1 - a_1/a_2)}} + V_1 \tag{10.74b}$$

For the case of the wave moving upstream it may be seen from Eq. (10.74*b*) that the difference is in the sign of the first term on the right side of the equation. When V_1 is zero, Eq. (10.74*a*) becomes

$$C_s = \sqrt{\frac{gD_2}{2D_1}(D_2 + D_1)} \tag{10.75a}$$

where C_s is the velocity of a translatory wave in still water. The expression for C_s may be written

$$C_s = \sqrt{\frac{gD_1}{2}}\sqrt{\frac{D_2}{D_1}\left(\frac{D_2}{D_1} + 1\right)} \tag{10.76a}$$

Equation (10.74*a*) can then be written as

$$C = C_s + V_1 \tag{10.77a}$$

When the wave is very small so that $D_2 \rightarrow D_1 \rightarrow D$, Eq. (10.74*a*) reduces to

$$C = \sqrt{gD} + V_1 \tag{10.78a}$$

In the corresponding equation for the general case [Eq. (10.78*aa*)] the mean depth D_m is introduced, which is the area divided by the top width, a/T.

Equation (10.78*a*) may be used to show why translatory waves which involve an increase in depth have abrupt faces similar to a hydraulic jump. If the total wave is considered as a series of small waves or steps, then, according to Eq. (10.78*a*), the highest wave has the largest velocity, thus tending to overtake the small waves at lower depths. The fact that a translatory wave is similar to a hydraulic jump may be shown by setting C equal to zero in Eq. (10.74*b*). Then

$$V_1 = \sqrt{\frac{gD_2}{2D_1}(D_1 + D_2)}$$

indicating that the wave is moving upstream at the same speed that the water is carrying it downstream. Rearrangement of this equation shows that it is identical with the one for a jump in rectangular channels, which was derived in Sec. 8 [Eq. (8.81)].

Lansford[45] has reported a series of tests conducted in a rectangular channel 1.5 m wide and 43 m long, in which sudden closure of a gate was made. His test results over a considerable range of discharges and depths indicated that Eq. (10.74*b*) gives reliable results. One of Lansford's test conditions will be used as a numerical example.

Example 10.6 In a rectangular channel 1.5 m wide, the depth D_1 is 75 cm and the discharge Q is 0.15 m³/s. A gate is suddenly closed, reducing the discharge to zero ($Q_2 = 0$ and $V_2 = 0$). Determine the velocity C at which the wave will travel upstream and the depth D_1 just after closure of the gate.

$$q_1 = \frac{Q_1}{b} = 0.15/1.5 = 0.1 \text{ m}^3/\text{s}\cdot\text{m}$$

$$v_1 = \frac{q_1}{D_1} = \frac{0.1}{0.75} = 0.133 \text{ m/s}$$

Then Eqs. (10.72*b*) and (10.77*b*) become

$$0 = \frac{0.75}{D_2}(0.133 - C) + C$$

and

$$C = -C_s + 0.133$$

These equations may be solved by trial in several ways. The method used here is to assume a value of D_2, compute C from Eqs. (10.76*b*) and (10.77*b*), and then insert these values in Eq. (10.72*b*), repeating until the right side of Eq. (10.72*b*) is zero.

Assume $D_2 = 0.78$. Then $D_2/D_1 = 1.04$, and C_s can be computed using Eq. (10.76*a*) as 2.8 m/s. Substitution in Eq. (10.77*b*) yields $C = -2.67$ m/s. Substitution of C and D_2 in Eq. (10.72*b*) gives

$$\frac{0.75}{0.78}(0.133 + 2.67) - 2.67 = 0.02 \approx 0$$

Additional iterations can be done to make the solution as close to zero as desired. ▲

Negative Waves

A decrease in depth is produced on the downstream side of a gate that is suddenly closed, or on the upstream side of a gate that is suddenly opened. If the resulting wave is considered as a series of small waves, Eq. (10.78*a*) or (10.78*b*) may be applied to the individual small waves by replacing V_1 with the local velocity at the depth D, as indicated in Fig. 10.25. The top wave will have the highest velocity, $c_{max} = \sqrt{gD_0} + V_0$. with the result that a rapid flattening of the wave will occur. Indeed, there is little or no appearance of a wave, but rather a gradual recession of the water surface.

The shape of a negative wave may be determined by a method suggested by Streeter.[46] If the momentum equation is written for a very small wave:

$$wy\,dy = \frac{w}{g}(C - V)y\,dV \qquad (10.79)$$

Inserting the value of C from Eq. (10.78*a*) and integrating yields

$$V = V_0 - 2\sqrt{gD_0} + 2\sqrt{gD} \qquad (10.80)$$

Combining Eqs. (10.80) and (10.78*a*),

$$C = V_0 + 3\sqrt{gD} - 2\sqrt{gD_0} \qquad (10.81)$$

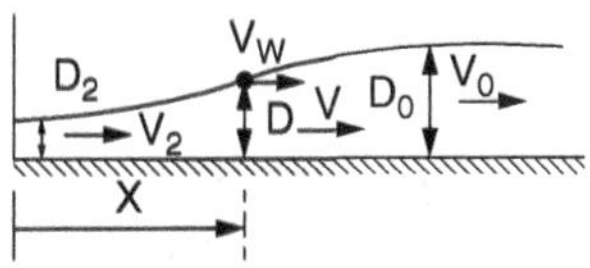

Figure 10.25 Negative wave.

Then, because $x = Ct$, the equation for the surface becomes:

$$x = (V_0 + 3\sqrt{gD} - 2\sqrt{gD_0})t \tag{10.82}$$

The depth D_2 at the gate depends on the discharge Q_2 after the gate has been partially closed. For total closure Q_2 and V_2 are zero. For partial closure Q_2 must be known and the equation $Q_2 = D_2V_2b$ may be combined with Eq. (10.82) to determine D_2.

References

1. U.S. Army Corps of Engrs., Beach Erosion Board Bull., spec. issue 2, 1953.
2. R. L. Wiegel, "Gravity Waves, Tables of Functions," Engineering Foundation, Council on Wave Research, Berkeley, CA, 1954.
3. G. B. Airy, "On Tides and Waves," in *Encyclopaedia Metropolitana*, vol. 5, p. 289, 1845.
4. A. S. Ramsey, *A Treatise on Hydromechanics*, pt. II, *Hydrodynamics*, Bell & Sons, London, 1957.
5. J. W. Johnson, M. P. O'Brien, and J. D. Isaacs, "Graphical Construction of Wave Refraction Diagrams," U.S. Navy Hydrographic Office, Publ. 605, 1948.
6. T. Saville, Jr., and K. Kaplan, "A New Method for the Graphical Construction of Wave Refraction Diagrams," U.S. Army Corps of Engrs., Beach Erosion Board Bull., vol. 6, no. 3, p. 23, 1952.
7. R. S. Arthur, W. H. Munk, and J. D. Isaacs, "The Direct Construction of Wave Rays," *Trans. Am. Geophys. Union*, Dec. 1952.
8. "Breakers and Surf, Principles in Forecasting," U.S. Navy Hydrographic Office, Publ. 234, Nov. 1944.
9. H. W. Iversen, "Laboratory Study of Breakers, Gravity Waves," *Natl. Bur. Standards* (U.S.), Circ. 521, 1952.
10. W. H. Munk, "The Solitary Wave Theory and Its Application to Wave Problems," *Ann. N. Y. Acad. Sci.*, vol. 51, p. 376, 1949.
11. R. R. Putz, "Statistical Distribution for Ocean Waves," *Trans. Am. Geophys. Union*, vol. 33, pp. 685–692, Oct. 1952.
12. R. L. Wiegel and J. Kukk, "Wave Measurements along the California Coast," *Trans. Am. Geophys. Union*, vol. 38, pp. 667–674, Oct. 1957.
13. R. L. Wiegel, *Oceanographical Engineering*, Prentice-Hall, Englewood Cliffs, NJ, 1964.
14. H. U. Sverdrup and W. H. Munk, "Wind Waves and Swell: Principles in Forecasting," U.S. Navy Hydrographic Office, Publ. 11275 misc., p. 275, 1944.
15. H. U. Sverdrup and W. H. Munk, "Wind, Sea, and Swell: Theory of Relations for Forecasting," U.S. Navy Hydrographic Office, Publ. 11, p. 275.
16. C. L. Bretschneider, "Revisions in Wave Forecasting: Deep and Shallow Water," *Proc. 6th Conf. on Coastal Eng.*, Engineering Foundation, Council of Wave Research, Berkeley, CA, 1958.
17. C. L. Bretschneider, "Revised Wave Forecasting Relationships," *Proc. 2d Conf. on Coastal Eng.*, Engineering Foundation, Council on Wave Research, Berkeley, CA, 1952.
18. C. O. Wisler and E. F. Brater, *Hydrology*, Wiley, New York, 1959, pp. 204–209.
19. C. L. Bretschneider, "Generation of Wind Waves over a Shallow Bottom," U.S. Army Corps of Engrs., Beach Erosion Board, Tech. Mem. 51, 1954.

20. *Shore Protection Manual*, U.S. Army Corps of Engrs., Coastal Engineering Research Center, 1973.
21. T. Saville, Jr., "The Effect of Fetch Width on Wave Generation," U.S. Army Corps of Engrs., Beach Erosion Board, Tech. Mem. 70, 1954.
22. "Wave Run-up and Overtopping Levee Sections, Lake Okeechobee, Fla.," U.S. Army Corps of Engrs. Waterways Experiment Station, Tech. Rep. 2. 449, 1957.
23. T. Saville, Jr., "Wave Run-up on Shore Structures," *Trans. ASCE*, vol. 123, pp. 139–150, 1958.
24. R. P. Savage, "Laboratory Data on Wave Run-up on Roughened and Permeable Slopes," U.S. Army Corps of Engrs., Beach Erosion Board, Tech. Mem. 109, 1959.
25. R. A. Jackson, "Design of Cover Layers for Rubble-Mound Breakwaters," U.S. Army Corps of Engrs. Waterways Experiment Station, Res. Rep. 2–11, 1968.
26. T. Saville, Jr., "Laboratory Data on Wave Run-up and Overtopping on Shore Structures," U.S. Army Corps of Engrs., Beach Erosion Board, Tech Mem. 64, 1955.
27. O. Sibul, "Model Study of Overtopping of Wind-Generated Waves on Levees with Slopes of 1:3 and 1:6," U.S. Army Corps of Engrs., Beach Erosies Board, Tech. Mem. 80, 1956.
28. R. R. Minikin, *Winds, Waves and Maritime Structures*, Charles Griffin, London, pp. 40–48, 1950.
29. R. Y. Hudson, "Wave Forces on Breakwaters," *Trans. ASCE*, vol. 118, p. 653, 1953.
30. C. W. Ross, "Laboratory Study of Shock Pressures of Breaking Waves," U.S. Army Corps of Engrs., Beach Erosion Board, Tech. Mem. 59, 1955.
31. L. A. Mueller, H. A. Knutson, and A. A. Koch, "Some Dynamic Aspects in the Design of Marine Structures on the Great Lakes," in *Proc 4th Conf. on Coastal Eng.*, Engineering Foundation, Council on Wave Research, Berkeley, CA, 1954.
32. D. D. Gaillard, "Wave Action in Relation to Engineering Structures," The Engineering School, Fort Belvoir, VA, 1935.
33. R. Y. Hudson, "Laboratory Investigation of Rubble-Mound Breakwaters," *Trans. ASCE*, vol. 126, p. 492, 1961.
34. P. Danel, "Tetrapods," in *Proc. 4th Conf. on Coastal Eng.*, Engineering Foundation, Council on Wave Research, Berkeley, CA, 1953.
35. J. R. Morison, "Design of Piling," in *Proc. 1st Conf. on Coastal Eng.*, Engineering Foundation, Council on Wave Research, Berkeley, CA, 1951.
36. J. R. Morison, J. W. Johnson, and M. P. O'Brien, "Experimental Study of Forces on Piles," in *Proc. 4th Conf. on Coastal Eng.*, Engineering Foundation, Council on Wave Research, Berkeley, CA, 1954.
37. E. F. Brater, J. S. McNown, and L. D. Stair, "Wave Forces on Submerged Structures," *Trans. ASCE*, vol. 126, pt. I, pp. 661–696, 1961.
38. E. F. Brater and R. Wallace, "Wave Forces on Submerged Pipe Lines," in *Proc. 13th Coastal Eng. Conf.*, ASCE, vol. III, pp. 1703–1722, 1972.
39. R. G. Folsom, "Measurement of Ocean Waves," *Trans. Am. Geophys. Union*, pp. 691–699, Oct. 1949.
40. D. Riabauchinski, "Sur la restance des fluides," in *Proc. Int. Cong. of Mathematics*, Strasbourg, France, pp. 568–585, 1920.
41. E. F. Brater and H. W. Baynton, "Extreme Lake Levels of Lake Erie near Monroe, Michigan," 1956, unpublished.
42. O. Sibul, "Laboratory Study of Wind Tides in Shallow Water," U.S. Army Corps of Engrs., Beach Erosion Board, Tech. Mem. 61, 1955.

43. E. G. Tickner, "Effect of Bottom Roughness on Wind Tide in Shallow Water," U.S. Army Corps of Engrs., Beach Erosion Board, Tech. Mem. 95, 1957.
44. T. Saville, Jr., "Wind Set-up and Waves in Shallow Water," U.S. Army Corps of Engrs., Beach Erosion Board, Tech. Mem. 27, 1952.
45. W. M. Lansford, Discussion of J. H. Wilkinson, "Translatory Waves in Natural Channels," *Trans. ASCE*, vol. 110, p. 1227, 1945.
46. V. L. Streeter, *Fluid Mechanics*, 3d ed., McGraw-Hill, New York, p. 511, 1962.

SECTION 11

Spatially Variable and Unsteady Flow

This section deals with selected aspects of unsteady flow. Because the same basic equations are required for problems in spatially variable flow, this topic is also treated in this section. The first three portions of this section are devoted to open-channel flow, and the fourth portion to water hammer problems.

Unsteady Open-Channel Flow

When the discharge entering a channel is gradually changing with respect to time, or water is being added or taken out throughout the length of the channel (spatially variable flow), it is necessary to deal with the basic equations of motion in determining the water-surface profile. The equations will first be derived for a general case of unsteady spatially variable flow. Thereafter, they will be simplified to apply to steady spatially variable flow and finally to flow in wide rectangular channels.

The derivation is carried out for a section of channel having a length dx and an area a, as shown in Fig. 11.1. The depth y is taken perpendicular to the bottom, although for small slopes this distance is virtually identical with the vertical distance from the bottom to the water surface. The derivation is based on the momentum relationship embodied in Newton's second law, which states that

$$F = \frac{dM}{dt} \tag{11.1}$$

In this equation, F is the sum of all external forces resolved in the direction of flow x, the positive sign indicating forces toward the right, as given by the equation

$$F = F_p + F_g + F_{fb} + F_{fa} \tag{11.2}$$

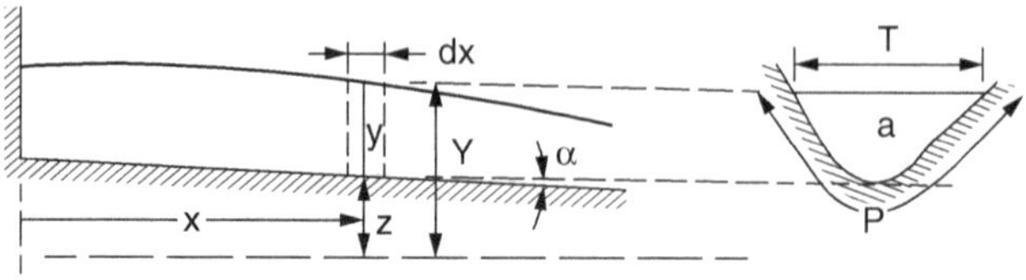

FIGURE 11.1 Definition of open-channel flow.

F_p is the pressure force, which, for parallel-walled channels, may be expressed as

$$F_p = -\rho g a \frac{\partial y}{\partial x} dx \tag{11.3}$$

The minus sign is introduced so that a negative value of $\partial y/\partial x$ will produce a positive force.

If the walls are not parallel, this force must include the component of pressure forces acting on the water in the direction of flow. For very steep slopes the hydrostatic force may be approximated by multiplying this term by cos α.

F_g is the component of the weight, which acts in the x directions. Its value is

$$F_g = \rho g a\, dx \sin \alpha \tag{11.4}$$

For small slopes,

$$\sin \alpha = \tan \alpha = s_0 \tag{11.5}$$

and

$$F_g = \rho g a\, dx\, s_0 \tag{11.6}$$

F_{fb} is the shear force on the walls as expressed by

$$F_{fb} = -\tau p\, dx \tag{11.7}$$

where τ is the average shear stress in pounds per square foot or newtons per square meter and p is the wetted perimeter. The shear stress may be evaluated from uniform flow conditions as follows. For uniform steady flow, the only forces acting in the direction of flow on a section of the water are the component of the gravitational force and the shear force. Therefore, from Eqs. (11.6) and (11.7),

$$F_{fb} = -\tau p\, dx = -\rho g a\, dx\, s \tag{11.8}$$

In this case, because flow is uniform, the slope s of the energy gradient may be used instead of the bottom slope s_0. It is assumed that the energy-gradient slope, as computed for uniform flow, will also give the correct shear force for nonuniform unsteady and spatially variable flow. The value of s is then obtained from the Manning equation [Eq. (7.46)] or from an equation of the form of Eq. (7.29). The roughness coefficients for spatially variable flow are usually larger, especially for small values of the Reynolds number, than for flow in which no such disturbance occurs.[1,2]

F_{fa} is the shear force due to the friction of the air on the water surface. This term will not be included here because it is usually very small compared with the other forces. However, for some special cases, as, for example, for wind tides (Sec. 10), it may be one of the most important terms.

The momentum term dM/dt can be expressed as

$$\frac{dM}{dt} = -\rho QV + \rho\left(Q + \frac{\partial Q}{\partial x}dx\right)(V + dV) \tag{11.9}$$

where

$$dV = \frac{\partial V}{\partial t}dt + \frac{\partial V}{dx}dx \tag{11.10}$$

Then, canceling terms and eliminating higher-order differentials,

$$\frac{dM}{dt} = +\rho Q\frac{\partial V}{\partial x}dt + pQ\frac{\partial V}{\partial x}dx + \rho\frac{\partial Q}{\partial x}dx\,V \tag{11.11}$$

Then substituting values from Eqs. (11.3), (11.6), (11.8), and (11.11) into Eq. (11.1),

$$\begin{aligned}\rho ga\frac{\partial y}{dx}dx + \rho ga\,dx\,s_0 - \rho ga\,dx\,s\\ = \rho aV\frac{\partial V}{\partial t}dt + \rho aV\frac{\partial V}{dx}dx + \rho\frac{\partial Q}{\partial x}dx\,V\end{aligned} \tag{11.12}$$

Dividing Eq. (11.12) by $\rho ga\ dx$ and letting the inflow per unit length $\partial Q/\partial x$ be denoted by q,

$$-\frac{\partial y}{\partial x} + s_0 - s = \frac{1}{g}\frac{\partial V}{\partial t} + \frac{V}{g}\frac{\partial V}{\partial x} + \frac{qV}{ga} \tag{11.13}$$

This is the dynamic equation for unsteady spatially variable flow.

For steady spatially variable flow, Eq. (11.13) becomes

$$-\frac{\partial y}{\partial x}+s_0-s=\frac{V}{g}\frac{\partial V}{\partial x}+\frac{qV}{ga} \tag{11.14}$$

and when flow is steady and not spatially variable, the expression is

$$-\frac{\partial y}{\partial x}+s_0-s=\frac{V}{g}\frac{\partial V}{\partial x} \tag{11.15}$$

Another useful form of Eq. (11.14) is obtained by noting in Fig. 11.1 that the following approximate arrangement exists:

$$Y=z+y \tag{11.16}$$

Then

$$\frac{dY}{dx}=\frac{dz}{dx}+\frac{dy}{dx}=-s_0+\frac{dy}{dx} \tag{11.17}$$

and Eq. (11.14) becomes

$$-\frac{dY}{dx}-s=\frac{V}{g}\frac{\partial V}{\partial x}+\frac{qV}{ga} \tag{11.18}$$

The continuity equation is derived as follows:

$$aV-\frac{\partial a}{\partial t}dx+q\,dx=\left(V+\frac{\partial V}{\partial x}dx\right)\left(a+\frac{\partial a}{\partial x}dx\right) \tag{11.19}$$

Cancellation of terms, eliminating a higher-order differential, and division by *dx* yield

$$-\frac{\partial a}{\partial t}+q=V\frac{\partial a}{\partial x}+a\frac{\partial V}{\partial x} \tag{11.20}$$

Noting that $\partial a = T\,\partial y$, the equation can be written

$$-\frac{\partial y}{\partial t}T+q=VT\frac{\partial y}{\partial x}+a\frac{\partial V}{\partial x} \tag{11.21}$$

For steady spatially variable flow, the continuity equation becomes

$$q=VT\frac{\partial y}{\partial x}+a\frac{\partial V}{\partial x} \tag{11.22}$$

For rectangular channels, $T = 1$ and $a = y$. Therefore,

$$q = V\frac{\partial y}{\partial x} + y\frac{\partial V}{\partial x} \tag{11.23}$$

When flow is steady and not spatially variable, the general continuity equation becomes

$$VT\frac{\partial y}{\partial x} + a\frac{\partial V}{\partial x} = 0 \tag{11.24}$$

which, for rectangular channels, reduces to the form

$$V\frac{\partial y}{\partial x} + y\frac{\partial V}{\partial x} = 0 \tag{11.25}$$

It may be noted that substitution of Eq. (11.24) into Eq. (11.15) yields the general equation for steady, gradually varied flow in open channels [Eq. (8.99)], and a similar combination of Eq. (11.25) with Eq. (11.15) gives Eq. (8.100), which is the corresponding equation for rectangular channels.

Steady Spatially Variable Flow

One of the applications of these equations most often encountered is to the case of steady spatially variable flow. Examples of this type of flow are flow in roof and street gutters, overland flow produced by rainfall, flow in side-channel spillways, and flow in channels with side-channel outlets or inlets.[1,3-5] An additional continuity equation can be stated for this type of flow,

$$Q = Q_0 + qx \tag{11.26}$$

where Q_0 is the discharge in the channel at $x = 0$. When $Q_0 = 0$,

$$q = \frac{Q}{x} = \frac{aV}{x} \tag{11.27}$$

The solution of spatially variable flow problems may be accomplished by changing the form of Eq. (11.14) so that the differentials become finite differences, as shown in the following equation, in which the subscripts 1 and 2 apply to the upstream and downstream ends, respectively, of the reaches of length x,

$$a_{av}(y_1 - y_2) + a_{av}\,\Delta x(s_0 - s_{av}) = \frac{a_1V_1(V_2 - V_1)}{g} + \frac{qV_2\Delta x}{g} \tag{11.28}$$

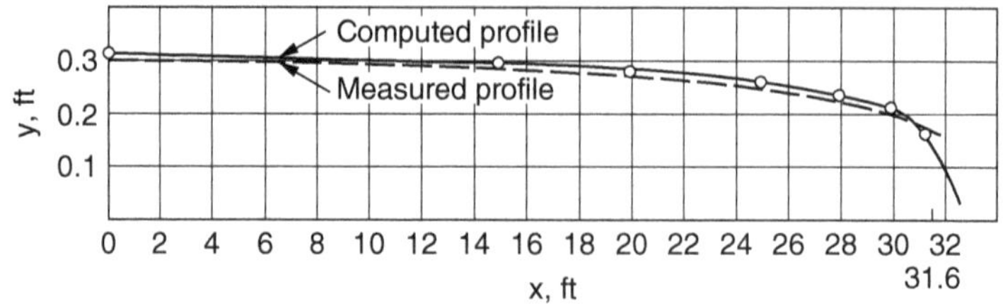

FIGURE 11.2 Spatially variable flow profile; after Beij.[4]

In Eq. (11.28) $a_{av} = (a_1 + a_2)/2$ and s_{av} is computed by using this average cross-sectional area in the reach. The solution for a water surface profile by means of Eq. (11.28) requires only that the depth be known at some point in the channel. For channels with a mild slope ($s_0 < s_c$) ending in a free fall, critical depth occurs near the outlet (Sec. 8), and computations can be started from this point. This is often the case with roof or street gutters and with overland flow. An illustration of the computation of a water profile for a horizontal channel ending in a downspout is given in the following example. The channel selected is one tested by Beij,[4] and the computed and measured profiles are plotted in Fig. 11.2.

Example 11.1 A smooth rectangular channel 0.5 ft wide and 31.6 ft long is supplied by a discharge q of 0.00608 ft³/s per foot of length. The channel is horizontal and ends in a downspout, which simulates conditions at a free fall. Based on the observed data, it was assumed that critical depth would occur 0.316 ft from the end of the channel. This is only approximately $2D_c$ upstream from the fall, which is a considerably shorter distance than that reported in Sec. 8 for steady, not spatially variable flow. Then the discharge

$$31.284 \times 0.00608 = 0.190 \text{ ft}^3/\text{s}$$

and critical depth can be computed from Eq. (8.26),

$$y_c = \sqrt[3]{\frac{\left(\frac{0.190}{0.5}\right)^2}{32.2}} = 0.165 \text{ ft}$$

Computations are then carried out for selected reaches Δx proceeding from y_c to the upstream end. The procedure is to assume a depth y_1 at the upstream end of the reach and solve for the terms in Eq. (11.28), repeating until this equation is satisfied. (Note that s_0 is zero for this example.) The final set of values for the first reach is shown in Table 11.1. The value of n was assumed to be 0.012. Values of s_{av} are computed by means of the Manning equation, Eq. (7.34). ▲

Table 11.1 Final Computation for First Reach in Fig. 11.2

	(1)	(2)	(3)	(4)	(5)	(6)	(7)	(8)	(9)	(10)	(11)	(12)	(13)	(14)	(15)
x, ft	Δx, ft	y, ft	Q, ft³/s	V, ft/s	y_{av}, ft	$\frac{y_{av}}{b}$	$\left(\frac{1}{K'}\right)^2$	Q_{av}, ft³/s	$\left(\frac{Q_{av}n}{b^{8/3}}\right)^2$	$-a_{av}s_{av}\Delta x$, ft³	$a_{av}(y_1 - y_2)$, ft³	$\frac{a_1V_1(V_2 - V_1)}{g}$, ft³	$qV_2\frac{\Delta x}{g}$, ft³	(10) + (11)	(12) + (13)
31.284		0.165	0.190	2.30											
	1.284				0.189	0.378	24.6	0.186	0.000205	−0.00061	0.0045	0.0033	0.00056	0.0039	0.0039
30		0.213	0.182	1.71											

For channels with a steep slope ($s_0 > s_c$), critical depth no longer occurs at the outlet, and the solution for the water-surface profile requires the establishment of the magnitude and location of a controlling depth in order to provide a starting point for the computations. If, for example, water were entering the channel at a known depth and discharge at the upper end of the spatially variable flow region, this would establish a control, and computations could be started from this point and carried downstream in steps in the same manner that they were carried upstream for channels with a mild slope.

In the discussion up to this point, it has been assumed that q represents inflow in cubic feet per second per foot or cubic meters per second per meter of channel. There are many cases where a uniform rate of outflow occurs. Examples are infiltration during overland flow and flow in conduits with side outlets. If all the momentum of the discharge leaving the channel is reduced to zero before the water leaves the channel, the term involving q would simply have its sign changed. This would probably apply to the case of infiltration during overland flow. However, for side outlets, the water leaving the channel usually retains part of its original momentum in the direction of flow after having left the channel. This situation might be handled by retaining some fractional part of the term involving q with a minus sign.

Unsteady Flow With Free Surface

The equations developed in this section form the basis for computation of the mode of travel of a gentle wave, such as a flood wave, down a channel. Such computations, called flood routing, have usually been carried out by less elegant methods because these equations are not readily adaptable to nonprismatic channels and because the computations, even for rectangular channels, were very tedious. However, the recognition that the so-called method of characteristics could be used to solve the equations and the adaptability of this method to solution by means of digital computers have made it possible to use these equations for flood routing.[6–10] The method is based on the fact that Eqs. (11.13) and (11.21), with the spatially variable flow term eliminated, can be arranged into the form of the following expressions for total derivatives:

$$\frac{dy}{dt} = \frac{\partial y}{\partial x}\frac{dx}{dt} + \frac{\partial y}{\partial t} \tag{11.29}$$

$$\frac{dV}{dt} = \frac{\partial V}{\partial x}\frac{dx}{dt} + \frac{\partial V}{\partial t} \tag{11.30}$$

to derive an equivalent set of equations,

$$\frac{dx}{dt} = V + C \tag{11.31a}$$

$$d(V + 2C) = g(s_0 - s)\,dt \tag{11.31b}$$

and

$$\frac{dx}{dt} = V - C \tag{11.32a}$$

$$d(V - 2C) = g(s_0 - s)\,dt \tag{11.32b}$$

where C is the celerity of a very small wave ($C = \sqrt{gy}$), as shown by Eq. (10.7). Consider a coordinates system having the coordinate x, the distance from some initial point in a channel, and t, the time elapsed after the introduction of a change in depth at $x = 0$, as, for example, the initial step in introducing a flood wave at the upper end of the channel. Then a series of lines on this coordinate system would have slopes given by Eqs. (11.31*a*) and (11.32*a*). For subcritical flow, which is the most usual situation, Eq. (11.31*a*) would give a positive slope, as, for example, *AB* in Fig. 11.3, and Eq. (11.32*a*) would give a negative slope, such as *BC* in Fig. 11.3. Inherent in the derivation of Eqs. (11.31) and (11.32) is the implication that Eq. (11.31*b*) must be satisfied along a line having a slope given by Eq. (11.31*a*) and that similar relations exist for Eq. (11.32). If y and V at some time t_1 (Fig. 11.3) are known at all points in a channel, then the slopes of lines *AB* and *CB* can be computed for points *A* and *C* from Eqs. (11.31*a*) and (11.32*a*). Although the slopes are not constant from *A* to *B* and from *C* to *B*, as a first trial they may be considered constant, and since Eqs. (11.31*b*) and (11.32*b*) must both hold at *B*, their simultaneous solution at that point will yield first-trial values of V_B, y_B, and C_B. The solution of Eqs. (11.31*b*) and (11.32*b*) is carried out for finite differences in t and x. Thus

$$dt \approx t_2 - t_1 = \Delta t \tag{11.33}$$

and the slope of *AB* should be the average of the slopes at *A* and *B*.

Thus

$$\frac{dx}{dt_{av}} = \frac{1}{2}(V_A + C_A + V_B - C_B) \tag{11.34}$$

For finite differences Eq. (11.31*b*) can be written

$$\Delta(V + 2C) = g(s_0 - s_{av})\,\Delta t \tag{11.35}$$

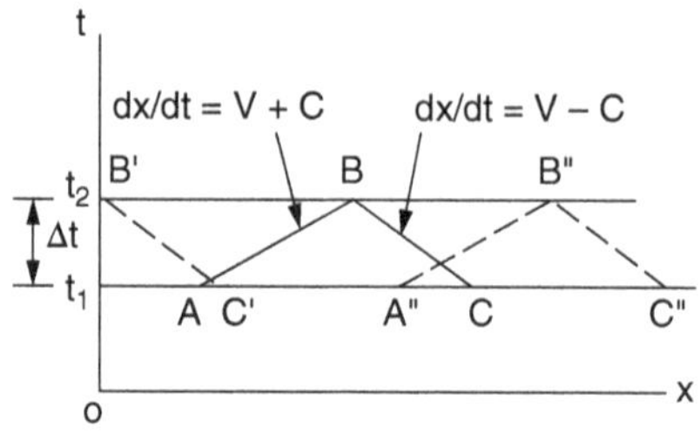

FIGURE 11.3 Method of characteristics.

or
$$(V_B + 2C_B) - (V_A + 2C_A) = g(s_0 - s_{av})\Delta t \tag{11.36}$$

where
$$s_{av} = \frac{s_A + s_B}{2} \tag{11.37}$$

The corresponding rearrangement of Eq. (11.32*b*) yields

$$(V_B - 2C_B) - (V_C - 2C_C) = g(s_0 - s_{av})\,\Delta t \tag{11.38}$$

where
$$s_{av} = \frac{s_B + s_C}{2} \tag{11.39}$$

The value of s_{av} may be obtained from the Manning equation in one of the following forms developed in Sec. 7:

$$s = \left(\frac{1}{K'}\right)^2 \left(\frac{Qn}{b^{8/3}}\right)^2 \tag{11.40}$$

Note that the unknowns in Eqs. (11.36) and (11.38) are V_B, C_B, and s_B. However, s_B is related to V_B and C_B ($C_B = \sqrt{gy_B}$) by means of Eq. (11.40). Thus there are three equations and three unknowns, and solutions can be obtained by successive approximations.

The solution is carried forward to time t_2 in the same manner at other selected values of *x*, so that *y* and *V* are known at points *B*, B_1, B_2, and so on. The solution for the channel entrance ($x = 0$) at time t_2, which is designated at *B′* in Fig. 11.3, requires only the single characteristic *C′B′* because *V* is the only unknown, *y* having been established by the nature of the flood wave that was introduced into the channel.

Lin[9] and Chow[11] give examples of graphic procedures that aid in the solution. Even with the graphic aids the solutions are tedious, and it is only by the utilization of digital computers that this procedure becomes practical.

Water Hammer

Assume a pipe with a valve at its outlet discharging from a reservoir. If the valve is suddenly closed, a dynamic pressure, in addition to the normal static pressure, is created within the pipe. This dynamic pressure is commonly called *water hammer*. It is caused by the sudden transformation of kinetic energy to pressure energy. If P is the pressure due to water hammer and M the mass of water in the pipe whose velocity is reduced the amount dV in the time interval dt, then

$$P = M\frac{dV}{dt} \tag{11.41}$$

If $dt = 0$, the pressure becomes infinite. Instantaneous change in velocity is not, however, possible. In the following nomenclature, all units are expressed in feet and seconds or meters and seconds except as noted:

b = thickness of pipe walls
d = inside diameter of pipe
A = cross-sectional area of pipe, ft^2 or m^2
e = modulus of elasticity of pipe walls, lb/ft^2 or N/m^2
e' = modulus of elasticity of water, lb/ft^2 or N/m^2
E = modulus of elasticity of pipe walls, lb/ft^2 or N/m^2
E' = modulus of elasticity of water, lb/ft^2 or N/m^2
g = acceleration of gravity
ΔH = head due to water hammer (in excess of static head)
H = total head producing discharge through valve
H_0 = original head in pipe
L = length of pipeline
T = time of closing valve
V_0 = mean velocity of water in pipe before closure of valve
α = velocity of pressure wave along pipe

Let $l_1, l_2, l_3, \ldots, l_n$ represent successive infinitesimally short sections of pipe beginning at the outlet. The instant the valve is closed, the water in section l_1 is brought to rest, its kinetic energy is transformed into pressure energy, the water is somewhat compressed, and the pipe wall with which it is in contact expands slightly. Because of the enlarged cross-sectional area of l_1 and the compressed condition of the water within it, a greater mass of water is contained within this section than before the closure. A small volume of water has flowed into l_1 after the valve was closed. An instant later, a similar procedure takes place in l_2 and then in l_3, causing a wave of increased pressure to travel up the pipe. The instant this wave reaches the reservoir, the entire pipe is expanded and the water within it is compressed. There is now no moving mass of water to maintain this high pressure, and the pipe begins to contract and the water to expand with a consequent return to normal static pressure. This process starts at the reservoir and travels as a wave to the lower end. During this second period some of the water stored within the pipe flows back into the reservoir, but on account of the inertia of this moving mass, an amount flows back greater than the excess amount stored at the end of the first period, so that the instant this second wave reaches the valve, the pressure at that point drops, not only to the normal static pressure but also below it. A third period now follows during which a wave of pressure less than normal static pressure passes up the pipe to the reservoir. When it reaches the reservoir, the entire pipe is under less than static pressure, but since all the water is again at rest, the pressure in l_n immediately returns to the normal static pressure due to the head of water in the reservoir. This starts a fourth period marked by a wave of normal static pressure moving down the pipe. When the valve is reached, the pressure there is normal, and for an instant the conditions throughout the pipe are similar to what they were when the valve was first closed. The velocity of the water (and the resultant water hammer) is now, however, somewhat less than it was at the time of closure because of friction and the imperfect elasticity of the pipe and the water.

Instantly another cycle begins, similar to the one described, and then another, and so on, each set of waves successively diminishing until the waves finally die out.

Equation (11.41) shows that, for instantaneous closure of the valve, the pressure created would be infinite if the water were incompressible and the pipe inelastic. It is impossible to close a valve instantaneously, but if the valve is completely closed before the first pressure wave has time to return to the valve as a wave of low pressure, or in other words, if T is less than $2L/\alpha$, the pressure will continue to increase up to the time of complete closure, and the resulting pressure is just the same as if the valve had been closed instantaneously. If T is greater than $2L/\alpha$, the earlier pressure waves return as waves of low pressure and tend to reduce the rise of pressure resulting from the final stages of valve closure.

Hence if T is equal to or less than $2L/\alpha$, h will be the same as for instantaneous closure, but if T is greater than $2L/\alpha$, h will be diminished as T increases.

The kinetic energy contained in the moving columns of water within the pipe is

$$\text{KE} = \frac{MV_0^2}{2} = wAL\frac{V_0^2}{2g} \tag{11.42}$$

This energy is used in the work of compressing the water and stretching the pipe walls. If the increase in pressure head is ΔH feet (meters), the compression of the water column absorbs $(w\ \Delta H)^2AL/2e'$ foot-pounds (meter-newtons) of energy because the average pressure force of the water is $w\ \Delta Ha/2$ and the distance through which this force moves is $w\ \Delta HL/e'$.

In a similar manner, the work done in stretching the pipe walls is $(w\ \Delta H)^2LAD/2eb$. Then, relating the energy of the water to the work done in stopping it,

$$wAL\frac{V_0^2}{2g} = \frac{(w\ \Delta H)^2 AL}{2e'} + \frac{(w\ \Delta H)^2 LAD}{2eb} \tag{11.43}$$

or

$$H = \frac{V_0}{g}\sqrt{\frac{1}{w/g(1/e' + D/eb)}} \tag{11.44}$$

As this pressure wave moves along the pipe with a velocity α, the force may be equated to the rate of change in momentum as follows:

$$w\ \Delta HA = \frac{w}{g}aAV_0 \tag{11.45}$$

or

$$\Delta H = \frac{a}{g}V_0 \tag{11.46}$$

and

$$a = \sqrt{\frac{1}{w/g(1e' + D/eb)}} \tag{11.47}$$

This derivation is approximate in that the stress in the longitudinal direction is neglected. When this is included, Eq. (11.47) becomes[12]

$$a = \sqrt{\frac{1}{w/g(1/e' + Dc/eb)}} \tag{11.48}$$

where c is related to Poisson's ratio and the manner in which the pipe is held in place. Values of c are somewhat less than unity[12]; however, it is on the safe side to use Eq. (11.47), in which c is assumed to be unity. The analysis presented here is credited to Professor N. Jankovsky of Moscow.[13] The analysis may also be carried out by writing the basic dynamic and continuity equations in a manner similar to the derivation presented for open channels in the first part of this section.[13,14]

Pressure When $T \leqq 2L/\alpha$

For this case, ΔH may be computed directly from Eq. (11.46) after α is determined from Eq. (11.47).

If, for example, α is 1400 m/s and V_0 is 3.5 m/s, closure of a valve in a time $T \leqq 2L/v_\omega$ will create an additional pressure head at the gate of

$$\Delta H = \frac{\alpha V_0}{g} = 500 \text{ m}$$

Pressure When $T \geqq 2L/\alpha$

For slow gate closure the computations may be carried out in steps by a method suggested by Gibson[15] or by means of a graphic solution. Both methods will be described and illustrated by means of a numerical example. In the step solution, Eq. (11.46) is written as

$$\Delta H = \frac{\alpha}{g} \Delta V \tag{11.49}$$

where ΔH is the increase in pressure head resulting from one step in the valve closure, which reduces the velocity in the pipe by the amount ΔV. It is convenient to write Eq. (11.39) in terms of the initial condition,

$$\frac{\Delta H}{H_0} = \frac{\alpha V_0}{g H_0} \frac{\Delta V}{V_0} \tag{11.50}$$

Relations between head and pipe velocity for various gate openings must next be developed. It is assumed that this relationship can be expressed as

$$Q = \alpha V = C_d A_g \sqrt{2gH} \tag{11.51}$$

where a is the area of the pipe, V the average velocity in the pipe, $C_d A_g$ the product of the coefficient of discharge and the area of the gate, and H the head at the gate. Then, for full gate opening,

$$Q = A V_0 = (C_d A_g)_0 \sqrt{2gH_0} \tag{11.52}$$

and, in general,

$$\frac{V}{V_0} = \frac{C_d A_g}{(C_d A_g)_0} \sqrt{\frac{H}{H_0}} \tag{11.53}$$

or, for simplicity,

$$\frac{V}{V_0} = \tau \sqrt{\frac{H}{H_0}} \tag{11.54}$$

where τ is $C_d A_g/(C_d A_g)_0$. Values of τ vary from 1.0 for a fully open gate to zero when the gate is closed. Solutions of slow closure problems consist of the step-by-step simultaneous solution of Eqs. (11.50) and (11.54), as illustrated in Example 11.2.

Example 11.2 Determine the rise in pressure in a penstock leading to a power plant if the turbine gates are closed in 2.1 s. $H_0 = 50$ m, $L = 250$ m, $V_0 = 3.6$ m/s, and $\alpha = 1410$ m/s. The solution of the problem will be carried forward in six time increments of 2.1/6 = 0.35 s, arbitrarily chosen equal to the travel time of the pressure wave from the gate to the reservoir and back to the gate,

$$\frac{2L}{a} = 0.35 \text{ s}$$

A sketch of the penstock is given in Fig. 11.4. The gate is closed such a manner that six uniform decrements of area occur during each of the six time intervals. Then τ has the values shown in column 2 of Table 11.2. In solving the problem, it helps to visualize that the head at the gate is increasing linearly during the time of uniform closure. Then at $t = 0.167$ s, the front of the pressure wave has just reached the reservoir and the pressure in the pipe is as shown by line 1–2 in

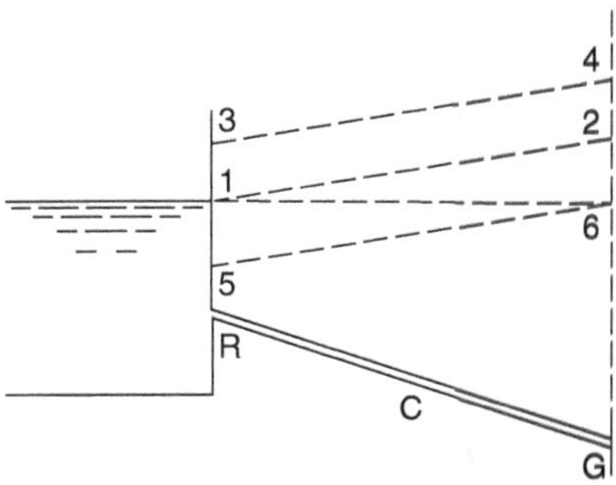

Figure 11.4 Definition of water hammer.

TABLE 11.2 Water Hammer Computation for Example 11.2

	(1)	(2)	(3)	(4)	(5)	(6)	(7)	(8)
Time interval	t, s	τ	$\frac{\Delta V}{V_0}$	$\frac{\Delta H}{H_0}$	$\frac{V}{V_0}$	V, m/s	$\frac{H}{H_0}$	H, m
0	0	1.00	0	0	1	3.60	1	50.0
1	0.35	0.833	0.033	0.34	0.967	3.48	1.34	67.0
2	0.70	0.667	0.067	0.69	0.867	3.10	1.69	85.1
3	1.05	0.500	0.095	0.98	0.705	2.52	1.98	99.7
4	1.40	0.333	0.117	1.19	0.493	1.77	2.19	110.4
5	1.75	0.167	0.123	1.29	0.253	0.91	2.29	115.2
6	2.10	0	0.130	1.32	0	0	2.32	116.5

Fig. 11.4. At $t = 0.35$ s, the pressure is the sum of the positive pressures indicated by line 3–4 and the negative pressures indicated by line 5–6. Thus at $t = 0.35$ s, the pressure wave has just returned to the gate G but has zero value at that point. Therefore, at time $t = 0.35$ s there is only an increase in pressure at G computed as follows, noting first that $V_1 = V_0 - \Delta V_1$, $H_1 = H_0 + \Delta H_1$, and $\tau = 0.833$, the subscripts indicating values at the end of the first time interval.

Equation (11.54) becomes

$$\frac{V_1}{V_0} = 1 - \frac{\Delta V_1}{V_0} = \tau\sqrt{\frac{H_1}{H_0}} = 0.833\sqrt{1 + \frac{\Delta H_1}{H_0}} \tag{11.54a}$$

Also, Eq. (11.50) becomes

$$\frac{\Delta H_1}{H_0} = \frac{v_w V_0}{gH_0}\frac{\Delta V_1}{V_0} = \frac{1410 \times 3.60}{9.8 \times 50.0}\frac{\Delta V_1}{V_0} = 10.36\frac{\Delta V_1}{V_0} \tag{11.50a}$$

These two equations are then solved by successive approximations. As a first trial assume $\Delta V_1/V_0 = 0.10$. Then, from Eq. (11.50a),

$$\frac{\Delta H_1}{H_0} = 10.36 \times 0.10 = 1.04$$

and Eq. (11.54*a*) becomes

$$1-\frac{\Delta V_1}{V_0}0.833\sqrt{1+\frac{\Delta H_1}{H_0}}$$

$$1-0.10=0.833\sqrt{1+1.04}$$

$$0.90\neq 1.19$$

Because this is not an equality, further trials must be carried out. The final solution for this time interval gave values of $\Delta V_1/V_0 = 0.033$ and $\Delta H/H_0 = 0.34$. Then $V_1/V_0 = 0.967$ and

$$V_1 = 0.967\times 3.60 = 3.48 \text{ m/s}$$

Similarly, $H_1/H_0 = 1.34$ and

$$H_1 = 1.34\times 50.0 = 67 \text{ m}$$

These values are recorded in Table 11.2.

During the second time interval the solution of Eq. (11.50*a*) is carried out as before, except for the subscripts, because this equation deals only with the pressure increment resulting from another reduction in velocity,

$$\frac{\Delta H_2}{H_0} = 10.36\frac{\Delta V_2}{V_0} \tag{11.50a}$$

However, Eq. (11.54*a*) must be revised because there is superimposed upon the positive pressure wave the negative wave returning from the reservoir. At $T = 0.70$ s, this negative wave will cause a reduction in velocity equal to the increment from the first interval denoted here as ΔV_1 ($\Delta V_1/V_0 = 0.033$) and a reduction in pressure ΔH_0 ($\Delta H_1/H_0 = 0.34$). These values must be included in the discharge relationship at the gate as given by Eq. (11.54*a*),

$$\begin{aligned}\frac{V_2}{V_0} &= \frac{V_1}{V_0} - \frac{\Delta V_1}{V_0} - \frac{\Delta V_2}{V_0}\\ &= \tau\sqrt{\frac{H_2}{H_0}} = \tau\sqrt{\frac{H_1}{H_0} - \frac{\Delta H_1}{H_0} + \frac{\Delta H_2}{H_0}}\end{aligned} \tag{11.54a}$$

Inserting numerical values,

$$0.967-0.033-\frac{\Delta V_2}{V_0} = 0.667\sqrt{1.34-0.34+\frac{\Delta H_2}{H_0}}$$

or $$0.934 - \frac{\Delta V_2}{V_0} = 0.667\sqrt{1 + \frac{\Delta H_2}{H_0}}$$

Solution yields $\Delta V_2/V_0 = 0.067$ and $\Delta H_2/H_0 = 0.693$. Then $V_2/V_0 = 0.867$ and $H_2/H_0 = 1.693$. Corresponding values of V_2 and H_2 are shown in Table 11.2. Computations were continued in this manner, and the results are tabulated. ▲

References

1. D. C. Woo and E. F. Brater, "Spatially Varied Flow from Controlled Rainfall," *J. Hydraul. Div., ASCE*, vol. 88, no. 6, Nov., p. 31, 1962.
2. G. H. Keuligan, "Determination of Critical Depth in Spatially Variable Flow," in *Proc. 2d Midwest. Conf. on Fluid Mech.* Ohio State University, College of Engineering, Columbus, OH, p. 67, 1952.
4. K. H. Beij, "Flow in Roof Gutters," U.S. Bureau of Standards Res. Paper RPG44, 1934.
5. G. H. Keuligan, "Spatially Variable Discharge over a Sloping Plane," *Trans. Am. Geophys. Union*, pt. VI, p. 956, 1944.
6. J. Massau, "Mémoire sur l'intégration graphique des équations aux deriviés particelles," *Ann. Assoc. Ingenieurs Sortis des Écoles Spéciales de Gand*, vol. 23, p. 95, 1900.
7. H. J. Putnam, "Unsteady Flow in Open Channels," *Trans. Am. Geophys. Union*, vol. 29, p. 227, 1948.
8. B. R. Gilcrest, "Flood Routing," in H. Rouse (ed.), *Engineering Hydraulics*, Wiley, New York, Chap. 10, p. 635, 1950.
9. P.-N. Lin, "Numerical Analysis of Unsteady Flow in Open Channels," *Trans. Am. Geophys. Union*, vol. 33, p. 226, Apr. 1952.
10. J. J. Stoker, "Numerical Solution of Flood Prediction and River Regulation Problems," Rep. 1, p. 200, Inst. Math. and Mech. New York University New York, 1953.
11. V. T. Chow, *Open Channel Hydraulics*, McGraw-Hill, New York, p. 587, 1959.
12. J. Parmakian, *Water Hammer Analysis*, Prentice-Hall, Englewood Cliffs, NJ, 1955.
13. O. Simin, "Water Hammer, with Special Reference to Researches of Professor N. Jankovsky," *Proc. Am. Water Works Assoc.*, vol. 24, 1904.
14. G. R. Rich, *Hydraulic Transients*, McGraw-Hill, New York, 1951.
15. N. R. Gibson, "Pressures in Penstocks Caused by the Gradual Closing of Turbine Gates," *Trans. ASCE*, vol. 83, 1919.
16. R. W. Angus, "Simple Graphical Solution for Pressure Rise in Pipes and Pump Discharge Lines," *J. Eng. Inst. Canada*, p. 72, Feb. 1935.

SECTION 12

Measurement of Flowing Water

In general, all methods of measuring flowing water may be classed in one of two divisions which, together with a list of the more important methods coming under each classification, are as follows:

1. *Velocity methods*
   a. Current meter
   b. Pitot tube
   c. Color
   d. Salt-velocity
2. *Direct-discharge methods*
   a. Gravimetric
   b. Volumetric
   c. Weirs
   d. Orifices
   e. Gibson time-pressure diagram
   f. Venturi meter
   g. Pipe orifice
   h. Standardized nozzle
   i. California pipe
   j. Venturi flume
   k. Critical-depth meter
   l. Contracted opening
   m. Chemical gauging
   n. Electromagnetic flow meter

In general, the current meter, the weir, the venturi flume, the contracted section, and the critical-depth meter are used for measuring the flow in open channels. The color method, the salt-velocity method, the Gibson method, the venturi meter, the pipe orifices, the standardized nozzle, and the California-pipe method are used for measuring the flow in pipes. Pitot tubes and chemical gauging are used for both open and closed conduits. Orifices are used only for measuring comparatively small quantities of water.

Velocity-area methods require the determination by actual measurement of the mean velocity of the water. The discharge is the product of the area and the mean velocity. The color and salt-velocity methods determine the mean velocity from a single observation. The current meter and pitot tube give velocities at one point for each observation, and floats give the mean velocity in the path traveled. The current meter is generally preferred for measuring velocities is open channels where extreme accuracy is not required, and it is used almost exclusively for gauging rivers.

Direct-discharge methods do not involve velocity measurements. In some of the methods the determination of velocity is a step in the proceedings, but no actual measurement of velocity is made. Gravimetric and volumetric methods of measuring water, which require the determination of the weight and volume, respectively, of water flowing in a given period of time, are adapted primarily to experimental work in laboratories and can be used only for measuring comparatively small flows. These methods are simple and require no explanation. The use of orifices and weirs is explained in other sections of this handbook. All the other methods listed are described in this section.

Current Meter

The actuating element of a current meter is a wheel, comprising a series of vanes or cups, which is impelled by the current. The rate at which the wheel revolves varies with the velocity of the water. There are several arrangements for determining the speed of the wheel. Usually, this is accomplished by means of a mechanism which, at each revolution or at a given number of revolutions, makes and breaks an electric circuit in which is included a telephone receiver or other suitable electrical indicating or recording device. The acoustic meter has an attachment that strikes a drum at a given number of revolutions, the sound being conveyed to the observer through the tube by which the meter is held. Other meters are arranged with mechanical recording devices. Current meters are suspended from a cable or attached to a rod. For gauging large streams the former arrangement is preferable. Meters attached to rods are convenient for gauging small streams.

There are two general types of current meters—the differential, or cup, type, consisting of a vertical axis carrying a series of cups which revolve by the excess pressure on the concave side over that on the convex side; and the direct, or propeller, type, having vanes set on a horizontal axis which revolve by the direct action of the current. The cup meter always registers the full velocity, regardless of the direction of the current or the direction in which the meter is pointed. Moving the meter vertically, either up or down, causes the wheel to move in a positive direction. There is therefore a tendency for the cup meter always to indicate too high a velocity. The propeller-type meter does not have this objectionable feature, at least not to the same degree as the cup meter. It is probable, however, that any of the standard makes of current meters, if used properly, under the conditions to which they are best suited, will prove satisfactory in ordinary stream-gauging work.[a]

Pitot Tube

The pitot tube, in its simplest form, is a pipe with open ends and a right-angled bend near one end. When held with the opening in the shorter portion directed against the current, as indicated in Fig. 12.1*a*, the velocity at the opening being v, water will rise in the tube a distance $v^2/2g$ above the surface of the stream. It has been proved experimentally that this relation holds rigidly regardless of the dimensions of the tube or the size of opening. In this simple form the pitot tube has little practical value, since the distance above the surface of the stream that the water in the tube will rise cannot be measured accurately.

The tubes in Fig. 12.1*b* and *c* are exactly like tube *a*, except that tube *b* is pointed downstream and tube *c* is pointed across the current. A suction is created in both, the amount of suction being measured by

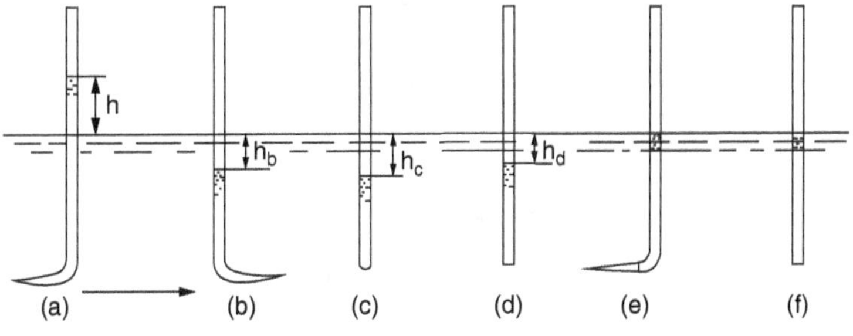

Figure 12.1 Principle of pitot tube.

[a]For a more complete discussion of the use of current meters, see Wisler and Brater.[1]

distances h_b and h_c by which the water surface in each tube is drawn below the surface of the stream. Approximate values of h_b and h_c, from experiments by Darcy are

$$h_b = 0.43\frac{v^2}{2g} \quad \text{and} \quad h_c = 0.68\frac{v^2}{2g}$$

Similarly, for tube *d*, a straight tube open at both ends, there is a depression h_d. Tube *e* is the same as tube *a*, except that the lower end is closed and there is a small hole in each sidewall. If the tube is pointed either upstream or downstream so that the two holes are aligned perpendicular to the direction of flow, the water surface in the tube will remain at about the same elevation as the surface of the stream. In other words, a tube of this type records very closely the true static pressure. Tube *f* is a straight pipe, closed at its lower end and with a small hole on each side. If this tube is held so that the two holes are aligned perpendicular to the current, it records approximately the static pressure, but probably not as closely as tube *e*.

A device designed by Darcy to measure velocities in open channels is shown in Fig. 12.2. A kinetic leg and a static leg, Fig. 12.1*a* and *b*, respectively, have glass tubes attached to their upper ends and are connected to an air suction pump through a common chamber containing a valve. There are two valves on one stem just below the glass tubes. To obtain a velocity measurement, the kinetic leg (either leg

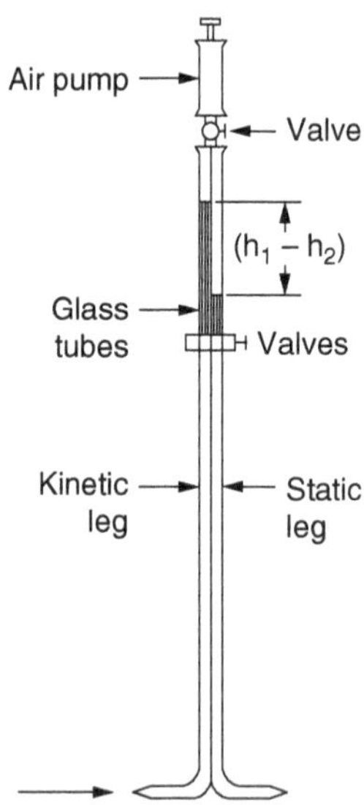

FIGURE 12.2 Darcy tube.

for this particular design) is held to point against the current. Then by operating the pump, air is rarefied sufficiently to draw both water columns into the glass tubes. The valves are then closed, and the instrument is lifted from the water and read. Any one of the forms of static legs *c, d, e*, or *f*, of Fig. 12.1 as well as other designs not shown could be substituted for leg *b*.

If $h_1 - h_2$ (Fig. 12.2) is the difference in height of the water columns in feet or meters and *c* is a coefficient, constant for each instrument the velocity is

$$v = c\sqrt{2g(h_1 - h_2)} \tag{12.1}$$

Approximate values of *c* for some of the forms of static leg shown in Fig. 12.1 are

For *b*,	$c = 0.84$
For *c*,	$c = 0.77$
For *e* and *f*,	$c = 1$

For accurate work each instrument should be rated by moving it through still water and *c* obtained for several velocities.

The pitot tube shown in Fig. 12.3 is a form of pipe contained in the Report of the Special Research Committee on Fluid Meters of the ASME.[2] With this arrangement the static head is taken in the wall of the pipe entirely separate from the kinetic leg. There should be

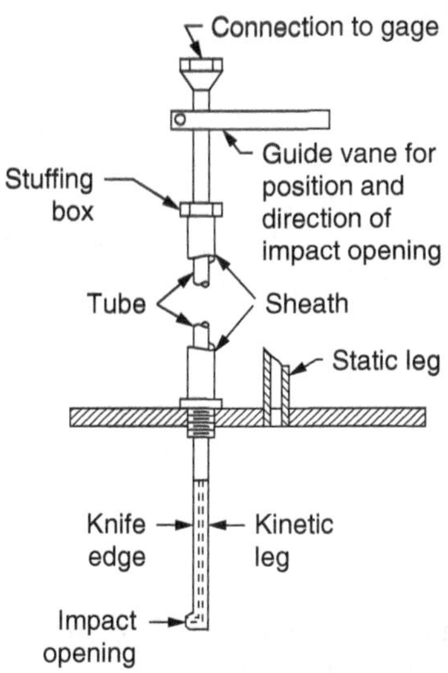

FIGURE 12.3 Single-opening pitot tube for pipes.

not less than two pairs of diametrically opposed holes to transmit the static head. These openings should not exceed ¼ in. (6 mm) in diameter, and their edges should be sharp. The tubes leading from them should be normal to the internal surface of the conduit wall. The difference in pressure in the two legs may be measured with a U tube.

Color Method

This method is used for determining the discharge in pipes or other closed conduits. It is not necessary that the conduit be of uniform or regular cross section. A small quantity of a concentrated solution of some powerful coloring matter is injected into the conduit, and the time it requires to travel to some point of observation is determined. The length of reach, that is, the distance between the point where the coloring matter is introduced and the point where it is observed, is measured.

Scobey[3] found that the coloring matter is contained in a prism having a length equal to about 10 percent of the distance traveled. He used three chemicals for coloring—fluorescein, congo red, and potassium permanganate. One teaspoonful of fluorescein dissolved in one-half liter of water gives sufficient coloring matter for four injections in a pipe carrying 60 ft^3/s (1.8 m^3/s) of water. Congo red and potassium permanganate each require about 10 times this amount of chemical to produce the same intensity in color.

The coloring matter can be introduced at the intake, or it can be injected by a force pump or gun through an opening in the wall of the conduit. The point of observation must be where the water can be seen, as at the outlet or when a valve can be opened. The length of reach should not be less than 200 times the velocity. Time observations should be made at the instant the coloring matter is introduced and at the first and last appearances of the coloring matter at the point of observation. The average of these time intervals is considered the time of travel. In a pipe or conduit of uniform cross-sectional area the length of reach divided by the time of travel equals the mean velocity, and the discharge is the product of this mean velocity and the cross-sectional area of the conduit. If the cross section of the conduit is not uniform, the volume of the reach must be determined. This volume divided by the time of travel equals the discharge.

Salt-Velocity Method

This is a modification of the color method and was used by Slichter[4] in 1901 to measure the rate of flow of ground water. Its application to the measurement of flow in pipes was described by King[5] in 1918 in the first edition of this handbook. Later the method was used by Allen and Taylor[6] to measure the discharge of penstocks in making turbine tests.

A concentrated salt brine is introduced at the intake or through openings in the conduit wall. The salt increases the electrical conductivity of the water. Electrodes, connected to an ammeter or electrical recording meter, are installed at one or more points of observation below the place where salt is introduced. An increase in electric current is indicated when the prism of water containing the salt passes the electrodes.

If an ammeter is used with but one point of observation, time observations are made at the instant the salt is introduced and also at the beginning and end of increased deflection of the ammeter needle. The average of these time intervals is considered the time of travel. If there are two or more points of observation, particularly if one of the points is near the place where the salt is introduced, it is not usual to record the time of introducing the salt. The time of travel between two points of observation is the average of the times between the beginnings and the ends of increased deflection. Readings at one point of observation, in addition to the two readings necessary for a discharge measurement, provide an independent check on the result.

If a recording meter is used, the time when the salt is considered to pass a point of observation is given by the center of gravity of the area included between the line of normal current extended and the line showing excess current. The method of computing the discharge is identical with that described for the color method. The volume of the conduit for the reach being considered must be determined. The discharge equals this volume divided by the time of travel.

Gibson Method

This can be used for measuring the discharge in a pipe or other closed conduit which has a valve, or some other means of gradually shutting off the water, at least 10 and preferably 20 diameters or more below the intake. The conduit need not be of uniform cross-sectional area. The method is particularly adapted to measuring the discharge in penstocks in which the flow of water is controlled by turbine gates.

A mercury U tube attached to the penstocks above the gates indicates the hydrostatic pressure. In making a measurement the gates are closed gradually, and a continuous photographic record showing equal time intervals and the height of the mercury in the U tube

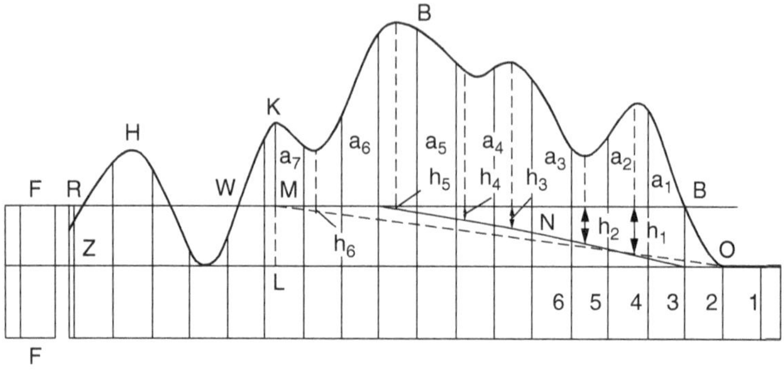

FIGURE 12.4 Gibson time-pressure diagram.

is made by the Gibson apparatus.[b] This is a time-pressure diagram which shows the total impulse necessary to bring the column of water to rest and from which the discharge can be computed.

Figure 12.4 is reproduced from one of Gibson's time-pressure diagrams. Vertical lines 1, 2, 3, etc., are equal time intervals, preferably 1 s. The curved line *OBBKH* is the locus of elevations of the surface of mercury in the U tube. The detached rectangle *FF* is a short record of pressure after the gates are closed and oscillations have ceased. The horizontal line *OZ* shows the pressure before the gates began to close, and the line *FB* is the hydrostatic pressure at zero velocity. The vertical distance between these lines is therefore the sum of the velocity head and the friction head when the gates are open. At *O* the gates began to close, and at *K* they are completely closed. The position of *K* is determined by drawing the ordinate *MK*, the point *M* being approximately one-fourth of a complete wavelength, ¼*RW*, to the right of *W*.

The line *ONM* eliminates the area produced by the recovery of friction and velocity heads, the area above this line being a measure of the rate of discharge in the conduit at the moment the gates began to close. Trial and error are used to locate the position of line *ONM*.

As a first assumption, the straight line *OM* may be taken, but usually it will be found easy to estimate by inspection the approximate position so that only one correction will be necessary.

A number of vertical lines are drawn from the pressure line *BBK* to intersect the line *OM*, and the areas a_1, a_2, a_3, …, enclosed by these

[b]Method and apparatus patented in the United States, Oct. 13 and Nov. 22, 1921. A brief description of the method is given here. For fuller details, the theory involved, minor corrections to be made, and a description of the Gibson apparatus, the reader is referred to Gibson.[7]

lines are measured by planimeter or by scaling the dimensions. The sum of these areas will be called A_1. The distance *LM*, the sum of the friction and velocity heads, is measured and called *Y*. Let

$$P_1 = \frac{a_1}{A_1}, \qquad P_2 = \frac{a_1 + a_2}{A_1}, \qquad p_3 = \frac{a_1 + a_2 + a_3}{A_1}, \qquad \ldots$$

and let

$$h_1 = Y(1 - P_1)^2, \qquad h_2 = Y(1 - P_2)^2, \qquad \ldots$$

Having determined the values of h_1, h_2, h_3, …, from these equations, a point is located on the line separating areas a_1 and a_2, or this line is produced by measuring down vertically from *FB* a distance equal to h_1. A distance equal to h_2 is measured down vertically from *FB* on the line separating areas a_2 and a_3. Similarly, a distance equal to each of the values h_3, h_4, …, is measured down from *FB* on the respective lines separating areas a_3, a_4, … The points so plotted are then joined, forming a line from *O* to *M*. If this line fails to coincide with the assumed line, a second trial is made using the new line in place of the one first assumed. This process is repeated until a base line *ONM* is found such that when h_1, h_2, etc., are computed as before, but using the new areas a_1, a_2, a_3, …, above line *ONM*, the line joining the ends of h_1, h_2, …, will coincide with *ONM*.

Referring to Fig. 12.4, let

A = area of surface *OBBKMNO*
r = vertical height, corresponding to 1 ft of pressure change in conduit
S = horizontal length, in., corresponding to 1 s of time

Then if $K = g/r$, g being the acceleration due to gravity and L the length of pipe in meters,

$$\upsilon = \frac{KA}{SL} \tag{12.2}$$

This is a convenient form of the final equation for a pipe of uniform cross-sectional area. The mean velocity υ is multiplied by the area to obtain the discharge.

When the conduit is not of uniform cross-sectional area throughout its length, but is made up of a series of different sections of lengths l_1, l_2, l_3, …, having average cross-sectional areas a_1, a_2, a_3, …, the discharge is given by the formula

$$Q = \frac{KA}{SF} \tag{12.3}$$

where $F = \Sigma(l/a)$, that is, the sum of the quotients obtained by dividing the length of each section by its average cross-sectional area.

This method can be modified by substituting for the simple U tube with a single piezometer connection a differential U tube with connections to the conduit at two points at least 20 diameters apart. The procedures in preparing and using the diagrams in the two cases are substantially the same.

Venturi Meter

The principle of the venturi meter was first stated in 1797 by J. B. Venturi, an Italian, and was first applied by Herschel[8] to the measurement of flow in pipes in 1887. Figure 12.5 shows a venturi meter reproduced from a drawing contained in the Report of the Special

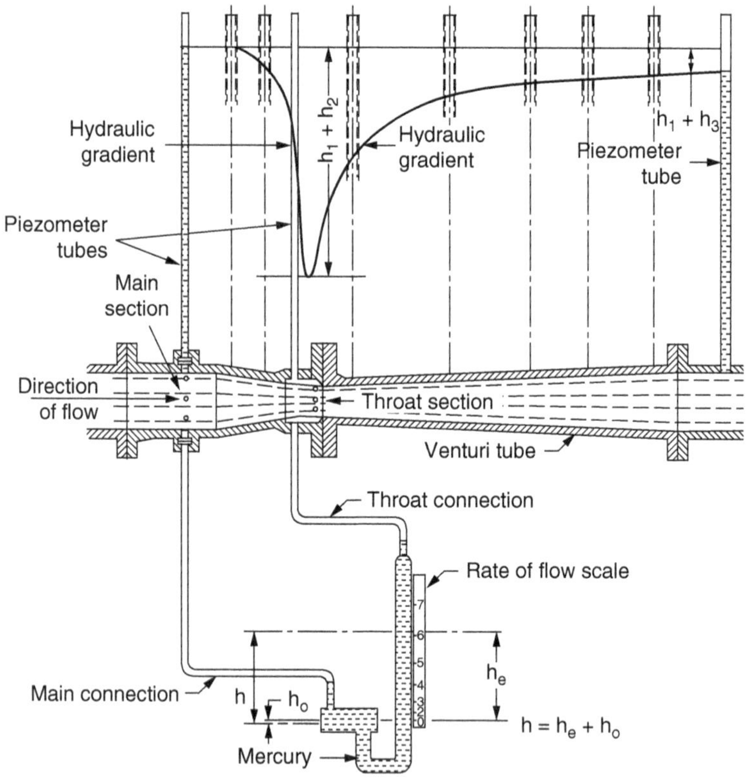

FIGURE 12.5 Section of venturi meter.

Committee on Fluid Meters of the ASME.[c] It consists of a constricted portion, which may be inserted between two flanges in a pipeline for the purpose of accelerating the water and lowering its static pressure. The proportions of the venturi meters now used are much the same as those originally adopted by Herschel and shown in Fig. 12.5. Section 3 discusses the basic principles involved in the use of the venturi meter as well as methods of computing the discharge at which cavitation will begin in a venturi.

Starting at the upstream flange, there is, first, a short cylindrical portion, which continues the pipeline but is machined inside or cast smooth so that its diameter may be accurately determined. In this portion a side hole is drilled through the wall, or several holes lead into a piezometer ring, so that a connection may be made for measuring the static pressure of the fluid before it enters the constriction.

Following this preliminary straight part, an entrance cone of about 21° total angle leads to the short cylindrical throat which is accurately machined and is provided with a side hole or holes for taking off the static pressure in the throat. The transition from the entrance cone into the straight throat is rounded off to an easy tangential curve to avoid the resistance caused by a sharp corner and to preclude the possibility that the fluid might break away from the wall at high speeds and so not fill the throat completely. The diameter of the throat is usually between one-half and one-fourth of the entrance or pipeline diameter, and there is usually a joint, a handhole, or in large sizes even a manhole at the throat which facilitates inspection of the condition of the throat and side holes and permits the accurate measurement of the throat diameter. The end of the throat leads, by another easy curve, into the exit cone or diffuser, which has a total angle of about 5° to 7°. This terminates in the downstream flange for connecting the venturi to the following pipeline. A pressure connection is sometimes provided at the downstream flange for the purpose of determining the overall loss of head caused by the resistance of the venturi as a whole.

Very small venturis—say, for pipelines less than 2 in. (5 cm) is diameter—are commonly made of brass or bronze and smoothly finished all over the inside to reduce resistance. Larger venturis are usually of cast iron, the throat and sometimes the straight entrance portion being lined with brass or bronze and machined to a smooth finish. Very large venturis, up to 20 ft (6 m) in diameter, have been

---

[c]The figure and most of the information contained in this description are taken from this report.[2]

made of smooth-surface concrete, only the throat being of machined bronze. Wood staves and steel plate have also been used for the cones of large venturis.

Venturi tubes, especially in large sizes, are usually set horizontally or nearly so, although this position is not at all essential to the operation of the meter. If reliable measurements are required, it is essential that the venturi should not follow immediately after a valve, elbow, or other irregularity in the pipeline, which would tend to cause great turbulence or set up permanent crosscurrents or whirls, but should be preceded by a length of 5 or more diameters of straight pipe in which the flow may steady down after any such cause of disturbance. There is probably never any appreciable advantage in accuracy to be gained by using more than 20 diameters of straight pipe ahead of the venturi. When it is impossible to provide a sufficient line of pipe, the flow should be steadied by straightening vanes. A sheet-metal cross, 1 or 2 diameters long, inserted in the pipe just ahead of the venturi will serve this purpose.

The function of the long diverging cone is to retard or decelerate the fluid smoothly and restore the pressure as nearly as possible to the value it had at the entrance. This restoration is imperfect because there is always some resistance to flow through the venturi, and some loss of head is unavoidable. The overall loss is 10 to 20 percent of the venturi head, or differential from entrance to throat. In other words between 80 and 90 percent of the fall of pressure at the throat is restored in the diverging cone. The percentage loss decreases as the speed of flow increases or as the meter is made larger.

Let h_1 and h_2 represent the height above the axis of the meter to which the water rises in the piezometer tubes at the entrance and throat of the meter, respectively, and let V_1, d_1, and a_1 and V_2, d_2, and a_2 be the corresponding velocities, diameters, and areas. Then, from Bernoulli's theorem, neglecting friction,

$$\frac{V_2^2 - V_1^2}{2g} = h_1 - h_2 \tag{12.4}$$

and since

$$Q = a_1 V_1 = a_2 V_2$$

the formula for discharge through a venturi meter, including the empirical coefficient c, becomes

$$Q = \frac{c a_1 a_2}{\sqrt{a_1^2 - a_2^2}} \sqrt{2g(h_1 - h_2)} \tag{12.5}$$

or, expressed in terms of diameter,

$$Q = \frac{c\pi d_1^2 d_2^2}{4\sqrt{d_1^4 - d_2^4}}\sqrt{2g(h_1 - h_2)} \tag{12.6}$$

which reduces to

$$Q = cKd_2^2\sqrt{h_1 - h_2} \tag{12.7}$$

where

$$K = \frac{\pi}{4}\sqrt{\frac{2g}{1-(d_2/d_1)^4}} \tag{12.8}$$

Values of K corresponding to different values of d_2/d_1 and values of c are given in Tables 12.1 and 12.2. Coefficients of discharge c were obtained from experiments by Ledoux[9] on 19 different venturi meters, in which the throat diameter varied from 0.4 to 18 in. (1 to 45 cm). Discharges were measured volumetrically, and heads were measured with hook gauges. It is believed that these experiments provide the most reliable data available for computing venturi coefficients. Values in Table 12.2 represent the approximate mean of the experimental values. Individual measurements vary by 1 percent or more from the tabulated values. The experiments include throat velocities of 1 to 30 ft/s (0.3 to 9 m/s) approximately. Values in Table 12.2 for the higher velocities as well as for the throat diameter of 48 in. (120 cm) were obtained by extending curves plotted from the experiments. The Ledoux experiments indicated values of the coefficient to be about 2 percent smaller for throat velocities of 5 ft/s (1.5 m/s) and about 1 percent smaller for the higher velocities than the values given by the Report of the Special Committee on Fluid Meters of the ASME.

The coefficients of discharge c are for temperatures of 15°C. The following are corrective factors for other temperatures:

Temperature, °C	4	15	27	38	49	66	93
f	1.4	1.0	0.8	0.6	0.5	0.4	0.3

To determine c for a temperature other than 15°C, multiply the actual throat velocity by f and take the coefficient corresponding to the equivalent throat velocity from Table 12.2.

Measurements by Ledoux on a 10 × 3.8-cm venturi meter with heavy deposits in the cone but clean throat, gave a coefficient before cleaning of about 2.5 percent less than the coefficient obtained for the same meter after cleaning. Snow[10] found that a venturi meter in a sewage main, after having been in service for 20 years, had its throat diameter reduced from 38 to 30 cm by the deposition of carbonate of lime, which came from trade wastes.

Table 12.1 Value of *K* in Formula for Venturi Meters

$\frac{d_2}{d_1}$	*K* (Metric Units)	*K* (English Units)
0.20	3.48	6.31
0.21	3.48	6.31
0.22	3.48	6.31
0.23	3.48	6.31
0.24	3.48	6.31
0.25	3.48	6.31
0.26	3.49	6.31
0.27	3.49	6.32
0.28	3.49	6.32
0.29	3.49	6.33
0.30	3.50	6.33
0.31	3.50	6.33
0.32	3.50	6.33
0.33	3.50	6.34
0.34	3.51	6.34
0.35	3.51	6.35
0.36	3.51	6.35
0.37	3.52	6.36
0.38	3.52	6.37
0.39	3.52	6.37
0.40	3.52	6.38
0.41	3.53	6.39
0.42	3.54	6.40
0.43	3.54	6.41
0.44	3.55	6.42
0.45	3.55	6.43
0.46	3.56	6.45
0.47	3.57	6.46
0.48	3.58	6.47
0.49	3.59	6.49
0.50	3.60	6.51
0.51	3.61	6.52
0.52	3.61	6.54

Table 12.1 Value of *K* in Formula for Venturi Meters (*Continued*)

$\frac{d_2}{d_1}$	*K* (Metric Units)	*K* (English Units)
0.53	3.62	6.56
0.54	3.63	6.59
0.55	3.64	6.61
0.56	3.66	6.64
0.57	3.67	6.66
0.58	3.69	6.69
0.59	3.71	6.72
0.60	3.73	6.75
0.61	3.75	6.79
0.62	3.77	9.82
0.63	3.79	6.86
0.64	3.81	6.91
0.65	3.84	6.95
0.66	3.87	7.00
0.67	3.90	7.05
0.68	3.93	7.11
0.69	3.96	7.17
0.70	3.99	7.23
0.71	4.03	7.30
0.72	4.07	7.37
0.73	4.11	7.45
0.74	4.16	7.53
0.75	4.21	7.62
0.76	4.26	7.72
0.77	4.32	7.82
0.78	4.39	7.94
0.79	4.46	8.06
0.80	4.53	8.20
0.81	4.61	8.35
0.82	4.70	8.51
0.83	4.80	8.69
0.84	4.91	8.89

TABLE 12.2 Coefficients of Discharge *c* for Venturi Meters

Dia. of Throat, in./cm	Throat Velocity, ft/s / Throat Velocity, m/s								
	3	**4**	**5**	**10**	**15**	**20**	**30**	**40**	**50**
	0.9	***1.2***	***1.5***	***3***	***4.5***	***6***	***9***	***12***	***15***
1/2.5	0.935	0.945	0.949	0.958	0.963	0.966	0.969	0.970	0.972
2/5.0	0.939	0.948	0.953	0.965	0.970	0.973	0.974	0.975	0.977
4/10.0	0.943	0.952	0.957	0.970	0.975	0.977	0.978	0.979	0.980
8/20	0.948	0.957	0.962	0.974	0.978	0.980	0.981	0.982	0.983
12/30	0.955	0.962	0.967	0.978	0.981	0.982	0.983	0.984	0.985
18/45	0.963	0.969	0.973	0.981	0.983	0.984	0.985	0.986	0.986
48/120	0.970	0.977	0.980	0.984	0.985	0.986	0.987	0.988	0.988

Pipe Orifice

The equation for discharge through a circular sharp-edged orifice was derived in Sec. 4. This equation may be restated in the following form:

$$Q = Ka\sqrt{2g\left(\frac{p_1}{w} - \frac{p_2}{w}\right)} \tag{12.9}$$

where K is defined as

$$K = \frac{C}{\sqrt{1 - C^2(a/a_1)^2}} \tag{12.10}$$

In Eqs. (12.9) and (12.10), Q is the discharge in ft³/s (or m³/s), a and a_1 are the areas of the orifice and the pipe in ft² (or m²), p_1 and p_2 are the pressures at the upstream and downstream sides of the orifice in lb/ft² (or newtons per square meter), and w is the unit weight of the liquid in lb/ft³ (or newtons per cubic meter). It will be noted from the derivation in Sec. 4 that C takes care of the energy loss and the contraction of the jet. It follows from Eq. (12.10) that the coefficient K includes these factors as well as the velocity of approach.

Values of K for square-edged pipe orifices have been presented by the ASME[11] for three different types of pressure taps. These are flange taps, vena contracta taps, and radius taps. Flange taps are located with the center of the high-pressure tap 1 in. (2.5 cm) from the upstream face of the orifice plate and the center of the low-pressure tap 1 in. (2.5 cm) from the downstream side of the plate. For vena contracta taps, the high-pressure tap is located 1 pipe diameter d_1 upstream from the face of the orifice plate, and the low-pressure tap is located at the vena contracta. The mean distance from the orifice plate to the vena contracta, in terms of pipe diameters, for various values of the ratio of orifice diameter to pipe diameter d/d_1 is given in Sec. 4. Radius taps are located 1 pipe diameter upstream and 12 pipe diameters downstream from the orifice plate.

Values of K are presented in terms of the Reynolds number,

$$R = \frac{dV}{\nu} \tag{12.11}$$

where d is the orifice diameter in meters, V the velocity at the orifice, and ν the kinematic viscosity of the fluid. Values of ν for water may be obtained from Table 1.2. Values of K to be used with flange taps for 6- and 10-in. (15- and 25-cm) pipe are shown in Fig. 12.6. These were taken from the ASME Power Test Codes.[11] The curves for 6 in. (15-cm) pipe are quite similar to those for 3- and 4-in. (7.5- and 10-cm) pipe, whereas those for 10-in. (25-cm) pipe are similar to those for 8- and 14-in. (20- and 35-cm) pipe. Values of K for 2-in. (5-cm) pipe are somewhat lower than those for 6-in. (15-cm) pipe for $R < 10^5$ and slightly higher for $R > 2 \times 10^5$. If an installation of a pipe orifice is to be made without a calibration, it is recommended that reference be made to the original data for the particular pipe size, as well as to detailed installation requirements.[11]

Curves giving values of K for vena contracta taps are shown in Fig. 12.7. While these curves are for 15-cm pipe,[11] they are nearly identical with those for other pipe diameters.

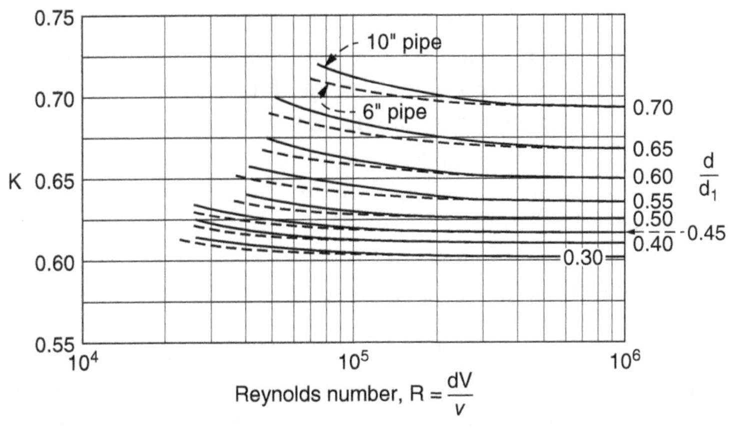

FIGURE 12.6 Pipe-orifice coefficients for flange taps.

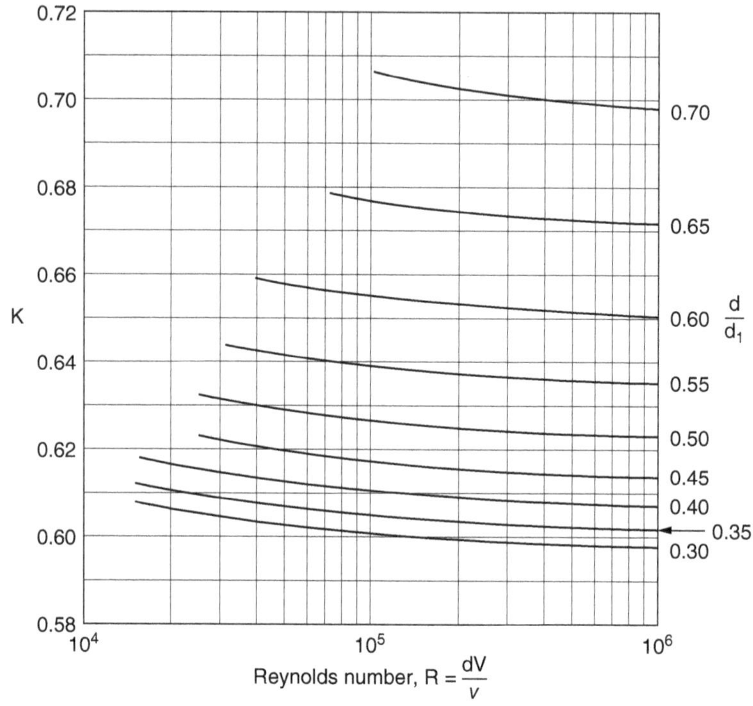

FIGURE 12.7 Pipe-orifice coefficients for vena contracta taps.

Values of K for a pipe orifice in any particular location may be expected to differ somewhat from the values given in Figs. 12.6 and 12.7 unless installation recommendations are carefully followed. Detailed instructions are available.[11] Some of the more important restrictions have to do with the location of pipe fittings in the vicinity of the orifice. For example, elbows or open valves must be located 5 pipe diameters upstream from the orifice for small values of d/d_1 and as many as 25 diameters upstream from the orifice for large values of d/d_1. No elbow should be placed nearer than 5 diameters from the downstream side of the orifice. Partially open valves should not be located upstream from the orifice. Where these restrictions cannot be followed, flow straighteners must be used or the orifice must be calibrated in place. Studies regarding the effect of pipe fittings on orifice flow have been conducted, and information is available.[12]

The ISA (International Federation of National Standardizing Associations) standard orifice is also a square-edged circular-pipe orifice. The pressure taps, however, are located immediately adjacent to the orifice plate. Coefficients for this type of taps have been published by the Verein Deutscher Ingenieure, translated into English by the ASME, and published by the National Advisory Committee for Aeronautics.[13]

Standardized Nozzles

This method of measuring the flow in pipes is the same as the pipe-orifice method, except that the orifice has rounded corners and prolonged sides (Fig. 12.8), which lead the water smoothly to the orifice so that it issues without contraction in a straight cylindrical jet of the same diameter as the orifice. In effect, the flow nozzle is a venturi meter with the long diffuser omitted. Discharge coefficients have been published for two standard types of nozzles. The ISA nozzle was developed in Germany. Discharge coefficients have been published in the United States by the ASME[11] and NACA.[13] The ASME long-radius nozzle,[11] as the name implies, is longer and constricts the flow more gently than the ISA nozzle. The installation requirements described earlier for pipe orifices also apply to standardized nozzles.

California-Pipe Method

This is a method, developed by Van Leer,[14] for determining the rate of flow from the open end of a partially filled horizontal pipe discharging freely into the air. It is particularly adapted to the measurement of comparatively small flows in pipes. It can also be used to measure flows in small open channels if the water can be diverted into a pipe which it does not completely fill and which discharges without any submergence of the outlet.

The discharge pipe should be level and at least 6 diameters long. If the pipe is flowing nearly full, there should be an air vent a few

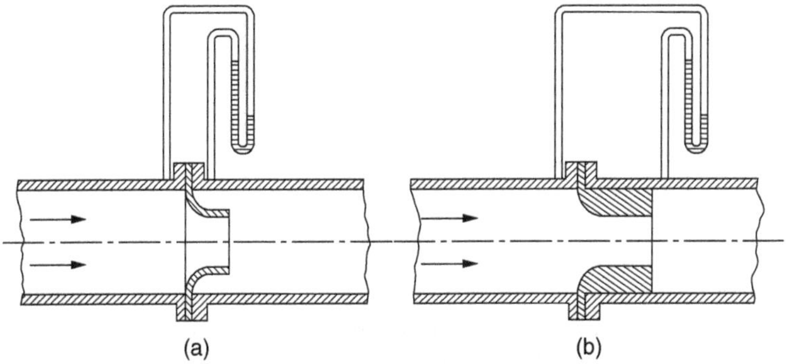

Figure 12.8 Flow nozzles. (*a*) Spun type. (*b*) Thick-plate type.

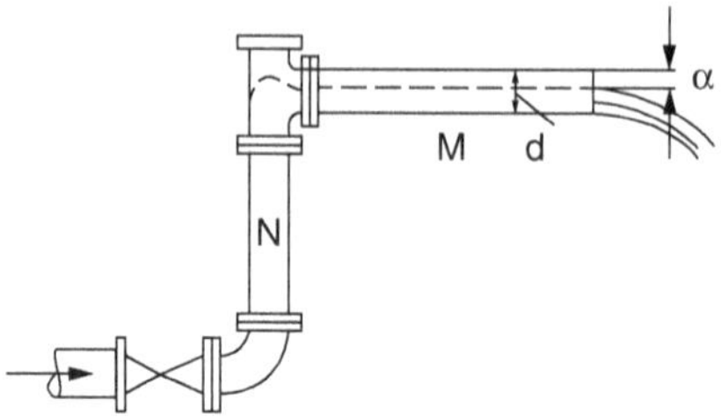

Figure 12.9 California-pipe method of measuring discharge.

diameters back from the outlet to provide for the free circulation of air in the unfilled portion of the discharge pipe. Water should enter the discharge pipe without excessive velocity of approach. The equipment required is simple, and it can be easily installed and used.

One arrangement for measuring the flow in a pipeline is shown in Fig. 12.9. The nipple N connects at its lower end to the supply pipe, and at the upper end it is connected by a tee to the discharge pipe M. The open end of the tee permits free circulation of air.

Let d equal the internal diameter of the pipe in meters and a the distance in meters, measured in the plane of the end of the pipe, from the top of the inside surface of the pipe to the water surface. This distance can be measured approximately with a rule or more accurately with calipers. The discharge in cubic feet per second is given by the formula

$$Q = 8.69\left(1 - \frac{a}{d}\right)^{1.88} d^{2.48}$$

The discharge in cubic meters per second is given by the formula

$$Q = 4.71\left(1 - \frac{a}{d}\right)^{1.88} d^{2.48} \tag{12.12}$$

This formula is based on experimental data for pipes 3 to 10 in. (7.5 to 25 cm) in diameter, and it applies accurately within these limits. There appears no reason to doubt, however, that it can be used with considerable assurance for larger pipes.

Venturi Flume

This is a method of adapting the venturi principle to the measurement of flow in open channels. A venturi flume[15–18] consists of a constricted portion, or throat, with a gradually contracting portion leading to the throat and a gradually expanding portion leading from it. It is illustrated in Fig. 12.10, with the exception that in the true venturi flume the floor is laid substantially to the canal grade, and stilling wells for measuring the head are provided at both the entrance and the throat. The difference in level in the two wells represents the change from potential head to velocity head plus friction and is a measure of the discharge. Figure 12.10 represents a flume having a trapezoidal section, but the section may be rectangular, triangular, or of any other regular form.

The venturi flume is particularly adapted to the measurement of irrigation water. It has advantages over the weir in that it does not back up materially the water in the channel, and its accuracy is not apt to be interfered with by the deposition of silt or floating matter.

The derivation of the formula for discharge is identical with that for venturi meters (see Sec. 3). If a_1 and a_2 are the respective cross-sectional areas of the stream and h is the difference in water levels at the entrance and the throat, the formula for discharge through the flume, including the empirical coefficient c, is

$$Q = cMa_2\sqrt{h} \tag{12.13}$$

where

$$M = \sqrt{\frac{2g}{1-(a_2/a_1)^2}} \tag{12.14}$$

Values of M corresponding to different values of a_2/a_1 are given in Table 12.3. The value of c will ordinarily lie between 0.95 and 1.00. It can be kept at 0.98 or higher if care is taken to have smooth surfaces faces and to round off all corners so as to lead the water to the throat of the flume without contractions or unnecessary turbulence.

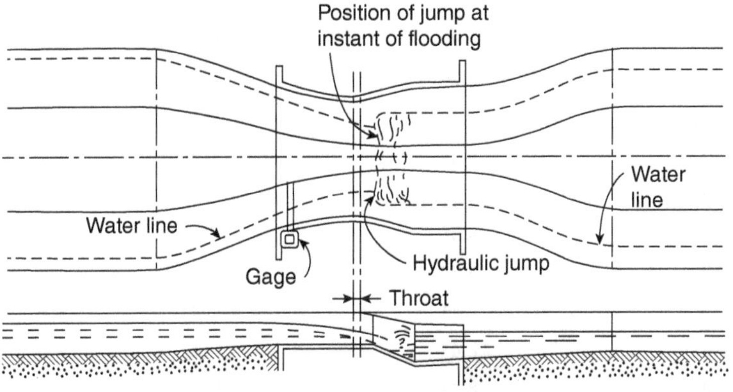

FIGURE 12.10 Critical-depth meter.

Table 12.3 Values of M for Venturi Flume and Contracted-Section Formulas

$\frac{a_2}{a_1}$	M English Units	M Metric Units
0.00	8.02	4.43
0.01	8.02	4.43
0.02	8.02	4.43
0.03	8.02	4.43
0.04	8.03	4.43
0.05	8.03	4.43
0.06	8.03	4.44
0.07	8.04	4.44
0.08	8.05	4.44
0.09	8.05	4.45
0.10	8.06	4.45
0.11	8.07	4.45
0.12	8.08	4.46
0.13	8.09	4.46
0.14	8.10	4.47
0.15	8.11	4.47
0.16	8.12	4.48
0.17	8.14	4.49
0.18	8.15	4.50
0.19	8.17	4.51
0.20	8.19	4.52
0.21	8.20	4.53
0.22	8.22	4.54
0.23	8.24	4.55
0.24	8.26	4.56
0.25	8.28	4.57
0.26	8.31	4.58
0.27	8.33	4.60
0.28	8.35	4.61
0.29	8.38	4.63
0.30	8.41	4.64
0.31	8.44	4.66
0.32	8.47	4.68

Table 12.3 Values of *M* for Venturi Flume and Contracted-Section Formulas (*Continued*)

$\frac{a_2}{a_1}$	*M* English Units	*M* Metric Units
0.33	8.50	4.70
0.34	8.53	4.72
0.35	8.56	4.73
0.36	8.60	4.75
0.37	8.63	4.77
0.38	8.67	4.79
0.39	8.71	4.81
0.40	8.75	4.83
0.41	8.79	4.85
0.42	8.84	4.87
0.43	8.88	4.90
0.44	8.93	4.93
0.45	8.98	4.96
0.46	9.03	4.99
0.47	9.09	5.02
0.48	9.14	5.05
0.49	9.20	5.08
0.50	9.26	5.11
0.51	9.32	5.14
0.52	9.39	5.18
0.53	9.46	5.22
0.54	9.53	5.26
0.55	9.60	5.30
0.56	9.68	5.34
0.57	9.76	5.39
0.58	9.84	5.44
0.59	9.93	5.49

Critical-Depth Meter

This is a modified venturi flume. The main feature is a constriction in the channel, so proportioned as to produce flow at critical depth. The principles, formulas, and tables which apply are described in Sec. 8. The adaptation of this device to the measurement of irrigation water was first described by Hinds.[19] Figure 12.10 shows a structure of trapezoidal form which was installed at the hydraulic laboratory of the U.S. Bureau of Reclamation at Boise, Idaho. The constriction is provided by drawing in the sides and raising the floor above the bed of the channel. A slight drop, to prevent interference from backwater, is provided just below the throat.

It is essential that flow at critical depth should occur without interference from backwater. A standing wave (hydraulic jump) usually forms in the lower channel. This will not affect the proper functioning of the device so long as the wave forms so far downstream that the control is not flooded. Figure 12.10 illustrates a test run which was made to determine the conditions under which flooding of the control will occur. By means of checks the water surface in the lower channel was slowly raised while conditions at the control were studied. It was found that there was no submergence (that is, the depth at the control was not greater than the critical depth) until the standing wave reached approximately the location of the control. The figure shows the characteristic appearance of the wave just before submergence. After submergence the wave usually changes to a series of ripples, which may be practically as rough as the last stage of the wave, but which have a distinctly different appearance. After the control is flooded, the device becomes in principle a venturi flume.

A large number of tests on different structures by the U.S. Bureau of Reclamation give an average coefficient of discharge of practically unity, that is, the theoretical formulas [Eqs. (8.35) and (8.63)] apply without any empirical modification. These formulas include no correction for friction or the effects of nonuniform velocities in the channel of approach. Since these corrections are of opposite sign, they tend to offset each other, and this may, to some extent, explain the close agreement between theoretical and measured results.

Until more experiments are available it is probable that the best that can be done is to use the theoretical formulas for determining discharges by the control meter. Even under these circumstances it appears to compare favorably in accuracy with the sharp-crested weir. It is at a disadvantage in that it may be flooded when the lower channel is obstructed. Otherwise it has all the advantages of the venturi flume and the additional advantage of requiring but a single head measurement.

Contracted Opening

This method of measuring discharge in open channels is particularly adapted to the measurement of floods. It can be used where a constriction in the channel produces a drop of 1 ft (0.3 m) or more in the water surface. The method is described in European literature. It was used to measure flood flows by the Miami Conservancy District. The details of these measurements were described by Houk.[20] The principle is that of the venturi flume, but any constriction, such as might be caused by a bridge or similar obstruction, is utilized.

The formula for discharge through a contracted opening, the same as for a venturi flume, is

$$Q = cMa_2\sqrt{h} \tag{12.13}$$

where

$$M = \sqrt{\frac{2g}{1-(a_2/a_1)^2}} \tag{12.14}$$

In these expressions h is the drop in water surface, and a_1 and a_2 are the effective cross-sectional area of the stream just above the beginning of the pronounced drop in water surface and the area at the smallest cross section of the opening. Values of M corresponding to different values of a_2/a_1 are given in Table 12.3. Following the analysis used for orifices,

$$c = C_c C_v$$

where C_c is the coefficient of contraction and C_v the coefficient of velocity. The latter corrects for all losses due to friction and turbulence.

A correction for contraction is required wherever there is a sharp-edged entrance such as the corner of a bridge pier or abutment. It is probable that for a vertical right-angled corner the Francis correction for end contractions (that is, reducing the length of the opening by 0.1 the depth of water for each contraction) will apply approximately. The contraction will be less for piers having rounded or pointed heads. Wherever practicable, the estimated contraction should be checked by field measurements.

For openings of fairly regular form which permit the water to approach the section of minimum area without undue turbulence, C_v will probably have a value of 0.90 or more. For extremely rough or irregular openings, C_v may be 0.50 or even less. The mean value of C_v from 23 measurements by the Miami Conservancy District was 0.86, 20 of the values being above 0.80. The minimum value was 0.45, the maximum 0.97.

Cross sections should be obtained just above the pronounced drop in water surface and at the smallest part of the opening. Elevations of water surface at these sections are also necessary. Their difference is the drop in water surface. Where the drop in water surface cannot be obtained by direct measurement, as occurs when it is desired to determine the maximum flood discharge after the crest has passed, high-water marks may be utilized.

Chemical Gauging

Chemical gauging consists in determining discharges by introducing a chemical at a known rate into flowing water and determining the quantity of the chemical in the stream at a section far enough downstream to ensure a thorough mixture of the chemical with the water. Common salt (NaCl) is the chemical usually employed. For convenience the salt is dissolved in water before being introduced into the stream.

Let Q represent the discharge of the stream in ft³/s (or m³/s). If w lb/s (or newtons per second) of salt is introduced and after thorough mixture a sample taken from the stream shows that 1 lb (or 1 newton) of water contains n lb (or newtons) of salt in addition to the salt that the water already contained, then

$$\frac{w}{62.4Q} = \frac{n}{1} \quad \text{(in lb)}$$

$$\frac{w}{9800Q} = \frac{n}{1} \quad \text{(in newtons)}$$

$$Q = \frac{w}{62.4n} \quad \text{(in lb)}$$

$$Q = \frac{w}{9800n} \quad \text{(in newtons)} \tag{12.15}$$

The determination of n requires precise laboratory techniques. The successful use of this method depends on thorough mixing, which requires a high degree of turbulence. Therefore, the method does not lend itself to the measurement of sluggish, slow-moving water. Detailed discussions of this method of measuring discharge have been presented by Groat[21] and Nagler.[22]

Electromagnetic Flowmeters

When a fluid which is an electric conductor moves across a magnetic field at 90°, as shown in Fig. 12.11, an electromotive force is induced in the fluid at right angles to both the flux of the magnetic field and

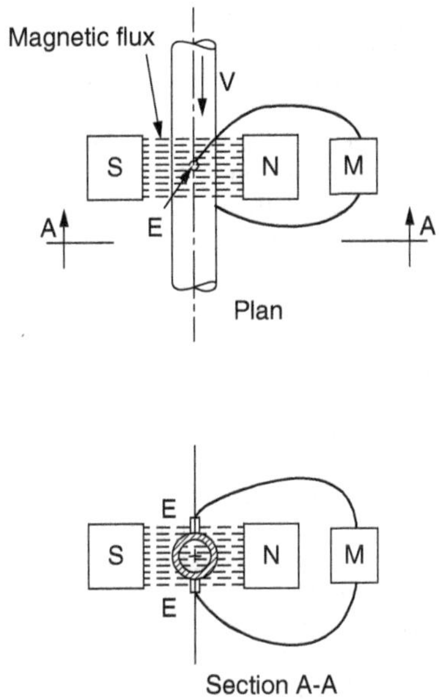

FIGURE 12.11 Electromagnetic flowmeter.

the velocity of the fluid. The induced voltage is proportional to the average velocity of the fluid V. In Fig. 12.11, S and N are magnetic poles, M is a voltmeter, and the electrodes are designated as E. If the pipe is a conductor, it is not necessary for the electrodes to penetrate the pipe walls.

This method is particularly useful where corrosive materials are being gauged or where it is undesirable to penetrate the conduit, as in the case of the flow of blood in blood vessels. The original development of this flow-measuring procedure has been ascribed to Faraday by Shirer, Shackelford, and Jochim.[23] Their article also provides a detailed description of the procedure and a historical account of the development of the method. Applications to the gauging of mercury have been made[24] which show a good relation between analytical consideration and test results.

References

1. C. O. Wisler and E. F. Brater, *Hydrology*, 2d ed., Wiley, New York, 1959.
2. C. Herschel, "Fluid Meters: Their Theory and Application," Report of Special Research Committee on Fluid Meters, ASME, 1924.
3. F. C. Scobey, "The Flow of Water in Concrete Pipe," U.S. Dept. of Agriculture, Bull. 852, 1920.

4. C. S. Slichter, "Motion of Underground Waters," U.S. Geol. Survey Water Supply and Irrigation Paper 67, 1902.
5. H. W. King, *Handbook of Hydraulics*, 1st ed., McGraw-Hill, New York, p. 247, 1918.
6. C. M. Allen and E. A. Taylor, "The Salt-Velocity Method of Water Measurement," *Trans. ASME*, vol. 86, 45, pp. 285–341, 1923.
7. N. R. Gibson, "The Gibson Method and Apparatus for Measuring the Flow of Water in Closed Conduits," *Trans. ASME*, vol. 86, pp. 343–376, 1923.
8. C. Herschel, "The Venturi Water Meter", *Trans. ASCE*, vol. 17, p. 228, 1887.
9. J. W. Ledoux, "Venturi Tube Characteristics," *Trans. ASCE*, vol. 91, pp. 1787–1796, 1927.
10. B. F. Snow, "Venturi Tube Characteristics," discussion, *Trans. ASCE*, vol. 91, p. 581, 1927.
11. "Flow Measurement by Means of Standardized Nozzles and Orifice Plates," Power Test Codes, Supplement on Instruments and Apparatus, pt. 5, Chap. 4, ASME, 1949. This supplement was prepared and published by ASME under the auspices of the Power Test Codes Committee.
12. "Investigation of Orifice Meter Installation Requirements," Interim Res. Rep., American Gas Assoc. and ASME, Mar. 1951.
13. "Standards for Discharge Measurement with Standardized Nozzles and Orifices," 4th ed., DIN 1952, NACA Tech. Mem. 952.
14. B. R. Van Leer, "The California-Pipe Method of Water Measurement," *Eng. News-Rec.*, Aug. 3, 1922; Aug. 21, 1924.
15. V. M. Cone, "The Venturi Flume," *J. Agr. Res.*, vol. 9, no. 4, 1917.
16. R. L. Parshall and C. Rohwer, "The Venturi Flume," Colo. Agricult. Experiment Station Bull. 265, 1921.
17. P. S. Wilson and C. A. Wright, "A Study of the Venturi Flume," *Eng. News-Rec.*, vol. 85, no. 10, pp. 452–457, Sep. 2, 1920.
18. R. L. Parshall, "The Improved Venturi Flume," *Trans. ASCE*, vol. 89, pp. 841–851, 1926.
19. J. Hinds, "Venturi Flume Data Throws Light upon 'Control Weir,'" *Eng. News-Rec.*, vol. 85, Dec. 23, 1920; also discussion, "The Improved Venturi Flume," *Trans. ASCE*, vol. 89, 1926.
20. I. E. Houk, "Calculations of Flow in Open Channels," Tech. Rep., pt. IV, Miami Conservancy District, Dayton, OH, 1918.
21. R. F. Groat, "Chemi-Hydrometry and Precise Turbine Testing," *Trans. ASCE*, vol. 80, pp. 951–1305, 1916.
22. F. A. Nagler, "Verification of Bazin Weir Formula by Hydrochemical Gaging," *Trans. ASCE*, vol. 83, pp. 105–156, 1919.
23. H. W. Shirer, R. B. Shackelford, and K. E. Jochim, "A Magnetic Flowmeter for Recording Cardiac Output," *Proc. IRE*, vol. 47, p. 1901, 1959.
24. H. G. Elrod, Jr., and R. R. Fouse, "An Investigation of Electromagnetic Flow Meters," *Trans. ASME*, vol. 24, p. 589, 1952.

SECTION 13

Computational Hydraulics

Hydraulics is one of several engineering disciplines that have experienced rapid advances in the past 20 years. Examples include the introduction of aeration devices for the control of cavitation-related problems, acoustic flow/current meters, finite-element methods for flow-field analysis, and the development of many advanced software packages. These advancements result mainly from our continuing effort to further our understanding of the basic principles of physics applied in hydraulics and, of equal importance, the rapid advancements in computer technology and numerical techniques. Today nearly all engineers have at their disposal relatively powerful personal computers. With increased computational power, hydraulic engineers are solving more and more complicated problems, such as one-dimensional hydraulic transient problems in complex pipe network systems and in open-channel systems.

With environmental concerns having become ever more important during the past decade, the approach to the design of hydraulic structures and water resources management practices has undergone significant changes. More sophisticated hydraulic/hydrodynamic and transport/mixing analyses are required. These analyses are usually two-dimensional. Numerous three-dimensional procedures also exist.

In the design of modern hydropower stations, hydraulic engineers have looked more to the application of sophisticated computational fluid dynamics (CFD) software packages for gaining insight into the complicated flow conditions inside the major components of machinery such as spiral cases, turbine runners, and draft tubes, which may be difficult to observe in conventional physical model studies. Recognizing the need for further development of computational hydraulics, specific task committees have been established in various professional societies.

The objectives of this section are to provide an introduction to some basic computational methods and to provide a number of examples in which simple hydraulic analyses are formulated.

Elementary Computational Concepts

The steps involved in developing a usable program to solve a particular problem include (1) a full understanding of the scope of the problem and the flow system to be analyzed, (2) establishment of the mathematical description or the governing equations, (3) selection of the computational method, such as the finite-difference or the finite-element method, (4) programming, which includes the actual coding, (5) program testing and verification, and (6) production and analysis of results.

It should be noted that an understanding of the problem and the flow system to be analyzed is required whether or not a computer is used as the computational tool. It is necessary to establish or define the problem scope prior to obtaining a solution by means of a computer. The predominant physical characteristics of the flow system need to be identified before the appropriate problem-solving approach can be selected, for example, whether the problem can be treated by a one- or two-dimensional approach, by steady-state or unsteady flow analysis, potential versus viscous flow analysis, and laminar versus potential flow analysis. The mathematical description may involve simple algebra, or it may involve a set of partial differential equations.

For problems that require solution of the governing partial differential equations, numerical procedures such as the finite-difference or the finite-element method can be applied to convert the governing equations into a set of algebraic equations with the initial and boundary conditions incorporated. A computational algorithm is then developed to enable the engineer to build the computer model for the flow system to be analyzed. Successful production runs are only possible after a complete testing of any newly developed program. The construction of a computer model is considered complete only after the model has gone through proper calibration and verification processes.

There are three computational methods that may be used in solving a technical problem—direct computation, iteration, and successive approximation. Direct computation and the iteration of a series of direct computations are well known to practicing engineers in their daily work. Successive approximation is less widely used and the hydraulic engineer may benefit from a description aimed specifically at hydraulic computation.

Solution by Successive Approximations

Iteration can be used to facilitate a trial-and-error solution of many complex equations. Programmed logical decisions are also helpful in arriving at the desired result.

As an illustration, if it were desired to find the normal depth associated with a particular flow in a trapezoidal channel, a direct solution of Manning's equation is not possible. Successive values of depth could substitute until the computed flow matches the actual flow, but this may involve many trials. Ideally, a systematic procedure should be established so that the normal depth can be determined within a specified degree of accuracy, and in a minimum number of trials. In any successive approximation scheme a trial value is assumed, substituted, and then improved upon until the answer is obtained. An important consideration is that a method should be used that seeks an improved value which ensures rapid convergence to the final result. The bisection and Newton's methods are used in the following examples.

Example 13.1 Determine the normal depth using the bisection method for a flow of 5 m^3/s for a channel in which $z = 1.5$, $n = 0.015$, $b = 10$ m, and $S_0 = 0.0002$.

The Manning formula is written in the following form, and the value of D is sought so that the function is zero,

$$f(D) = 1 - \frac{Qn}{AR^{2/3}S^{1/2}} \tag{13.1}$$

The bisection or half-interval method is used to solve for normal depth. An initial guess of the depth is used to start. Two bracketing values maxdepth and mindepth are then computed. Successive values of the variable function are computed and compared to zero. The bracketing values are then adjusted and function is recomputed. The process is repeated a number of times, which should be sufficient to give a good approximation to the correct value of the depth, 3.72 m. ▲

Example 13.2 Determine the critical depth using Newton's method for the conditions given in Example 13.1, in which $Q = 400$ m^3/s, $z = 1.5$, $n = 0.015$, $b = 10$ m, and $S_0 = 0.0002$.

The critical-depth relationship (see "Critical Depth—General Case," Sec. 8) is written in the following form:

$$f(D) = 1 - \frac{Q^2 T}{gA^3} \tag{13.2}$$

Newton's method is used in the program to solve Eq. (13.2). Alternatively, the bisection method could be used with equal success.

In Newton's method the derivative of $f(D)$ [Eq. (13.2)] is needed,

$$\frac{df(D)}{dD} = -\frac{Q^2}{gA^3}\left(2z - \frac{3T^2}{A}\right) \tag{13.3}$$

According to Newton's method, the correction to the current depth D_k at the kth iteration is

$$\Delta D = \frac{-f(D)}{df(D)/dD} \tag{13.4}$$

and the newly corrected depth is

$$D_{k+1} = D_k + \Delta D \tag{13.5}$$

The computed depth for this example is 1.96 m. ▲

Steady Nonuniform Flow in Prismatic Open Channels

The computation of gradually varied water-surface profiles is readily adaptable to an automated solution. Several different approaches may be applied, any of which might be fully satisfactory, in most situations. One approach is to start with the basic differential equation, Eq. (8.101),

$$\frac{dD}{dx} = \frac{S_0 - S}{1 - Q^2T/gA^3} \tag{13.6}$$

and use a standard numerical integration procedure. If Eq. (13.6) is inverted, it can be expressed as

$$x = \int_{D_0}^{D_1} f(D)\, dD \tag{13.7}$$

since the right side is a function of depth only. At $x = 0$, the depth is D_0 and at any distance x the depth is D_1. The problem with this method is that the distance between the specified depths is computed forcing the user to interpolate to determine the depth at the required location.

One of the most direct approaches is to program the hand calculation proposed in Sec. 8, namely, the solution of Eq. (8.105b). Determine the water-surface profile, the energy grade line, and sufficient information so that the location of the hydraulic jump can be determined. The equation

$$WSEL_2 + \frac{V_2^2}{2g} = WSEL_1 + \frac{V_1^2}{2g} + L\overline{S}_f \tag{13.8}$$

is solved for depth, where

$$\overline{S}_f = \frac{S_{f_1} + S_{f_2}}{2}, \quad S_f = \left(\frac{Qn}{AR^{2/3}}\right)^2 \tag{13.9}$$

Example 13.3 Determine the water-surface profile, energy grade line, and location of the hydraulic jump or jumps for the open channel with a sluice gate at the upstream end of the channel to control the flow and a reservoir at the downstream end.

The computational procedure is divided into several steps:

1. Compute slopes, stations, and inverts at each section.
2. Compute critical depth at each section using Newton's method as in Example 13.3.
3. Compute normal depth at each section using the bisection method as in Example 13.2.
4. Starting at the downstream end where the water-surface elevation is given, compute the subcritical water-surface profile. At each successive cross-section upstream a depth is computed that balances Eq. (13.8). The procedure is to estimate the depth at the section and compute a new depth from the equation. Using this new depth the procedure is repeated until the depth does not change appreciably between iterations. If the computed depth is below the critical depth, then the depth is set equal to the critical depth and the computations move to the next section.
5. Starting at the upstream end compute the supercritical water surface profile. The solution procedure is similar to that used for the subcritical profile, except that a different procedure is used to solve for the depth. The correct value of depth is bracketed and then approached from both directions using an extrapolation of the energy grade line and the energy computed using the previously computed value of depth. This is necessary because of the rapid change in energy with depth.
6. Compare the value of force plus momentum at each section to determine the correct flow regime and the location of any hydraulic jumps. ▲

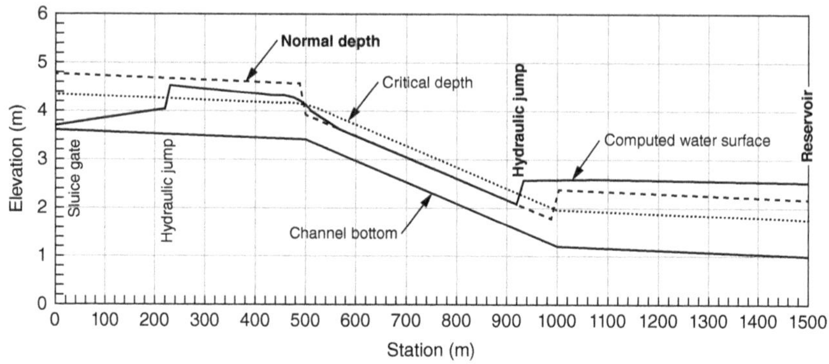

FIGURE 13.1 Water surface profile.

Steady Flow in Pipes

Problems associated with the flow of liquids in pipelines can be separated into three types, namely, determination of (1) the head loss, (2) the discharge, or (3) the pipe size. The Darcy-Weisbach equations [Eq. (6.19)] is normally used to relate the loss to the discharge in a straight section of a particular pipe,

$$h_f = \frac{fL}{D}\frac{V^2}{2g} \tag{13.10}$$

When flow is turbulent, the friction factor f from the Moody diagram is dependent on the relative roughness ε of the pipe and the Reynolds number R (see "Turbulent Flow in Pipes," Sec. 6). When a computer is used, it is convenient to express the information from the Moody diagram as an explicit relationship between relative roughness, Reynolds number, and friction factor. Wood[1] recommends the use of the following expressions:

$$f = a + bR^{-c} \tag{13.11}$$

where

$$a = 0.094\left(\frac{\varepsilon}{D}\right)^{0.225} + 0.53\left(\frac{\varepsilon}{D}\right)$$

$$b = 88.0\left(\frac{\varepsilon}{D}\right)^{0.44} \tag{13.12}$$

$$c = 1.62\left(\frac{\varepsilon}{D}\right)^{0.134}$$

If these expressions are used, a program to compute the head loss of a particular liquid flowing in a given pipe in which discharge, length, diameter, roughness, and fluid viscosity are given is required. The procedure is to compute the Reynolds number, the relative roughness, and the friction factor; then the head loss can be found directly from the Darcy-Weisbach equation. This procedure is illustrated in Example 6.1 using the Moody diagram (Figure 6.4) and Table 6.1.

The second type of pipe-flow problem in which the discharge is to be computed involves an iterative solution. In this case, since the discharge is unknown, the friction factor cannot be explicitly evaluated, f is corrected, and the process is repeated until the new value of f is within an acceptable tolerance of the previous value.

Example 13.4 Determine the flow of water in a 1-ft-diameter commercial steel pipe 5000 ft long if the head loss is 30 ft, $\nu = 0.0001$ ft^2/s, and $\varepsilon = 0.001$ ft.

Equation (13.10) is iteratively solved for velocity, as shown in Eq. (13.10*a*).

$$v = \sqrt{\frac{Dh_f 2g}{fL}} \tag{13.10a}$$

Note that the computation continues until the change in f is less than some tolerance value. If this never happens the solution does not converge to a satisfactory solution. This is a possibility if unrealistic data have been used. The computed result shows the discharge to be 3.34 ft^3/s. Inasmuch as the friction-factor equations are valid for turbulent flow only, they should not be used for a condition of laminar flow in the pipeline.

The third type of pipe-flow problem is the determination of a suitable pipe size to transport a certain flow to a specific destination within a limited pressure drop. In addition to the diameter, the variables velocity, relative roughness, and friction factor are also unknown. A typical problem is solved in the following example. ▲

Example 13.5 Determine the size of cast-iron pipe necessary to handle a flow of 2.05 m^3/s of water with a head loss of not more than 30.5 m. The length of the pipeline is 4000 m and $\nu = 0.00001$ m^2/s.

The computational procedure is to assume a friction factor, compute a trial diameter by use of the Darcy-Weisbach equation, then determine relative roughness, the Reynolds number, and a new f value. The procedure is repeated until the newly calculated f does not change within a specified tolerance. The computed pipe size is then used to select a nominal pipe from a list of commercially available sizes. The head loss corresponding to the given flow through the actual pipe can also be computed. The solution to this example shows that a 1-m-diameter pipe will handle the 2.05-m^3/s flow with a head loss of 30.63 m. ▲

Experimental Measurements—Least-Squares Analysis

The calibration of many fluid metering devices such as weirs, venturi meters, and orifice meters requires gathering data involving at least two variables. The calibration of a weir along with statistical tests of the derived equation will be used as an example. In the case of a weir, these data are the flow rates Q at various levels H of fluid above the weir notch. The calibration of the device is completed by correlating the data in some manner so that an explicit relationship is available between the variables.

Flow over a notched weir (see "V-Notch Weirs," Sec. 5) is expressed by an equation in the form

$$Q = CH^m \tag{13.13}$$

where C and m are the unknown weir coefficients to be determined by calibration. Taking the logarithm of Eq. (13.13) yields

$$\ln Q = m \ln H + \ln C \tag{13.14}$$

which has the form of the straight-line equation

$$y = mx + b \tag{13.15}$$

The values of m and b are desired so that the line best matches the data points. The process of arriving at these values is known as the method of least squares, and the straight line is known as the regression line of y on x, where y and x correspond to values of ln H and ln Q, respectively. A line fitted to the data (x_i, y_i), $i = 1, 2, \ldots, n$, by the method of least squares is the one that minimizes the sum of the squares of the deviations of the measured values y_i from corresponding points on the line y. The sum of the squares of the deviations is a function of the unknowns m and b and is given by

$$\sum_{i=1}^{n} (y_i - y)^2 = \sum_{i=1}^{n} (y_i - mx_i - b)^2 \tag{13.16}$$

By taking the separate partial derivatives of this equation with respect to each unknown and setting them equal to zero, two equations are obtained,

$$\sum_{y_i} - m\sum x_i^2 - nb = 0 \tag{13.17}$$

$$\sum x_i y_i - m\sum x_i^2 - b\sum x_i = 0$$

It is understood that the summations are for i varying from 1 to n. By denoting the mean values of the variables by x and y, defined by $ny = \Sigma y_i$ and $nx = \Sigma x_i$, the solutions for the unknowns can be written,

$$m = \frac{\sum x_i y_i - n\bar{x}\,\bar{y}}{\sum x_i^2 - n\bar{x}^2} \tag{13.18}$$

$$b = \bar{y} - m\bar{x} \tag{13.19}$$

These values of m and b define the best-fit straight line on the log-log graph of Q versus H. The value of m is the exponent of H in the weir equation [Eq. (13.13)], and $C = e^b$.

The scatter of the data points in the y direction is measured by the sample standard deviation of the regression given by

$$s = \sqrt{\frac{\Sigma(y_i - y)^2}{n-2}} \tag{13.20}$$

This quantity is a measure of the failure of the line to fit the data.

The sample linear correlation coefficient is defined by

$$r = \frac{\Sigma xy - n\bar{x}\,\bar{y}}{\sqrt{\left(\Sigma x^2 - n\bar{x}^2\right)\left(\Sigma y^2 - n\bar{y}^2\right)}} \tag{13.21}$$

and always lies between −1 and +1. If all points lie on the regression line, then $r = \pm 1$. If $r = 0$, no linear relation exists between the two variables x and y. The sign of r merely agrees with the sign of the slope of the regression.

Unsteady Flow in Pipes and Open Channels

The method of characteristics can be used to solve the unsteady flow equations.[2] The equations of motion and continuity are combined and then, using the method of characteristics, two new equations are derived,

$$H_P - H_A + \frac{a}{gA}(Q_P - Q_A) + \frac{f\,\Delta x}{2gDA^2} Q_A |Q_A| = 0 \tag{13.22}$$

$$H_P - H_B - \frac{a}{gA}(Q_P - Q_B) - \frac{f\,\Delta x}{2gDA^2} Q_B |Q_B| = 0 \tag{13.23}$$

where $\Delta t = \Delta x/a$ and a is the wave speed. These two equations are solved at each time step for the variables H_p and Q_p. The solution of the equations can be seen in Fig. 13.2. The two lines are known as the characteristic lines C^+ and C^-, along which Eqs. (13.22) and (13.23) are valid.

In open-channel transients a pair of equations similar to Eqs. (13.22) and (13.23) can be written,

$$V_P - V_R + \frac{g}{c_R}(y_P - y_R) + g(S_R - S_0)\,\Delta t + \frac{q(V_R - c_R)}{A_R}\Delta t = 0 \tag{13.24}$$

$$V_P - V_s - \frac{g}{c_s}(y_P - y_S) + g(S_s - S_0)\,\Delta t + \frac{q(V_s - c_s)}{A_s}\Delta t = 0 \tag{13.25}$$

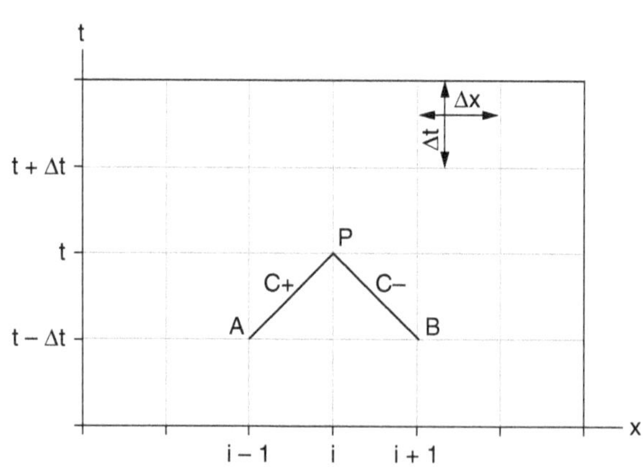

FIGURE 13.2 Characteristic lines in x–t plane for pipeline problem.

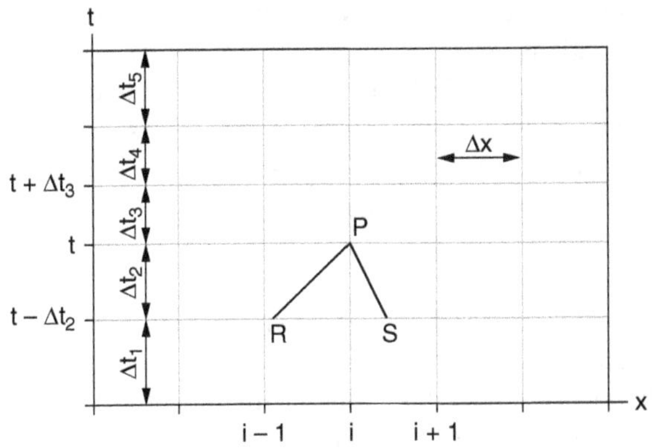

FIGURE 13.3 Characteristic lines for open-channel problem.

where

$$\Delta_t = \frac{x_P - x_R}{V_R + C_R} \qquad \text{or} \qquad \Delta t = \frac{x_P - x_s}{V_s - c_s}$$

In pipe-flow transients the wave speed is nearly constant for a particular pipe throughout the transient, and so the value of Δt is also constant. In open-channel transients the value of Δt changes from time step to time step and from location to location as the depth in the channel, and, thus the wave celerity c changes. Thus Δt is different everywhere along the channel. A diagram similar to Fig. 13.2 can be drawn for this case, as shown in Fig. 13.3. For subcritical flow the point P falls between points R and S; for supercritical flow point P is downstream, that is, to the right of both points R and S.

Advanced Numerical Methods

In order to solve two- and three-dimensional hydraulic problems more advanced computational techniques must be used. Two methods are described next.

Finite-Difference Method

The method of finite differencing calls for the governing flow equations to be transformed into a form that allows them to be solved numerically. The numerical approximation is achieved by discretization of the domain through a one-, two-, or three-dimensional grid, and the approximation of the differential equations by difference equations which form a set of algebraic equations. These equations are then solved for the values of the field variables. The rates of change, or the "derivatives," in the equations are replaced with algebraic expressions involving the finite quantities dx and dt. The original equations, which are defined at every point in space and time, thus become algebraic equations, accurate only at the finite number of grid points. These equations are easily solved, and by solving them over a fine enough grid, the resulting numerical solution can approach the original, exact solutions.

Finite-Element Method

The finite-element method divides a complex flow domain into a mesh of geometrically simple subdomains called finite elements. Like the finite-difference method, the finite-element method transforms the original equations, expressed in terms of derivatives, into algebraic equations. These transformed equations, called element equations, can be applied to each finite element separately. Terms in the element equations are evaluated numerically for each element, and the values obtained are used to form a larger, global set of algebraic equations, which are solved for the problem at hand. It provides great flexibility in handling variable-size elements of different properties. A method of weighted residuals is most commonly used. A shape function (trial function) substituted into the differential equation generally does not equal the exact solution, the difference being the residual. This residual, multiplied by a weighting function, yields a weighted residual. The finite-element method attempts to force the weighted residuals to zero in an average sense. The particular choice of the weighting function provides a different weighted residual. One of these choices, the Galerkin method, requires the weighting function to be equal to the shape function. The finite-element method can be viewed as the inverse of the finite-difference method. Finite differencing uses a global, regular coordinate system to divide a computational domain into small segments, whereas the finite-element method views each small segment as a basic unit and connects them together to assemble the computational domain. The finite-element method has found wide acceptance in a variety of flow problems. Examples include analysis of shallow, wide channels with irregular boundary geometry and variable bottom characteristics and seepage flow through zoned earth-fill dams. However, its success in computing hydraulic transients has been limited.

References

1. D. J. Wood, "An Explicit Friction Factor Relationship," *Civil Eng., ASCE*, pp. 60–61, 1966.
2. V. L. Streeter and E. B. Wylie, *Fluid Transients*, Feb Press, Ann Arbor, MI, 1982.

SECTION 14

Physical and Mathematical Modeling of Hydraulic Structures[a]

Introduction

Large hydraulic structures such as spillways, stilling basins, large water intakes, and outlet works are special design challenges in that they must be designed to work within their unique physical environment. Although these structures are designed and analyzed using the hydraulic principles discussed in the previous chapters of this book, hydraulic modeling is often the only tool available to confirm that the structures function as intended within their unique surroundings.

For many years physical hydraulic modeling has been a standard design tool for hydraulic engineering of dams and their appurtenant structures. An engineer's ability to gather and analyze data from physical models continues to improve with the development of more sophisticated electronic instrumentation, faster computers and advanced software. These same advancements in computer technology have enabled the development of more advanced mathematical modeling techniques and several powerful two-dimensional and

[a] This section was written by Steven Barfuss.

three-dimensional software packages are commercially available as options to hydraulic engineers.

It is becoming more and more common for physical modeling to utilize numerical modeling as part of the modeling package. For example, it is not uncommon for the design engineer to utilize a numerical model to help develop a proposed hydraulic structure design. Numerical modeling can be cost effective and accurate when based upon quantifiable field or physical modeling data. When a numerical model is based upon sound prototype or physical modeling information, the numerical model can be used for years on end as a valuable design and operational tool. However, modelers need to recognize that numerical modeling is limited when it comes to multiphase flows (air/water mixtures), highly turbulent flows with vortices and flows where scour and/or sedimentation are a concern.

The current trend for modeling hydraulic structures is to utilize both a numerical and physical model in parallel during a study (also known as composite modeling). Composite modeling is the effective utilization of both physical and numerical modeling, used either in series or in parallel with one another to solve difficult hydraulic problems.

Although physical modeling is the proven standard for modeling hydraulic structures and has successfully been used for decades, it comes with limitations and constraints. Similarly, numerical modeling, which is relatively new, also carries with it a number of limitations and constraints as well as benefits not found in physical modeling. However, when the two modeling techniques are used together during a hydraulic study, many of the limitations of one technique can be complemented by the other and vice versa. Accordingly, when researchers understand the benefits and limitations of each modeling technique, they can develop a research plan that utilizes composite modeling approaches where both modeling techniques are used together to increase the efficiency and effectiveness of the modeling process.

For example, a numerical model may first be utilized to determine a proposed design for the structure. A physical model is then commonly constructed of the proposed design. Data is collected from the physical model and that data is used to calibrate the numerical model to ensure its accuracy. The physical model is then used to optimize hydraulic efficiency and minimize construction costs of the original design. With a calibrated numerical model, it (the numerical model) can then be utilized for many years to check flow conditions or operational procedures even after the physical model has been dismantled. The important thing to remember with this approach is that numerical modelers need real data either from a flood condition through the prototype or from a physical model study. The physical model study is critical to the success and accuracy of the numerical model. Currently, the trend in modeling is to utilize the benefits of both for design.

Composite Modeling

Physical Modeling Overview

Physical models of hydraulic structures are typically modeled using Froude scaling, which is based on the ratio of inertia to gravity forces. The similarity parameter used to operate the model is the Froude number, defined as

$$Fr = \frac{V}{\sqrt{gy}} \tag{14.1}$$

where y is a characteristic linear dimension, g is the gravitational acceleration constant, and V is a characteristic velocity. Similarity is achieved by operating the physical model so that the Froude number in the model is the same as the Froude number in the prototype. This ensures that the inertial and gravity forces are scaled properly. The proper length, discharge, velocity, pressures, time, and rotational speed for the model in relationship to the prototype can be calculated using Froude relationships.

For a model to properly simulate hydraulic conditions in the prototype it is necessary to maintain geometric, kinematic, and dynamic similitude. Geometric similitude is achieved by carefully building the model to a selected scale ratio with all geometric components scaled appropriately.

The selection of the model scale has a direct impact on its ability to maintain kinematic and dynamic similitude. Physical models are always built as large as possible to minimize the scaling effects attributed to other forces, such as viscous and surface tension forces. When a large physical model is constructed (small scaling ratio), then flow phenomena such as super elevation, flow surging, wave action, hydraulic jumps, vortices, flow separation, local scour and erosion, air entrainment, standing waves, complex and composite roughness, etc., can be accurately modeled. In addition to being able to physically simulate complex flow conditions, the physical model provides a hands-on working simulation that most engineering teams find essential to the design process. Some of the more significant benefits and limitations of physical modeling are listed and summarized below.

Benefits of Physical Modeling

- As long as the scale ratio is properly selected and the model is constructed as large as possible to avoid potential size scale effects, physical modeling is a trusted and reliable method that will produce accurate results.
- Small changes can be made in the physical model very easily using temporary structures of wood, plastic, mortar, sandbags, etc., so that the resulting hydraulic changes can be seen immediately. It is time consuming to simulate a wall or a raised embankment into the grid of a numerical model, especially if the model is complex and the mesh resolution is fine, but it only takes a matter of minutes to see the effect of minor geometric changes in the physical model.
- Multiple flow rates can be tested on a single model configuration in a very short time. For example, a flood routing or flood hydrograph can be simulated in a matter of minutes. Hydraulic design configurations can experience a hysteresis of flow conditions with changing flow rates. Physical modeling can easily determine the presence and magnitude of such hysteresis effects.
- Specific hydraulic problems are immediately apparent in an operating physical model, and many times, there are multiple problems occurring simultaneously. Problems like vortices, separation zones, scour potential, poor energy dissipation characteristics, wall overtopping, aeration effects, or wave action are noted immediately in a physical model.
- Once constructed, a physical model can quickly determine the flow capacity of a flow control structure by increasing supply or inlet flow to the model until the given pool elevation is reached. Depending upon the complexity of the geometry of the control structure and the approach channel, the numerical modeling effort to determine flow capacity can be significantly more time consuming.
- The effects of tailwater or submergence on a hydraulic structure are easily and quickly observed in a physical model by simply varying the height of water in the downstream channel.
- Three dimensional effects, aeration effects, highly turbulent regions, standing waves and super-elevation conditions are easily simulated in a physical model. However, one must be wary of scale effects when evaluating aeration.

- Physical models facilitate "team" engineering. Geotechnical, structural and hydraulic engineers can get immediate feedback and make contributory suggestions to the design of the hydraulic structure as they witness flowing water through the physical model.
- Physical models will allow the researcher to videotape or photograph the moving water so that modeled flow conditions can be compared to prototype conditions providing a valuable tool in quality control as well as public awareness.

Limitations of Physical Modeling

- The physical model can be limited by the required structure cost and size, and the time required to build and acquire data. The cost and time is usually proportional to the amount of information that is required and the time it takes to collect the data. It is often difficult to physically model the full extent of the flow region or channel and the approach conditions, due to limited laboratory floor space. Major changes to a physical model are often time/labor intensive and expensive.
- Pressure, flow rate, and velocity measurements taken in the physical model introduce some uncertainties as a function of the accuracy of the instrumentation being used for the measurements. It can be difficult to make detailed pressure measurements in a physical model due to the installation of pressure taps, pressure transducers, and the presence of undissolved air in the flow.
- If a physical model is not scaled properly, then the data collected or the cost of the model is affected. If the model scale is too large (small-sized model), then size scaling effects can influence the accuracy of the results. If the model scale is too small (large-sized model), then the cost of the model will increase. It should be emphasized, however, that it is very important to construct any physical model as large as possible to reduce scaling problems.
- It is difficult to vary the boundary roughness in a physical model and often, a separate sensitivity study is needed to determine appropriate methods for simulating roughness as well as evaluating the effect of roughness in the physical model.
- Utilizing Froude similitude for the physical model study can be problematic in that full velocity profiles are not developed

along the boundary because the model is not operating at the correct Reynolds number when operating at a Froude-scale velocity.

- Project schedule and limited budget usually minimize the number of possible design configurations or modifications that can be tested.
- The physical model has to be dismantled at some point in time, and may then not be available for other design changes resulting from future changes in construction or other constraints.
- There are real hydrodynamic processes, which are either poorly simulated or practically impossible to simulate on the physical models. Among these is, for instance, formation of a water-air-ice cloud in the spillway tailrace with a high degree of flow aeration (for instance, overhang spillways with jet ejection).

Numerical Modeling Overview

Although there are multiple approaches to numerical modeling such as 1D and 2D methods, the full 3D method commonly referred to as computational fluid dynamics (CFD) is the type of numerical modeling referred to in this chapter. The most common CFD programs solve the Reynolds-averaged Navier-Stokes (RANS) equations using a variety of approaches. Most of the leading commercial CFD programs on the market have the ability to track free surface flows commonly associated with hydraulic structures with the volume-of-fluid (VOF) method or some variation thereof being the most common algorithm for tracking the free surface interface.

CFD codes solve the RANS equations using a finite volume method applied to a computational domain. The computational domain defines the region of interest as discretized (gridded) into smaller cells. Gridding techniques are usually classified as boundary (body) fitted coordinates (BFC) or a porosity technique. Solid objects are defined in BFC coordinates by wrapping the grids around the object whereas in porosity methods, the solid is imported into the grid and the solid blocks the fluid flow. Each method has its advantages and disadvantages; BFC tend to be more computationally efficient, especially in resolving boundary layer hydraulics, but grid construction can be time consuming whereas porosity methods are easier to grid but not as computationally efficient. Additionally, the porosity method lends itself for importing survey data (site geometry) into the model.

One of the difficulties in numerically solving flow in a hydraulic structure is the presence of a free surface, in which the location must be solved as part of the solution and may be transient in time. This is especially difficult when the water surface is rapidly changing with a high degree of curvature, such as when the flow changes from subcritical flow to supercritical flow as it flows over a spillway crest.

Another common technique used in CFD modeling is to approximate curved surfaces with a straight approximation. It is easy to see that the use of smaller grids and thereby more straight segments, provides a better approximation to a curved surface. Therefore, in order to improve the approximation, variable grid spacing can be used with a tighter spacing or smaller grid size located near high curvature areas on the obstacles and in locations where a rapidly varying water surface was expected. The downside to smaller grids is the increase in computational time.

To numerically solve the RANS equations across the gridded domain, there are a variety of numerical approaches that can be used and each one may have multiple parameters. Primary inputs include defining the boundary conditions, the turbulence model and its parameters, model roughness, and the numerical approach for solving the equations (implicit versus explicit; first, second, or third order; etc.). Some parameters such as the boundary conditions are critical for properly simulating prototype conditions. Other parameters provide more subtle changes. Understanding the effect of the various methods, approaches, and parameters are crucial in making sure that the model represents the true physical flow.

Peer-reviewed papers have validated the use of CFD to model free surface flow for many different applications especially for subcritical flows. However, CFD modeling has not been validated for all types of flows, and like any numerical method the computed results can be no more accurate than the data and assumptions on which the computational techniques and methods are based—that is, accept the values as a decent approximation, but not the absolute truth. Some of the more significant benefits and limitations of numerical modeling are listed and summarized below.

Benefits of Numerical Modeling

- A numerical model is a good choice for large changes to physical topography or to the location of a control structure.
- Numerical modeling provides the option of testing a configuration without fabricating a physical structure; thus allowing multiple configurations to be investigated at the same time using multiple computers. Numerical modeling allows for the automated multivariate parametric optimization of design elements.

- Data produced from numerical modeling results allows for velocities, pressures, turbulence parameters, and water surface profiles to be determined at each time step and at any location in the flow domain.
- Numerical modeling helps the researcher see the "big picture" or, in other words, where problem areas may be located. Flow patterns such as velocity profiles and vortices can be easily mapped using vector plots, streamlines, and flow ribbons.
- The numerical model can eliminate the need for physical sectional models [example: section of a spillway constructed at a smaller scale (larger model) so that spillway surface pressures can be evaluated]. The numerical model does a good job of evaluating fluctuating pressures as well as the magnitude of negative pressures along a fluid boundary.
- RANS CFD models do struggle to simulate pressure fluctuations associated with dynamic eddy structures that are small and therefore smoothed out (i.e., pressure fluctuations present in a hydraulic jump). RANS CFD models are good at surface pressures if the flow lines are parallel near the surface and relatively steady with time. Large eddy simulation (LES) and detached eddy simulation (DES) CFD models largely eliminate this issue but are computationally expensive and are not (yet) widely used for free surface flows.
- Significant changes to a modeling configuration are simple to construct on the computer and only have labor and software licensing costs associated with them.
- Numerical modeling costs are often less than a physical model.
- Numerical models can be stored and maintained for future use/changes.
- Animations of the flow, including the free surface can be constructed using a variable of interest such as the velocity or pressure.

Limitations of Numerical Modeling

- A numerical model benefits from having prototype or physical modeling data for calibration but this is not always available or required.
- Numerical models are limited in the selection of input and operating parameters when there are not prototype data or physical modeling data to use as a basis. Additionally, numerical modelers who have not been properly trained or do not

have vast experience in specific applications introduce the possibility for data errors.

- Numerical models are limited in their application of geometric roughness and the development of the velocity profile at boundaries (law of the wall), especially in high turbulence supercritical flows. Depending on the grid size of the numerical model, the true near-the-wall velocities may not be accurate since the law-of-the-wall approximations are dependent on numerical grid size and other factors.
- Numerical modeling is often constrained by gridding/computational time for models with large spatial and/or time domains. To reduce computational time some CFD codes include hybrid solutions which allow 2D shallow water approximations for large areas where vertical gradients are not critical to be combined with full 3D hydraulic analysis in local regions.
- The number of runs required to evaluate the routing of a hydrograph will be very time consuming for a numerical study. Similarly, simulating the routing of a complete hydrograph can require simulation times that are unrealistic.
- Hysteresis effects between increasing flows of a hydrograph as compared to the decreasing flows of a hydrograph can happen and be significant. Numerical models are limited in their ability to analyze this type of flow relationship.
- It is not possible to videotape or photograph the moving water of a numerical model, so that modeled flow conditions can be compared to prototype conditions like can be done with a physical model. This type of analysis is extremely valuable as a quality control to ensure that the modeling is accurate. However, it is possible to create visual simulations using numerical output in the form of short video clips that aid in the comparison with prototype conditions.
- A numerical model is limited in its ability to analyze the hydraulic stability and effect of unsteady flow downstream of a control structure.
- Flow conditions like a hydraulic jump can be very sensitive to changes in numerical boundary roughness and there is wide latitude in the application of the theory if prototype or physical model data is not available.
- A numerical model is limited by size of the computational domain both spatially and temporarily. A large physical

domain may require a 2D solution rather than a 3D solution, or a hybrid solution as noted above, thereby potentially reducing the accuracy of the solution.

- Numerical models are limited due to the time it takes to construct the numerical grid and run the model when complex flow conditions are being simulated.
- Shallow flows can be computationally intensive, requiring a small grid spacing to accurately simulate a velocity profile. An example would be an extremely low flow over an ogee crest. Fortunately, most design concerns are based on larger flows.
- Complex flow conditions from turbulent flow, recirculating flow, flow separation, and flow super-elevation around a flow bend may not be accurately calculated in a numerical model due to turbulent modeling limits. Although turbulence models work well for certain documented cases, they may not be appropriate for all flow conditions.
- Incorrect or improperly applied gridding techniques have the potential to introduce numerical errors. Grid convergence testing is necessary to determine the adequacy of the selected grid sizing.
- There is a nonproportional time requirement to simulate and analyze a very small change in a numerical model as compared to the time it takes in a physical model.

Composite Modeling Overview

Composite modeling is the effective utilization of both physical and numerical modeling, used either in series or in parallel to solve difficult hydraulic problems. Despite weaknesses in the two modeling techniques when they are utilized independently, composite modeling has proven to be extremely effective and efficient. A properly applied composite modeling approach will improve the accuracy of the flow data and measurements by focusing on the benefits of both physical and numerical modeling as they are used together. It should be noted, however, that composite modeling will not always reduce overall cost over using one modeling technique or the other. The value of composite modeling is normally in the quality of flow data and the confidence in the proposed hydraulic design and not necessarily in the cost. The composite approach allows for more design options to be evaluated and therefore the design to be optimized.

Composite modeling can be viewed in three components: (1) Numerical modeling performed before the physical model is constructed and tested. (2) Numerical modeling performed at the same time or in parallel with the physical model. (3) Numerical modeling performed after the physical model study is complete.

Numerical Modeling Performed before the Physical Model Is Constructed and Tested

- Prior to the construction of a physical model, numerical modeling results will provide information so that the approach geometry (i.e., the physical laboratory head box) of the physical model can be properly aligned with the approach streamlines in the flow, and can be minimized so that the head box that supplies water to the model is not constructed any larger than necessary. This reduces construction costs for the physical model and allows a larger-sized model (smaller scale) to be constructed, thereby reducing the Reynolds scale effects especially in the region of interest.
- The numerical model can help reduce the number of physical modeling configurations necessary by performing pre-runs. The numerical model can test design conditions and alternative design concepts before the physical modeling to make decisions about the physical model and to obtain a general acceptance for design.
- The pre-runs from the numerical model will help engineers understand the general hydraulic problems—identifying problematic regions where flow separation, excessive velocities, or nonuniform flow may occur.

Numerical Modeling Performed in Parallel with the Physical Model

- One of the most significant benefits of composite modeling is "modeling the model," which means that the exact geometry of the initial or baseline physical model is numerically modeled at a 1:1 scale so that numerical modeling errors can be evaluated and corrected if possible. This quality control effort is effective in reducing or eliminating the uncertainties of the numerical model.
- It is also possible for the calibrated numerical model to then be rescaled to model the prototype to evaluate potential physical modeling scale effects. Again, these scale effects should be minimal for the physical model if the physical model is constructed large enough to minimize the scale effects.

- Once the numerical model has been calibrated or adjusted to the results of the physical model, it can be used to evaluate a myriad of discharges, velocities, water surface profiles, and detailed surface pressures and shear stresses that are labor intensive and expensive to perform in the physical model. Adjustments to the numerical model to match the physical model are normally done by adjusting surface roughness, turbulence model parameters, and grid cell resolution until the correlation between the two modeling types seems reasonable.
- With a calibrated numerical model, any proposed structural or geometric change that requires major modifications to the physical model can be simulated in the numerical model to evaluate design acceptability as compared to the initial baseline calibrated numerical modeling results. The results of these numerical modeling runs are then carefully compared to determine the appropriate trial or final geometric configuration that should be physically modeled.
- The trial or final physical model configuration is then constructed in order that detailed hydrograph and rating curves can be tested. A physical model can quickly and accurately determine the rating curve of a hydraulic structure, while a numerical model will take a significantly longer time to generate the necessary runs for a rating curve. Again, however, once the noted physical model configuration is constructed, the numerical model can be used to provide an infinite number of data points, streamlines, and/or flow ribbons within the flow domain that would take considerable effort to collect in the physical model.
- The physical model can be photographed and the flow conditions and problems associated with the structural design can be videotaped so while the data associated with both the physical and numerical models are reviewed, the recorded flow conditions can be reviewed. These recordings are also extremely valuable in publicizing the design to the general public and to funding agencies.
- During this phase of the composite model, regions in the flow domain in which the numerical model simulations are questionable are documented. This may include areas of highly turbulent flow, areas within the larger model in which extremely small depths and high-velocity flows occur, flow conditions that are near critical depth, transient surges or

wave oscillations, super-elevation as a result of supercritical flows rounding a tight curve, regions where scour is a potential problem, and flow conditions in which the water is highly aerated.

Numerical Modeling Performed after the Physical Model Study Is Complete

After every physical model study, it is necessary to dismantle the model so that the space that was utilized for the physical model in the laboratory can be used for another physical model study. However, the numerical model does not take up any physical space and does not need to be discarded. The benefits of maintaining a calibrated numerical model are

- The numerical model can be used indefinitely after the physical model is dismantled.
- Additional information can be collected within the flow field (pressure and velocities recorded at every cell) if needed after the physical model has been dismantled.
- Minor modifications to the geometric design can be made and tested in the calibrated numerical model with confidence that the results will be accurate.
- The calibrated numerical model can be used for training and operational decisions after the physical model has been dismantled.

Summary and Conclusions

To enhance the hydraulic modeling effort, using both a physical and a numerical model together as a composite model has proven to improve modeling accuracy and reduce modeling uncertainty. Composite modeling has been found to be an effective tool for hydraulic structure design.

Composite modeling provides a unique opportunity for researchers and engineers to understand the uncertainties and limitations of both the physical model and the numerical model, since their parallel operation allows for direct comparison and calibration. When complex three-dimensional flow conditions are being modeled, numerical models are often limited in their ability to simulate the flow field in all regions as compared with the physical model. This would include flow conditions with water surface profiles in which transverse super-elevation or nonsymmetrical hydraulic jumps occur. This also includes flow conditions where standing waves in flows that are near critical depth are present, where upstream subcritical flows are being merged across a control structure to flows that are rapidly changing downstream of the control structure, and where highly turbulent and/or mixed flow regimes occur.

In the past, numerical modeling has been proven to be very effective and accurate when applied to two-dimensional and some three-dimensional flow applications, but many of the hydraulic problems associated with hydraulic structures are complex, highly turbulent, and three-dimensional, and numerical modeling can be questionable for these types of flow conditions. Therefore, composite modeling allows for the verification and validation of flow rates, water surface profiles, and point velocities within the flow domain and consequently determines specific regions within the numerical model that are not being simulated accurately.

Composite modeling can produce a calibrated numerical model to collect very large and detailed amounts of data from the flow domain and produce a model that can be effectively used long after the physical model is dismantled. Composite modeling provides an opportunity for the numerical model to be calibrated to both the prototype structure at a 1:1 scale and the physical model at a 1:1 scale. This is all very important to the quality control component of the study.

The use of numerical modeling as an integrated part of the modeling process has proven to be both cost effective and time beneficial for the design hydraulic engineer. Using a numerical model (before the physical model) to define the physical domain of the head pool and the approach flow requirements in the physical model is a valuable component for the modeling process. This pre-modeling provides important information so that the physical model can be constructed as large as possible, thereby minimizing possible scaling effects. This pre-modeling also provides information about how the water supply to the physical model should be designed to ensure appropriate approach flows are simulated.

Additionally, the numerical pre-modeling provides information about potential problems associated with the theoretical design or proposed design changes to the structure, thereby reducing the number of physical modeling configurations necessary during the physical modeling portion of the study. This pre-modeling effort saves time and money once the physical modeling has commenced. With a calibrated numerical model, the numerical model is a valuable tool in noting hydraulic roughness sensitivity and other relative changes that are not so easy to modify in the physical model. As previously mentioned, the numerical model provides great benefits for quality control since it can be scaled to model the physical model directly, thereby producing validation of geometry and operating conditions in the physical model. Finally, the numerical model is a very important tool for maintenance and operation of the hydraulic structure for use after the physical model has been dismantled.

This section is based on a chapter in a new bulletin on spillway hydraulics that was written by Barfuss and Lindell.[1] Information in several technical publications contributed to the formulation of that chapter, including Willey et al.,[2] Erpicum et al.,[3] Rahmeyer et al.,[4] Paxson et al.,[5] and Savage et al.[6]

References

1. ICOLD Committee on Hydraulics for Dams, *Proposed Bulletin on Technical Advancements in Spillway Design,* submitted for approval to the July 2017 General Assembly of ICOLD.
2. James Willey, Tom Ewing, Bob Wark, and Eric Lesleighter, *Complimentary Use of Physical and Numerical Modeling Techniques in Spillway Design Refinement,* 25th Congress on Large Dams, Kyoto, Japan, June 2012.
3. Sébastien Erpicum, Benjamin J. Dewals, Jean-Marie Vuillot, Pierre Archambeau, and Michel Pirotton, *Coupling Physical and Numerical Models: Example of the Taoussa Project (Mali),* 4th IAHR International Symposium on Hydraulic Structures, Porto, Portugal, Feb. 9–11, 2012.
4. William Rahmeyer, Steve Barfuss, and Bruce Savage, *Composite Modeling of Hydraulic Structures,* Dam Safety 2011, National Harbor, MD, Sep. 25–29, 2011.
5. Greg Paxson, Brian Crookston, Bruce Savage, Blake Tullis, and Frederick Lux III, *The Hydraulic Design Toolbox: Theory and Modeling for the Lake Townsend Spillway Replacement Project,* Association of State Dam Safety Officials (ASDSO), Indian Wells, CA, Sep. 2008.
6. Bruce M. Savage, Michael C. Johnson, and Brett Towler, *Hydrodynamic Forces on a Spillway: Can We Calculate Them?,* Dam Safety 2009, Hollywood, FL, Oct. 2009.

Index

F

G

H

N

O

P

www.ingramcontent.com/pod-product-compliance
Lightning Source LLC
LaVergne TN
LVHW020630300926
849161LV00004B/657